W9-AWR-134

STATA DATA MANAGEMENT
REFERENCE MANUAL
RELEASE 10

A Stata Press Publication
StataCorp LP
College Station, Texas

Stata Press, 4905 Lakeway Drive, College Station, Texas 77845

Copyright © 1985–2007 by StataCorp LP
All rights reserved
Version 10
Typeset in TEX
Printed in the United States of America

10 9 8 7 6 5 4 3 2 1

ISBN-10: 1-59718-030-0
ISBN-13: 978-1-59718-030-6

This manual is protected by copyright. All rights are reserved. No part of this manual may be reproduced, stored in a retrieval system, or transcribed, in any form or by any means—electronic, mechanical, photocopying, recording, or otherwise—without the prior written permission of StataCorp LP.

StataCorp provides this manual "as is" without warranty of any kind, either expressed or implied, including, but not limited to, the implied warranties of merchantability and fitness for a particular purpose. StataCorp may make improvements and/or changes in the product(s) and the program(s) described in this manual at any time and without notice.

The software described in this manual is furnished under a license agreement or nondisclosure agreement. The software may be copied only in accordance with the terms of the agreement. It is against the law to copy the software onto CD, disk, diskette, tape, or any other medium for any purpose other than backup or archival purposes.

The automobile dataset appearing on the accompanying media is Copyright © 1979 by Consumers Union of U.S., Inc., Yonkers, NY 10703-1057 and is reproduced by permission from CONSUMER REPORTS, April 1979.

Stata and Mata are registered trademarks and NetCourse is a trademark of StataCorp LP.

Other product names are registered trademarks or trademarks of their respective companies.

For copyright information about the software, type help copyright within Stata.

The suggested citation for this software is

StataCorp. 2007. *Stata Statistical Software: Release 10*. College Station, TX: StataCorp LP.

Table of Contents

Cross-Referencing the Documentation

When reading this manual, you will find references to other Stata manuals. For example,

[U] **26 Overview of Stata estimation commands**

[R] **regress**

[XT] **xtreg**

The first is a reference to chapter 26, *Overview of Stata estimation commands* in the *Stata User's Guide*, the second is a reference to the `regress` entry in the *Base Reference Manual*, and the third is a reference to the `xtreg` entry in the *Longitudinal/Panel-Data Reference Manual*.

All the manuals in the Stata Documentation have a shorthand notation:

[GSM]	*Getting Started with Stata for Macintosh*
[GSU]	*Getting Started with Stata for Unix*
[GSW]	*Getting Started with Stata for Windows*
[U]	*Stata User's Guide*
[R]	*Stata Base Reference Manual*
[D]	*Stata Data Management Reference Manual*
[G]	*Stata Graphics Reference Manual*
[P]	*Stata Programming Reference Manual*
[XT]	*Stata Longitudinal/Panel-Data Reference Manual*
[MV]	*Stata Multivariate Statistics Reference Manual*
[SVY]	*Stata Survey Data Reference Manual*
[ST]	*Stata Survival Analysis and Epidemiological Tables Reference Manual*
[TS]	*Stata Time-Series Reference Manual*
[I]	*Stata Quick Reference and Index*
[M]	*Mata Reference Manual*

Detailed information about each of these manuals may be found online at

http://www.stata-press.com/manuals/

Title

> **intro** — Introduction to data-management reference manual

Description

This entry describes this manual and what has changed since Stata 9. See the next entry, [D] **data management**, for an introduction to Stata's data-management capabilities.

Remarks

This manual documents most of Stata's data-management features and is referred to as the [D] manual. Some specialized data-management features are documented in such subject-specific reference manuals as [TS] *Stata Time-Series Reference Manual*, [ST] *Stata Survival Analysis and Epidemiological Tables Reference Manual*, and [XT] *Stata Longitudinal/Panel-Data Reference Manual*.

Following this entry, [D] **data management** provides an overview of data management in Stata and Stata's data-management commands. The other parts of this manual are arranged alphabetically. If you are new to Stata's data-management features, we recommend that you read the following first:

> [D] **data management** — Introduction to data-management commands
>
> [U] **12 Data**
>
> [U] **13 Functions and expressions**
>
> [U] **11.5 by varlist: construct**
>
> [U] **21 Inputting data**
>
> [U] **22 Combining datasets**
>
> [U] **23 Dealing with strings**
>
> [U] **25 Dealing with categorical variables**
>
> [U] **24 Dealing with dates and times**
>
> [U] **16 Do-files**

You can see that most of the suggested reading is in [U]. That is because [U] provides overviews of most Stata features, whereas this is a reference manual and provides details on the usage of specific commands. You will get an overview of features for combining data from [U] **22 Combining datasets**, but the details of performing a match-merge (merging the records of two files by matching the records on a common variable) will be found here, in [D] **merge**.

Stata is continually being updated, and Stata users are always writing new commands. To ensure that you have the latest features, you should install the most recent official update; see [R] **update**.

What's new

This section is intended for previous Stata users. If you are new to Stata, you may as well skip it.

1. Stata 10 has new date/time variables, so you can now record values like 14jun2007 09:42:41.106 in one variable. They are called %tc and %tC variables. The first is unadjusted for leap seconds; the second is adjusted.

What used to be called "daily variables" are now called %td variables. This is just a jargon change; daily (%td) variables continue to work as they did before—0 means 01jan1960, 1 means 02jan1960, and so on.

%tc and %tC variables work similarly: 0 means 01jan1960 00:00:00. Here, however, 1 means 01jan1960 00:00:00.001, 1000 means 01jan1960 00:00:01.000, and 02jan1960 08:00:00 is 115,200,000. The underlying values are big—so it is important you store them as doubles—but the %tc and %tC formats make the values readable, just as the %td format makes daily (%td) values readable.

There are many new functions to go along with this new value type. clock(), for instance, converts strings such as "02jan1960 08:00:00" (or even "8:00 a.m., 1/2/1960") to their numeric equivalents. dofc() converts a %tc value (such as 115,200,000, meaning 02jan1960 08:00:00) to its %td equivalent (namely, 1, meaning 02jan1960). cofd() does the reverse (the result would be 86,400,000, meaning 02jan1960 00:00:00).

See [D] **dates and times**.

2. The previously existing date() function, which converts strings to %td values, is now smarter. In addition to being able to convert strings such as "21aug2005", "August 21, 2005", it can convert "082105", "08212005", "210805", and "21082005". See [D] **dates and times**.

3. New command datasignature allows you to sign datasets and later use that signature to determine whether the data have changed. An early version of the command was made available during the Stata 9 release. That command is now called _datasignature and was used as the building block for the new, improved datasignature. See [D] **datasignature** and [P] **_datasignature**.

4. Existing command clear now clears data and value labels only. Type clear all to clear everything. This change will bite you the first few times you type clear expecting it to clear all. The problem was that new users were surprised when clear by itself cleared everything, whereas use *filename*, clear loaded new data and value labels but left everything else in place. The new users were right.

   clear now has the following subcommands:

   a. clear all clears everything from memory.

   b. clear ado clears automatically loaded ado-file programs.

   d. clear programs clears all programs, automatically loaded or not.

   c. clear results clears saved results.

   d. clear mata clears Mata functions and objects from memory.

   See [D] **clear**.

5. Stata for Unix now supports unixODBC [*sic*], making it easier to connect to databases such as Oracle, MySQL, and PostgreSQL; see [D] **odbc**.

6. Existing command describe now allows option varlist that was previously allowed only by describe using. Existing command describe using *filename* now allows option simple that was previously allowed only by describe. Option varlist saves the variable names in r(varlist), and option simple displays the variable names in a compact format. See [D] **describe**.

7. Existing command collapse now supports four additional *stat*s: first, the first value; last, the last value; firstnm, the first nonmissing value; and lastnm, the last nonmissing value. See [D] **collapse**.

8. Existing command cf (compare files) now provides a detailed listing of observations that differ when the verbose option is specified. Setting version to less than 10.0 restores the earlier behavior. See [D] **cf**.

9. Existing command codebook has new option compact that produces more compact output. See [D] **codebook**.

10. Existing command insheet has new option case that preserves the case of variable names when importing data; see [D] **insheet**.

11. Existing command outsheet has new option delimiter() that specifies an alternative delimiter; see [D] **outsheet**.

12. Existing commands infile and infix can now read up to 524,275 characters per line; the previous limit was 32,765. See [D] **infile** and [D] **infix (fixed format)**.

13. Existing commands icd9 and icd9p have now been updated to use the V24 codes; see [D] **icd9**.

14. New function itrim() returns the string with consecutive, internal spaces collapsed to one space; see *String functions* in [D] **functions**.

15. New functions lnnormal() and lnnormalden() provide the natural logarithm of the cumulative standard normal distribution and of the standard normal density; see *Probability distributions and density functions* in [D] **functions**.

16. New functions for calculating cumulative densities are now available:

binomial(n, k, p)	lower tail of the binomial distribution
ibetatail(a, b, x)	reverse (upper tail) of the cumulative beta distribution
gammaptail(a, x)	reverse (upper tail) of the cumulative gamma distribution
invgammaptail(a, p)	inverse reverse of the cumulative gamma distribution
invibetatail(a, b, p)	inverse reverse of the cumulative beta distribution
invbinomialtail(n, k, p)	inverse of right cumulative binomial

See *Probability distributions and density functions* in [D] **functions**.

17. Existing function Binomial(n, k, p) has been renamed binomialtail(n, k, p), thus making its name consistent with the naming convention for probability functions. The accuracy of the function has also been improved for very large values of n. At the other end of the number line, the function now returns the appropriate 0 or 1 value when $n = 0$, rather than returning missing. Binomial() continues to work as a synonym for binomialtail().

18. The behavior and accuracy of the following probability functions have been improved:

a. F(n_1, n_2, f) and Ftail(n_1, n_2, f) are more accurate for small values of n_1 and large values of n_2. Also, F() is more accurate for large f where n_1 and n_2 are less than 1.

b. gammap(a, x) is more accurate when a is large and x is near a.

c. ibeta(a, b, x) now is more accurate when x is near $a/(a + b)$ and a or b is large.

d. invbinomial(n, k, p), invchi2(n, p), invchi2tail(n, p), invF(n_1, n_2, p), and invgammap(a, p) are more accurate for small values of p or for returned values close to zero.

e. invFtail(n_1, n_2, p) and invibeta(a, b, p) are more accurate for small values of p or for returned values close to zero.

f. invttail(n, p) is more accurate for small values of p or for returned values close to zero.

g. ttail(n, t) is more accurate for exceedingly large values of n.

19. Existing function invbinomial(n, k, p) now returns the probability of a success on one trial such that the probability of observing k or *fewer* successes in n trials is p. The previous behavior of invbinomial() is restored under version control.

20. New function fmtwidth() returns the display width of a *%fmt* string; see *Programming functions* in [D] **functions**.

21. The maximum length of a *%fmt* has increased from 12 to 48 characters; see [D] **format**. (This change was necessitated by the new date/time variables.)

22. Existing commands corr2data and drawnorm now allow singular correlation (or covariance) structures. New option forcepsd modifies a matrix to be positive semidefinite and thus to be a proper covariance matrix. See [D] **corr2data** and [D] **drawnorm**.

23. Existing command hexdump, analyze now saves the number of \r\n characters in r(Windows) rather than in r(DOS). r(DOS) is still set when version is less than 10. See [D] **hexdump**.

For a complete list of all the new features in Stata 10, see [U] **1.3 What's new**.

Also See

[U] **1.3 What's new**

[R] **intro** — Introduction to base reference manual

Title

data management — Introduction to data-management commands

Description

This manual, called [D], documents Stata's data-management features.

Data management for statistical applications refers not only to classical data management—sorting, merging, appending, and the like—but also to data reorganization because the statistical routines you will use assume that the data are organized in a certain way. For example, statistical commands that analyze longitudinal data, such as xtreg, generally require that the data be in long rather than wide form, meaning that repeated values are recorded not as extra variables, but as extra observations.

Here are the basics everyone should know:

[D] **use**	Use Stata dataset
[D] **save**	Save datasets
[D] **describe**	Describe data in memory or in file
[D] **inspect**	Display simple summary of data's attributes
[D] **codebook**	Describe data contents
[D] **data types**	Quick reference for data types
[D] **missing values**	Quick reference for missing values
[D] **dates and times**	Date and time (%t) values and variables
[D] **list**	List values of variables
[D] **edit**	Edit and list data with Data Editor
[D] **rename**	Rename variable
[D] **format**	Set variables' output format
[D] **label**	Manipulate labels

You will need to create and drop variables, and here is how:

[D] **generate**	Create or change contents of variable
[D] **functions**	Functions
[D] **egen**	Extensions to generate
[D] **drop**	Eliminate variables or observations
[D] **clear**	Clear memory

(*Continued on next page*)

For inputting or importing data, see

[D] **use**	Use Stata dataset
[D] **sysuse**	Use shipped dataset
[D] **webuse**	Use dataset from Stata web site
[D] **input**	Enter data from keyboard
[D] **insheet**	Read ASCII (text) data created by a spreadsheet
[D] **infile**	Overview of reading data into Stata
[D] **infile (fixed format)**	Read ASCII (text) data in fixed format with a dictionary
[D] **infile (free format)**	Read unformatted ASCII (text) data
[D] **infix (fixed format)**	Read ASCII (text) data in fixed format
[D] **hexdump**	Display hexadecimal report on file
[D] **odbc**	Load, write, or view data from ODBC sources
[D] **xmlsave**	Save and use datasets in XML format
[D] **fdasave**	Save and use datasets in FDA (SAS XPORT) format
[D] **icd9**	ICD-9-CM diagnostic and procedure codes

and for exporting data, see

[D] **outfile**	Write ASCII-format dataset
[D] **outsheet**	Write spreadsheet-style dataset
[D] **fdasave**	Save and use datasets in FDA (SAS XPORT) format
[D] **odbc**	Load, write, or view data from ODBC sources

The ordering of variables and observations (sort order) can be important; see

[D] **order**	Reorder variables in dataset
[D] **sort**	Sort data
[D] **gsort**	Ascending and descending sort

To reorganize or combine data, see

[D] **merge**	Merge datasets
[D] **append**	Append datasets
[D] **reshape**	Convert data from wide to long form and vice versa
[D] **collapse**	Make dataset of summary statistics
[D] **fillin**	Rectangularize dataset
[D] **expand**	Duplicate observations
[D] **expandcl**	Duplicate clustered observations
[D] **stack**	Stack data
[D] **joinby**	Form all pairwise combinations within groups
[D] **xpose**	Interchange observations and variables
[D] **cross**	Form every pairwise combination of two datasets

In the above list, we particularly want to direct your attention to [D] **reshape**, a useful command beginners often overlook.

For random sampling, see

[D] **sample**	Draw random sample
[D] **drawnorm**	Draw sample from multivariate normal distribution

For file manipulation, see

[D] **type**	Display contents of a file
[D] **erase**	Erase a disk file
[D] **copy**	Copy file from disk or URL
[D] **cd**	Change directory
[D] **dir**	Display filenames
[D] **mkdir**	Create directory
[D] **rmdir**	Remove directory
[D] **cf**	Compare two datasets
[D] **filefilter**	Convert ASCII text or binary patterns in a file
[D] **checksum**	Calculate checksum of file

The entries above are important. The rest are useful when you need them:

[D] **datasignature**	Determine whether data have changed
[D] **type**	Display contents of a file
[D] **notes**	Place notes in data
[D] **label language**	Labels for variables and values in multiple languages
[D] **labelbook**	Label utilities
[D] **encode**	Encode string into numeric and vice versa
[D] **recode**	Recode categorical variable
[D] **impute**	Fill in missing values
[D] **ipolate**	Linearly interpolate (extrapolate) values
[D] **destring**	Convert string variables to numeric variables and vice versa
[D] **mvencode**	Change missing values to numeric values and vice versa
[D] **pctile**	Create variable containing percentiles
[D] **range**	Generate numerical range
[D] **by**	Repeat Stata command on subsets of the data
[D] **statsby**	Collect statistics for a command across a by list
[D] **compress**	Compress data in memory
[D] **recast**	Change storage type of variable

(Continued on next page)

[D] **assert**	Verify truth of claim
[D] **clonevar**	Clone existing variable
[D] **compare**	Compare two variables
[D] **contract**	Make dataset of frequencies and percentages
[D] **corr2data**	Create dataset with specified correlation structure
[D] **count**	Count observations satisfying specified conditions
[D] **duplicates**	Report, tag, or drop duplicate observations
[D] **isid**	Check for unique identifiers
[D] **lookfor**	Search for string in variable names and labels
[D] **memory**	Memory size considerations
[D] **obs**	Increase the number of observations in a dataset
[D] **separate**	Create separate variables
[D] **shell**	Temporarily invoke operating system
[D] **split**	Split string variables into parts

There are some real jewels in the above, such as [D] **notes**, [D] **compress**, and [D] **assert**, which you will find particularly useful.

Also See

[D] **intro** — Introduction to data-management reference manual

[R] **intro** — Introduction to base reference manual

Title

> **append** — Append datasets

Syntax

> <u>append</u> using *filename* [, *options*]

You may enclose *filename* in double quotes and must do so if *filename* contains blanks or other special characters.

options	description
<u>keep</u>(*varlist*)	keep specified variables from appending dataset
<u>nol</u>abel	do not copy value label definitions from dataset on disk
<u>nonotes</u>	do not copy notes from dataset on disk

Description

append appends a Stata-format dataset stored on disk to the end of the dataset in memory. If *filename* is specified without an extension, .dta is assumed.

Stata can also join observations from two datasets into one; see [D] **merge**. See [U] **22 Combining datasets** for a comparison of append, merge, and joinby.

Options

keep(*varlist*) specifies the variables to be kept from the using dataset. If keep() is not specified, all variables are kept.

The *varlist* in keep(*varlist*) differs from standard Stata varlists in two ways: variable names in *varlist* may not be abbreviated, except by the use of wildcard characters, and you may not refer to a range of variables, such as price-weight.

nolabel prevents Stata from copying the value label definitions from the disk dataset into the dataset in memory. Even if you do not specify this option, label definitions from the disk dataset never replace definitions already in memory.

nonotes prevents notes in the using dataset from being incorporated into the result. The default is to incorporate notes from the using dataset that do not already appear in the master data.

Remarks

The disk dataset must be a Stata-format dataset; that is, it must have been created by save (see [D] **save**).

▷ Example 1

We have two datasets stored on disk that we want to combine. The first dataset, called even.dta, contains the sixth through eighth positive even numbers. The second dataset, called odd.dta, contains the first five positive odd numbers. The datasets are

```
. use even
(6th through 8th even numbers)
. list
```

	number	even
1.	6	12
2.	7	14
3.	8	16

```
. use odd
(First five odd numbers)
. list
```

	number	odd
1.	1	1
2.	2	3
3.	3	5
4.	4	7
5.	5	9

We will append the even data to the end of the odd data. Since the odd data are already in memory (we just used them above), we type append using even. The result is

```
. append using even
. list
```

	number	odd	even
1.	1	1	.
2.	2	3	.
3.	3	5	.
4.	4	7	.
5.	5	9	.
6.	6	.	12
7.	7	.	14
8.	8	.	16

Since the variable number is in both datasets, the variable was extended with the new data from the file even.dta. Since there is no variable called odd in the new data, the additional observations on odd were forward-filled with *missing*. Since there is no variable called even in the original data, the first observations on even were back-filled with *missing*.

◁

▷ Example 2

The order of variables in the two datasets is irrelevant. Stata always appends variables by name:

```
. use http://www.stata-press.com/data/r10/odd1
(First five odd numbers)

. describe
Contains data from http://www.stata-press.com/data/r10/odd1.dta
  obs:            5                          First five odd numbers
  vars:           2                          9 Jan 2007 08:41
  size:          60 (99.9% of memory free)
```

variable name	storage type	display format	value label	variable label
odd	float	%9.0g		Odd numbers
number	float	%9.0g		

```
Sorted by:  number
. describe using http://www.stata-press.com/data/r10/even
Contains data from http://www.stata-press.com/data/r10/even
  obs:            3                          6th through 8th even numbers
  vars:           2                          9 Jan 2007 08:43
  size:          27
```

variable name	storage type	display format	value label	variable label
number	byte	%9.0g		
even	float	%9.0g		Even numbers

```
Sorted by:  number
. append using http://www.stata-press.com/data/r10/even

. list
```

	odd	number	even
1.	1	1	.
2.	3	2	.
3.	5	3	.
4.	7	4	.
5.	9	5	.
6.	.	6	12
7.	.	7	14
8.	.	8	16

The results are the same as those in the first example.

◁

When Stata appends two datasets, the definitions of the dataset in memory, called the *master* dataset, override the definitions of the dataset on disk, called the *using* dataset. This extends to value labels, variable labels, characteristics, and date–time stamps. If there are conflicts in numeric storage types, the more precise storage type will be used regardless of whether this storage type was in the *master* dataset or the *using* dataset. If a variable is stored as a str# in one dataset and a numeric storage type in the other, the definition in the *master* dataset will prevail. If a variable is stored as a string in one dataset that is longer than in the other, the longer str# storage type will prevail.

❏ Technical Note

If a variable is a string in one dataset and numeric in the other, Stata issues a warning message and then appends the data. If the using dataset contains the string variable, the combined dataset will have numeric missing values for the appended data on this variable; the contents of the string variable in the using dataset are ignored. If the using dataset contains the numeric variable, the combined dataset will have null strings for the appended data on this variable; the contents of the numeric variable in the using dataset are ignored.

❏

▷ Example 3

Since Stata has five numeric variable types—byte, int, long, float, and double—you may attempt to append datasets containing variables with the same name but of different numeric types; see [U] **12.2.2 Numeric storage types**.

Let's describe the datasets in the example above:

```
. describe using http://www.stata-press.com/data/r10/odd

Contains data from http://www.stata-press.com/data/r10/odd
  obs:           5                        First five odd numbers
 vars:           2                        9 Jan 2007 08:50
 size:          60

              storage  display      value
variable name    type   format      label      variable label

number          float   %9.0g
odd             float   %9.0g                   Odd numbers

Sorted by:
```

```
. describe using http://www.stata-press.com/data/r10/even

Contains data from http://www.stata-press.com/data/r10/even
  obs:           3                        6th through 8th even numbers
 vars:           2                        9 Jan 2007 08:43
 size:          27

              storage  display      value
variable name    type   format      label      variable label

number          byte    %9.0g
even            float   %9.0g                   Even numbers

Sorted by:  number
```

```
. describe using http://www.stata-press.com/data/r10/oddeven

Contains data from http://www.stata-press.com/data/r10/oddeven
  obs:           8                        First five odd numbers
 vars:           3                        9 Jan 2007 08:53
 size:         128

              storage  display      value
variable name    type   format      label      variable label

number          float   %9.0g
odd             float   %9.0g                   Odd numbers
even            float   %9.0g                   Even numbers

Sorted by:
```

The variable number was stored as a float in odd.dta but as a byte in even.dta. Since float is the more precise storage type, the resulting dataset, oddeven.dta, had number stored as a float. Had we, instead, appended odd.dta to even.dta, number would still have been stored as a float:

```
. use http://www.stata-press.com/data/r10/even, clear
(6th through 8th even numbers)
. append using http://www.stata-press.com/data/r10/odd
number was byte now float
. describe
Contains data from http://www.stata-press.com/data/r10/even.dta
  obs:           8                          6th through 8th even numbers
  vars:          3                          9 Jan 2007 08:43
  size:        128 (99.8% of memory free)
```

variable name	storage type	display format	value label	variable label
number	float	%9.0g		
even	float	%9.0g		Even numbers
odd	float	%9.0g		Odd numbers

```
Sorted by:
     Note:  dataset has changed since last saved
```

◁

▷ Example 4

Suppose that we have a dataset in memory containing the variable educ, and we have previously given a label variable educ "Education Level" command so that the variable label associated with educ is "Education Level". We now append a dataset called newdata.dta, which also contains a variable named educ, except that its variable label is "Ed. Lev". After appending the two datasets, the variable educ is still labeled "Education Level". See [U] **12.6.2 Variable labels**.

◁

▷ Example 5

Assume that the values of the variable educ are labeled with a value label named educlbl. Further assume that in newdata.dta, the values of educ are also labeled by a value label named educlbl. Thus there is one definition of educlbl in memory and another (although perhaps equivalent) definition in newdata.dta. When you append the new data, you will see the following:

```
. append using newdata
label educlbl already defined
```

If one label in memory and another on disk have the same name, append warns you of the problem and sticks with the definition currently in memory, ignoring the definition in the disk file.

◁

(*Continued on next page*)

❏ Technical Note

When you `append` two datasets that both contain definitions of the same value label, the codings may not be equivalent. That is why Stata warns you with a message like "label educlbl already defined". If you do not know that the two value labels are equivalent, you should convert the value-labeled variables into string variables, append the data, and then construct a new coding. `decode` and `encode` make this easy:

```
. use newdata, clear
. decode educ, gen(edstr)
. drop educ
. save newdata, replace
. use basedata
. decode educ, gen(edstr)
. drop educ
. append using newdata
. encode edstr, gen(educ)
. drop edstr
```

See [D] **encode**.

You can specify the `nolabel` option to force `append` to ignore all the value label definitions in the incoming file, whether or not there is a conflict. In practice, you will probably never want to do this.

❏

Also See

[D] **save** — Save datasets

[D] **use** — Use Stata dataset

[D] **cross** — Form every pairwise combination of two datasets

[D] **joinby** — Form all pairwise combinations within groups

[D] **merge** — Merge datasets

[U] **22 Combining datasets**

Title

assert — Verify truth of claim

Syntax

<u>as</u>sert *exp* $\big[$ *if* $\big]$ $\big[$ *in* $\big]$ $\big[$, <u>r</u>c0 <u>n</u>ull $\big]$

by is allowed; see [D] **by**.

Description

assert verifies that *exp* is true. If it is true, the command produces no output. If it is not true, assert informs you that the "assertion is false" and issues a return code of 9; see [U] **8 Error messages and return codes**.

Options

rc0 forces a return code of 0, even if the assertion is false.

null forces a return code of 8 on null assertions.

Remarks

assert is seldom used interactively since it is easier to use inspect, summarize, or tabulate to look for evidence of errors in the dataset. These commands, however, require you to review the output to spot the error. assert is useful because it tells Stata not only what to do but what also you can expect to find. Groups of assertions are often combined in a do-file to certify data. If the do-file runs all the way through without complaining, every assertion in the file is true.

```
. do myassert
. use trans, clear
(xplant data)
. assert sex=="m" | sex=="f"
. assert packs==0 if !smoker
. assert packs>0 if smoker
. sort patient date
. by patient: assert sex==sex[_n-1] if _n>1
. by patient: assert abs(bp-bp[_n-1]) < 20 if bp< . & bp[_n-1]< .
. by patient: assert died==0 if _n!=_N
. by patient: assert died==0 | died==1 if _n==_N
. by patient: assert n_xplant==0 | n_xplant==1 if _n==_N
. assert inval==int(inval)
.
.
end of do-file
```

▷ Example 1

You receive data from Bob, a coworker. He has been working on the dataset for some time, and it has now been delivered to you for analysis. Before analyzing the data, you (smartly) verify that the data are as Bob claims. In Bob's memo, he claims that (1) the dataset reflects the earnings of 522 employees, (2) the earnings are only for full-time employees, (3) the variable female is coded 1 for female and 0 otherwise, and (4) the variable exp contains the number of years, or fraction thereof, on the job. You assemble the following do-file:

```
use frombob, clear
assert _N==522
assert sal>=6000 & sal<=125000
assert female==1 | female==0
gen work=sum(female==1)
assert work[_N]>0
replace work=sum(female==0)
assert work[_N]>0
drop work
assert exp>=0 & exp<=40
```

Let's go through these assertions one by one. After using the data, you assert that _N equals 522. Remember, _N reflects the total number of observations in the dataset; see [U] **13.4 System variables (_variables)**. Bob said it was 522, so you check it. Bob's second claim was that the data are for only full-time employees. You know that everybody in your company makes a salary between $6,000 and $125,000, so you check that the salary figures are within this range. Bob's third assertion was that the female variable was coded zero or one.

You add something more. You know that your company employs both males and females, so you check that there are some of each. You create a variable called work equal to the running sum of female observations and then verify that the last observation of this variable is greater than zero. You then repeat the process for males and discard the work variable. Finally, you verify that the exp variable is never negative and is never larger than 40.

You save the above file as check.do, and here is what happens when you run it:

```
. do check

. use frombob, clear
(5/21 data)

. assert _N==522

. assert sal>6000 & sal<=125000
14 contradictions in 522 observations
assertion is false
r(9);

end of do-file
r(9);
```

Everything went fine until you checked the salary variable, when Stata told you that there were 14 contradictions to your assertion and stopped the do-file. Seeing this, you now interactively summarize the sal variable and discover that 14 people have missing salaries. You dash off a memo to Bob asking him why these data are missing.

◁

▷ Example 2

Bob responds quickly. There was a mistake in reading the salaries for the consumer relations division. He says it's fixed. You believe him but check with your do-file again. This time you type run instead of do, suppressing all the output:

```
. run check

. _
```

Even though you suppressed the output, if there had been any contradictions, the messages would have printed. check.do ran fine, so all its assertions are true.

◁

❏ Technical Note

assert is especially useful when you are processing large amounts of data in a do-file and wish to verify that all is going as expected. The error in this case may not be in the data but in the do-file itself. For instance, your do-file is rolling along, and it has just merged two datasets that it created by subsetting some other data. If everything has gone right so far, every observation should have merged. When you are performing merge interactively, use tabulate _merge to verify that the expected happened. In a do-file, include the line

```
assert _merge==3
```

to verify the correctness of the merge. If all the observations did not merge, the assertion will be false, and your do-file will stop.

As another example, you are combining data from many sources, and you know that after the first two datasets are combined, every individual's sex should be defined. So, you include the line

```
assert sex< .
```

in your do-file. Experienced Stata users include many assertions in their do-files when they process data.

❏

❏ Technical Note

assert is smart in how it evaluates expressions. When you type something like assert _N==522 or assert work[_N]>0, assert knows that the expression need be evaluated only once. When you type assert female==1 | female==0, assert knows that the expression needs to be evaluated once for each observation in the dataset.

Here are some more examples demonstrating assert's intelligence.

```
by female:  assert _N==100
```

asserts that there should be 100 observations for every unique value of female. The expression is evaluated once per by-group.

```
by female:  assert work[_N]>0
```

asserts that the last observation on work in every by-group should be greater than zero. It is evaluated once per by-group.

```
        by female:  assert work>0
```

is evaluated once for each observation in the dataset and, in that sense, is formally equivalent to `assert work>0`.

❏

Also See

[P] **capture** — Capture return code

[P] **confirm** — Argument verification

[U] **16 Do-files**

Title

> **by** — Repeat Stata command on subsets of the data

Syntax

by *varlist* : *stata_cmd*

by**sort** *varlist* : *stata_cmd*

The above diagrams show by and bysort as they are typically used.
The full syntax of the commands is

by *varlist₁* [(*varlist₂*)] [, s̲ort rc0] : *stata_cmd*

by**sort** *varlist₁* [(*varlist₂*)] [, rc0] : *stata_cmd*

Description

Most Stata commands allow the by prefix, which repeats the command for each group of observations for which the values of the variables in *varlist* are the same. by without the sort option requires that the data be sorted by *varlist*; see [D] **sort**.

Stata commands that work with the by prefix indicate this immediately following their syntax diagram by reporting, for example, "by is allowed; see [D] **by**" or "bootstrap, by, etc. are allowed; see [U] **11.1.10 Prefix commands**".

by and bysort are really the same command; bysort is just by with the sort option.

The *varlist₁* (*varlist₂*) syntax is of special use to programmers. It verifies that the data are sorted by *varlist₁ varlist₂* and then performs a by as if only *varlist₁* were specified. For instance,

        by pid (time): gen growth = (bp - bp[_n-1])/bp

performs the generate by values of pid but first verifies that the data are sorted by pid and time within pid.

Options

sort specifies that if the data are not already sorted by *varlist*, by should sort them.

rc0 specifies that even if the *stata_cmd* produces an error in one of the by-groups, then by is still to run the *stata_cmd* on the remaining by-groups. The default action is to stop when an error occurs. rc0 is especially useful when *stata_cmd* is an estimation command, and some by-groups have insufficient observations.

Remarks

▷ Example 1

```
. use http://www.stata-press.com/data/r10/autornd
(1978 Automobile Data)

. keep in 1/20
(54 observations deleted)

. by mpg: egen mean_w = mean(weight)
not sorted
r(5);

. sort mpg

. by mpg: egen mean_w = mean(weight)

. list
```

	make	weight	mpg	mean_w
1.	Cad. Eldorado	4000	15	3916.667
2.	AMC Pacer	3500	15	3916.667
3.	Chev. Impala	3500	15	3916.667
4.	Buick Electra	4000	15	3916.667
5.	Cad. Deville	4500	15	3916.667
6.	Buick Riviera	4000	15	3916.667
7.	Buick LeSabre	3500	20	3350
8.	Chev. Monte Carlo	3000	20	3350
9.	Buick Skylark	3500	20	3350
10.	Buick Century	3500	20	3350
11.	AMC Spirit	2500	20	3350
12.	AMC Concord	3000	20	3350
13.	Buick Regal	3500	20	3350
14.	Chev. Malibu	3000	20	3350
15.	Chev. Nova	3500	20	3350
16.	Cad. Seville	4500	20	3350
17.	Buick Opel	2000	25	2500
18.	Chev. Monza	3000	25	2500
19.	Chev. Chevette	2000	30	2000
20.	Dodge Colt	2000	30	2000

by requires that the data be sorted. In the above example, we could have typed by mpg, sort: egen mean_w = mean(weight) or bysort mpg: egen mean_w = mean(weight) rather than the separate sort; all would yield the same results.

◁

For more examples, see [U] **11.1.2 by varlist:**, [U] **11.5 by varlist: construct**, and [U] **27.2 The by construct**. For an extended introduction with detailed examples, see Cox (2002).

Reference

Cox, N. J. 2002. Speaking Stata: How to move step by: step. *Stata Journal* 2: 86–102.

Also See

[D] **sort** — Sort data

[D] **statsby** — Collect statistics for a command across a by list

[P] **byable** — Make programs byable

[P] **foreach** — Loop over items

[P] **forvalues** — Loop over consecutive values

[P] **while** — Looping

[U] **11.1.2 by varlist:**

[U] **11.1.10 Prefix commands**

[U] **11.4 varlists**

[U] **11.5 by varlist: construct**

[U] **27.2 The by construct**

Title

> **cd** — Change directory

Syntax

Stata for Windows

    cd

    cd ["] *directory_name* ["]

    cd ["] *drive*: ["]

    cd ["] *drive*:*directory_name* ["]

    pwd

Stata for Macintosh

    cd

    cd ["] *directory_name* ["]

    pwd

Stata for Unix

    cd ["] *directory_name* ["]

    pwd

If your *directory_name* contains embedded spaces, remember to enclose it in double quotes.

Description

Stata for Windows: cd changes the working directory to the specified drive and directory. pwd is equivalent to typing cd without arguments; both display the name of the current working directory. Note: you can shell out to a DOS window; see [D] **shell**. However, typing !cd *directory_name* does not change Stata's current directory; use the cd command to change directories.

Stata for Macintosh and Stata for Unix: cd (synonym chdir) changes the current working directory to *directory_name* or, if *directory_name* is not specified, the home directory. pwd displays the path of the current working directory.

Remarks

Remarks are presented under the following headings:

> *Stata for Windows*
> *Stata for Macintosh*
> *Stata for Unix*

Stata for Windows

When you start Stata for Windows, your working directory is set to the *Start in* directory specified in **Properties**. If you want to change this, see [GSW] **C.1 The Windows Properties Sheet**. You can always see what your working directory is by looking at the status bar at the bottom of the Stata window.

Once you are in Stata, you can change your directory with the cd command.

```
. cd
c:\data
. cd city
c:\data\city
. cd d:
D:\
. cd kande
D:\kande
. cd "additional detail"
D:\kande\additional detail
. cd c:
C:\
. cd data\city
C:\data\city
. cd \a\b\c\d\e\f\g
C:\a\b\c\d\e\f\g
. cd ..
C:\a\b\c\d\e\f
. cd ...
C:\a\b\c\d
. cd ....
C:\a
```

When we typed cd d:, we changed to the current directory of the D drive. We navigated our way to d:\kande\additional detail with three commands: cd d:, then cd kande, and then cd "additional detail". The double quotes around "additional detail" are necessary because of the space in the directory name. We could have changed to this directory in one command: cd "d:\kande\additional detail".

Notice the last three cd commands in the example above. You are probably familiar with the cd .. syntax to move up one directory from where you are. The last two cd commands above let you move up more than one directory. cd ... is shorthand for 'cd ..\..'. cd .... is shorthand for 'cd ..\..\..'. These shorthand cd commands are not limited to Stata; they will work in your DOS windows under Windows as well.

❏ Technical Note

When you type cd d: to change to the current directory of the D drive, Windows changes to the current directory of the process that started Stata.

A better way to understand this is to think about starting Stata from a DOS window under Windows. Say that in your DOS window, the current directory on the C drive is C:\WINDOWS, and the current directory on the D drive is D:\KANDE. Imagine starting Stata from the DOS prompt by typing c:\stata\wstata.exe. Inside Stata, you could type cd "d:\kande\additional detail" to change your current directory. You could then type cd c: to change to the current directory on the C drive—C:\WINDOWS. If you then typed 'cd d:', you would switch back to what Windows remembers as the current directory on the D drive—D:\KANDE.

No matter where you cd to inside Stata, if you cd away from the D drive and then cd back to the D drive without specifying a path, your current directory will be the directory of the D drive in DOS where you were before you started Stata.

❏

Stata for Macintosh

Read [U] **11.6 File-naming conventions** for a description of how filenames are written in a command language before reading this entry.

Invoking an application and then changing folders is an action foreign to most Macintosh users. If it is foreign to you, you can ignore cd and pwd. However, they can be useful. You can see the current folder (where Stata saves files and looks for files) by typing pwd. You can change the current folder by using cd or by selecting **File > Change Working Directory...**. Stata's cd understands '~' as an abbreviation for the home directory, so you can type things like cd ~/data.

```
. pwd
/Users/bill/proj
. cd "~/data/city"
/Users/bill/data/city
. _
```

If you now wanted to change to "/Users/bill/data/city/ny", you could type cd ny. If you wanted instead to change to "/Users/bill/data", you could type 'cd ..'.

Stata for Unix

cd and pwd are equivalent to Unix's cd and pwd commands. Like csh, Stata's cd understands '~' as an abbreviation for the home directory $HOME, so you can type things like cd ~/data; see [U] **11.6 File-naming conventions**.

```
. pwd
/usr/bill/proj
. cd ~/data/city
/usr/bill/data/city
. _
```

If you now wanted to change to /usr/bill/data/city/ny, you could type cd ny. If you wanted instead to change to /usr/bill/data, you could type 'cd ..'.

Also See

[D] **shell** — Temporarily invoke operating system

[D] **copy** — Copy file from disk or URL

[D] **dir** — Display filenames

[D] **erase** — Erase a disk file

[D] **mkdir** — Create directory

[D] **rmdir** — Remove directory

[D] **type** — Display contents of a file

[U] **11.6 File-naming conventions**

Title

cf — Compare two datasets

Syntax

cf *varlist* using *filename* $\left[\, , \, \underline{a}ll \; \underline{v}erbose \right]$

Description

cf compares *varlist* of the dataset in memory (the master dataset) with the corresponding variables in *filename* (the using dataset). cf returns nothing (i.e., a return code of 0) if the specified variables are identical and a return code of 9 if there are any differences. Only the variable values are compared. Variable labels, value labels, notes, characteristics, etc., are not compared.

Options

all displays the result of the comparison for each variable in *varlist*. Unless all is specified, only the results of the variables that differ are displayed.

verbose gives a detailed listing, by variable, of each observation that differs.

Remarks

cf produces messages having the following form:

```
varname: does not exist in using
varname: ___ in master but ___ in using
varname: ___ mismatches
varname: match
```

An example of the second message is "str4 in master but float in using". Unless all is specified, the fourth message does not appear—silence indicates matches.

▷ Example 1

We think the dataset in memory is identical to mydata.dta, but we are unsure. We want to understand any differences before continuing:

```
. cf _all using mydata

. _
```

All the variables in the master dataset are in mydata.dta, and these variables are the same in both datasets. We might see instead

```
. cf _all using mydata
             mpg:  2 mismatches
        headroom:  does not exist in using
    displacement:  does not exist in using
       gear_ratio:  does not exist in using
    r(9);
```

25

Two changes were made to the mpg variable, and the variables headroom, displacement, and gear_ratio do not exist in mydata.dta.

To see the result of each comparison, we could append the all option to our command.

```
. cf _all using mydata, all
        make:  match
       price:  match
         mpg:  2 mismatches
       rep78:  match
    headroom:  does not exist in using
       trunk:  match
      weight:  match
      length:  match
        turn:  match
displacement:  does not exist in using
  gear_ratio:  does not exist in using
     foreign:  match
r(9);
```

For more details on the mismatches, we can use the verbose option.

```
. cf _all using mydata, verbose
         mpg:  2 mismatches
               obs  1. 22 in master; 33 in using
               obs  2. 17 in master; 33 in using
    headroom:  does not exist in using
displacement:  does not exist in using
  gear_ratio:  does not exist in using
r(9);
```

This example shows us exactly which two observations for mpg differ, as well as the value stored in each dataset.

◁

▷ Example 2

We want to compare a group of variables in the dataset in memory against the same group of variables in mydata.dta.

```
. cf mpg headroom using mydata
         mpg: 2 mismatches
    headroom: does not exist in using
r(9);
```

◁

Saved Results

cf saves the following in r():

> Macros
> r(Nsum)   number of differences

Methods and Formulas

cf is implemented as an ado-file.

If you are using Small Stata, you may get the error "too many variables" when you stipulate _all and have many variables in your dataset. (This will not happen if you are using Stata/MP, Stata/SE, or Stata/IC.) If this happens, you will have to perform the comparison with groups of variables. See example 2 for details about how to do this.

Acknowledgment

Speed improvements in cf were based on code written by David Kantor.

Reference

Gleason, J. R. 1995. dm36: Comparing two Stata datasets. *Stata Technical Bulletin* 28: 10–13. Reprinted in *Stata Technical Bulletin Reprints*, vol. 5, pp. 39–43.

Also See

[D] **compare** — Compare two variables

Title

> **checksum** — Calculate checksum of file

Syntax

> checksum *filename* [, *options*]
>
> <u>set</u> checksum { on | off } [, <u>perm</u>anently]

options	description
save	save output to *filename*.sum; default is to display a report
replace	may overwrite *filename*.sum; use with save
<u>sav</u>ing(*filename2* [, replace])	save output to *filename2*; alternative to save

Description

checksum creates *filename*.sum files for later use by Stata when it reads files over a network. These optional files are used to reduce the chances of corrupted files going undetected. Whenever Stata reads file *filename.suffix* over a network, whether it be by use, net, update, etc., it also looks for *filename*.sum. If Stata finds that file, Stata reads it and uses its contents to verify that the first file was received without error. If there are errors, Stata informs the user that the file could not be read.

set checksum on tells Stata to verify that files downloaded over a network have been received without error.

set checksum off, which is the default, tells Stata to bypass the file verification.

❏ Technical Note

checksum calculates a CRC checksum following the POSIX 1003.2 specification and displays the file size in bytes. checksum produces the same results as the Unix cksum command. Comparing the checksum of the original file with the received file guarantees the integrity of the received file.

When comparing Stata's checksum results with those of Unix, do not confuse Unix's sum and cksum commands. Unix's cksum and Stata's checksum use a more robust algorithm than that used by Unix's sum. In some Unix operating systems, there is no cksum command, and the more robust algorithm is obtained by specifying an option with sum.

❏

Options

save saves the output of the checksum command to the ASCII file *filename*.sum. The default is to display a report but not create a file.

replace is for use with save; it permits Stata to overwrite an existing *filename*.sum file.

saving(*filename2* [, replace]) is an alternative to save. It saves the output in the specified filename. You must supply a file extension if you want one, as none is assumed.

permanently specifies that, in addition to making the change right now, the checksum setting be remembered and become the default setting when you invoke Stata.

Remarks

▷ Example 1

Say that you wish to put a dataset on your homepage so that colleagues can use it over the Internet by typing

```
. use http://www.myuni.edu/department/~joe/mydata
```

mydata.dta is important, and, even though the chances of the file mydata.dta being corrupted by the Internet are small, you wish to guard against that. The solution is to create the checksum file named mydata.sum and place that on your homepage. Your colleagues need type nothing different, but now Stata will verify that all goes well. When they use the file, they will see either

```
. use http://www.myuni.edu/department/~joe/mydata
(important data from joe)
```

or

```
. use http://www.myuni.edu/department/~joe/mydata
file transmission error (checksums do not match)
http://www.myuni.edu/department/~joe/mydata.dta not downloaded
r(639);
```

To make the checksum file, change to the directory where the file is located and type

```
. checksum mydata.dta, save
Checksum for mydata.dta = 263508742, size = 4052
file mydata.sum saved
```

◁

▷ Example 2

Let's use checksum on the auto dataset that is shipped with Stata. We will load the dataset and save it to our current directory.

```
. use http://www.stata-press.com/data/r10/auto
(1978 Automobile Data)
. save auto
file auto.dta saved
. checksum auto.dta
Checksum for auto.dta = 2043611174, size = 5949
```

We see the report produced by checksum, but we decide to save this information to a file.

```
. checksum auto.dta, save
. type auto.sum
1 5949 2043611174
```

The first number is the version number (possibly used for future releases). The second number is the file's size in bytes, which can be used with the checksum value to ensure that the file transferred without corruption. The third number is the checksum value. Although two different files can have the same checksum value, two files with the same checksum value almost certainly could not have the same file size.

This example is admittedly artificial. Typically, you would use `checksum` to verify that no file transmission error occurred during a web download. If you want to verify that your own data are unchanged, using `datasignature` is better; see [D] **datasignature**.

◁

Saved Results

`checksum` saves the following in `r()`:

Scalars
r(version)	checksum version number
r(filelen)	length of file, in bytes
r(checksum)	checksum value

Also See

[D] **use** — Use Stata dataset

[R] **net** — Install and manage user-written additions from the Internet

[D] **datasignature** — Determine whether data have changed

Title

clear — Clear memory

Syntax

```
clear

clear [ mata | results | programs | ado ]

clear [ all | * ]
```

Description

clear, by itself, removes data and value labels from memory and is equivalent to typing

```
. version 10
. drop _all              (see [D] drop)
. label drop _all        (see [D] label)
```

clear mata removes Mata functions and objects from memory and is equivalent to typing

```
. version 10
. mata: mata clear       (see [M-3] mata clear)
```

clear results eliminates saved results from memory and is equivalent to typing

```
. version 10
. return clear           (see [P] return)
. ereturn clear          (see [P] return)
. sreturn clear          (see [P] return)
. _return drop _all      (see [P] _return)
```

clear programs eliminates all programs from memory and is equivalent to typing

```
. version 10
. program drop _all      (see [P] program)
```

clear ado eliminates all automatically loaded ado-file programs from memory (but not programs defined interactively or by do-files). It is equivalent to typing

```
. version 10
. program drop _allado   (see [P] program)
```

clear all and clear * are synonyms. They remove all data, value labels, matrices, scalars, constraints, clusters, saved results, sersets, and Mata functions and objects from memory. They also close all open files and postfiles, clear the class system, close any open Graph windows and dialog boxes, drop all programs from memory, and reset all timers to zero. They are equivalent to typing

```
. version 10
. drop _all              (see [D] drop)
. label drop _all        (see [D] label)
. matrix drop _all       (see [P] matrix utility)
. scalar drop _all       (see [P] scalar)
. constraint drop _all   (see [R] constraint)
. cluster drop _all      (see [MV] cluster utility)
. file close _all        (see [P] file)
. postutil clear         (see [P] postfile)
```

31

. _return drop _all	(see [P] **_return**)
. discard	(see [P] **discard**)
. program drop _all	(see [P] **program**)
. timer clear	(see [P] **timer**)
. mata: mata clear	(see [M-3] **mata clear**)

Remarks

You can clear the entire dataset without affecting macros and programs by typing `clear`. You can also type `clear all`. This command has the same result as `clear` by itself but also clears matrices, scalars, constraints, clusters, saved results, sersets, Mata, the class system, and programs; closes all open files and postfiles; closes all open Graph windows and dialog boxes; and resets all timers to zero.

▷ Example 1

We load the `bpwide` dataset to correct a mistake in the data.

```
. use http://www.stata-press.com/data/r10/bpwide
(fictional blood pressure data)
. list in 1/5
```

	patient	sex	agegrp	bp_bef~e	bp_after
1.	1	Male	30-45	143	153
2.	2	Male	30-45	163	170
3.	3	Male	30-45	153	168
4.	4	Male	30-45	153	142
5.	5	Male	30-45	146	141

```
. replace bp_after = 145 in 3
(1 real change made)
```

We made another mistake. We meant to change the value of `bp_after` in observation 4. It is easiest to begin again.

```
. clear
. use http://www.stata-press.com/data/r10/bpwide
(fictional blood pressure data)
```

◁

Methods and Formulas

`clear` is implemented as an ado-file.

Also See

[D] **drop** — Eliminate variables or observations

[U] **11 Language syntax**

[U] **13 Functions and expressions**

Title

clonevar — Clone existing variable

Syntax

clonevar *newvar* = *varname* $\left[\textit{if} \right]$ $\left[\textit{in} \right]$

Description

clonevar generates *newvar* as an exact copy of an existing variable, *varname*, with the same storage type, values, and display format as *varname*. *varname*'s variable label, value labels, notes, and characteristics will also be copied.

Remarks

clonevar has various possible uses. Programmers may desire that a temporary variable appear to the user exactly like an existing variable. Interactively, you might want a slightly modified copy of an original variable, so the natural starting point is a clone of the original.

▷ Example 1

We have a dataset containing information on modes of travel. These data contain a variable named mode that identifies each observation as a specific mode of travel: air, train, bus, or car.

```
. use http://www.stata-press.com/data/r10/travel

. describe mode

              storage  display   value
variable name  type    format    label    variable label

mode           byte    %8.0g     travel   travel mode alternatives
. label list travel
travel:
          1 air
          2 train
          3 bus
          4 car
```

To create an identical variable identifying only observations that contain air or train, we could use clonevar with an if qualifier.

```
. clonevar airtrain = mode if mode == 1 | mode == 2
(420 missing values generated)

. describe mode airtrain

              storage  display   value
variable name  type    format    label    variable label

mode           byte    %8.0g     travel   travel mode alternatives
airtrain       byte    %8.0g     travel   travel mode alternatives
```

33

. list mode airtrain in 1/5

	mode	airtrain
1.	air	air
2.	train	train
3.	bus	.
4.	car	.
5.	air	air

The new variable `airtrain` has the same storage type, display format, value label, and variable label as mode. If mode had any characteristics or notes attached to it, they would have been applied to the new `airtrain` variable, too. The only differences in the two variables are their names and values for bus and car.

◁

❑ Technical Note

The `if` qualifier used with the `clonevar` command in example 1 referred to the values of mode as 1 and 2. Had we wanted to refer to the values by their associated value labels, we could have typed

. `clonevar airtrain = mode if mode == "air":travel | mode == "train":travel`

For more details, see [U] **13.9 Label values**.

❑

Methods and Formulas

`clonevar` is implemented as an ado-file.

Acknowledgments

`clonevar` was written by Nicholas J. Cox, Durham University, who, in turn, thanks Michael Blasnik, M. Blasnik & Associates, and Ken Higbee, StataCorp, for very helpful comments on a precursor of this command.

Also See

[D] **generate** — Create or change contents of variable

[D] **separate** — Create separate variables

Title

> **codebook** — Describe data contents

Syntax

codebook [*varlist*] [*if*] [*in*] [, *options*]

options	description
Options	
<u>a</u>ll	print complete report without missing values
<u>h</u>eader	print dataset name and last saved date
<u>n</u>otes	print any notes attached to variables
<u>m</u>v	report pattern of missing values
<u>t</u>abulate(#)	set tables/summary statistics threshold; default is tabulate(9)
<u>p</u>roblems	report potential problems in dataset
<u>d</u>etail	display detailed report on the variables; only with problems
<u>c</u>ompact	display compact report on the variables
dots	display a dot for each variable processed; only with compact
Languages	
<u>lang</u>uages [(*namelist*)]	use with multilingual datasets; see [D] **label language** for details

Description

codebook examines the variable names, labels, and data to produce a codebook describing the dataset.

Options

⌐ Options ⌐

all is equivalent to specifying the header and notes options. It provides a complete report, which excludes only performing mv.

header adds a header to the top of the output, which lists the dataset name, the date that the dataset was last saved, etc.

notes lists any notes attached to the variables; see [D] **notes**.

mv specifies that codebook search the data to determine the pattern of missing values. This is a CPU-intensive task.

tabulate(#) specifies the number of unique values of the variables to use to determine whether a variable is categorical or continuous. Missing values are not included in this count. The default is 9; when there are more than nine unique values, the variable is classified as continuous. Extended missing values will be included in the tabulation.

problems specifies that a summary report is produced describing potential problems that have been diagnosed:

- Variables that are labeled with an undefined value label
- Incompletely value-labeled variables
- Variables that are constant, including always missing
- Trailing, trimming, and embedded spaces in string variables
- Noninteger-valued date variables

detail may be specified only with the problems option. It specifies that the detailed report on the variables not be suppressed.

compact specifies that a compact report on the variables be displayed. compact may not be specified with any options other than dots.

dots specifies that a dot be displayed for every variable processed. dots may be specified only with compact.

⌐ Languages ⌐

languages [(*namelist*)] is for use with multilingual datasets; see [D] **label language**. It indicates that the codebook pertains to the languages in *namelist* or to all defined languages if no such list is specified as an argument to languages(). The output of codebook lists the data label and variable labels in these languages and which value labels are attached to variables in these languages.

Problems are diagnosed in all these languages, as well. The problem report does not provide details in which language problems occur. We advise you to rerun codebook for problematic variables; specify detail to produce the problem report again.

If you have a multilingual dataset but do not specify languages(), all output, including the problem report, is shown in the "active" language.

Remarks

codebook, without arguments, is most usefully combined with log to produce a printed listing for enclosure in a notebook documenting the data; see [U] **15 Printing and preserving output**. codebook is, however, also useful interactively, as you can specify one or a few variables.

▷ Example 1

codebook examines the data in producing its results. For variables that codebook thinks are continuous, it presents the mean, the standard deviation, and the 10th, 25th, 50th, 75th, and 90th percentiles. For variables that it thinks are categorical, it presents a tabulation. In part, codebook makes this determination by counting the number of unique values of the variable. If the number is nine or fewer, codebook reports a tabulation; otherwise, it reports summary statistics.

codebook distinguishes the standard missing values (.) and the extended missing values (.a/.z, denoted by .*). If extended missing values are found, codebook reports the number of distinct missing value codes that occurred in that variable. Missing values are ignored with the tabulate option when determining whether a variable is treated as continuous or categorical.

```
. use http://www.stata-press.com/data/r10/educ3
(ccdb46, 52-54)

. codebook fips division, all
                    Dataset:  http://www.stata-press.com/data/r10/educ3.dta
                 Last saved:  6 Mar 2007 22:20

                      Label:  ccdb46, 52-54
        Number of variables:  42
     Number of observations:  956
                       Size:  149,136 bytes ignoring labels, etc.

_dta:
  1.  confirmed data with steve on 7/22
```

fips					state/place code
type:	numeric (long)				
range:	[10060,560050]			units:	1
unique values:	956			missing .:	0/956
mean:	256495				
std. dev:	156998				
percentiles:	10%	25%	50%	75%	90%
	61462	120426	252848	391360	482530

division			Census Division
type:	numeric (int)		
label:	division		
range:	[1,9]	units:	0
unique values:	9	missing .:	4/956
unique mv codes:	2	missing .*:	2/956

```
              tabulation:  Freq.   Numeric  Label
                             69         1   N. Eng.
                             97         2   Mid Atl
                            202         3   E.N.C.
                             78         4   W.N.C.
                            115         5   S. Atl.
                             46         6   E.S.C.
                             89         7   W.S.C.
                             59         8   Mountain
                            195         9   Pacific
                              4         .
                              2        .a
```

Since division has nine unique nonmissing values, codebook reported a tabulation. If division had contained one more unique nonmissing value, codebook would have switched to reporting summary statistics, unless we had included the tabulate(#) option.

◁

▷ Example 2

The mv option is useful. It instructs codebook to search the data to determine patterns of missing values. Different kinds of missing values are not distinguished in the patterns.

```
. use http://www.stata-press.com/data/r10/citytemp
(City Temperature Data)

. codebook cooldd heatdd tempjan tempjuly, mv
```

---

cooldd   Cooling degree days

```
                type:   numeric (int)

               range:   [0,4389]                     units:  1
       unique values:   438                    missing .:  3/956

                mean:   1240.41
            std. dev:   937.668

         percentiles:        10%       25%       50%       75%       90%
                             411       615       940      1566      2761

     missing values:        heatdd==mv <-> cooldd==mv
                            tempjan==mv --> cooldd==mv
                           tempjuly==mv --> cooldd==mv
```

---

heatdd   Heating degree days

```
                type:   numeric (int)

               range:   [0,10816]                    units:  1
       unique values:   471                    missing .:  3/956

                mean:   4425.53
            std. dev:   2199.6

         percentiles:        10%       25%       50%       75%       90%
                            1510      2460      4950      6232      6919

     missing values:        cooldd==mv <-> heatdd==mv
                            tempjan==mv --> heatdd==mv
                           tempjuly==mv --> heatdd==mv
```

---

tempjan   Average January temperature

```
                type:   numeric (float)

               range:   [2.2,72.6]                    units:  .1
       unique values:   310                    missing .:  2/956

                mean:   35.749
            std. dev:   14.1881

         percentiles:        10%       25%       50%       75%       90%
                            20.2      25.1      31.3      47.8      55.1

     missing values:       tempjuly==mv <-> tempjan==mv
```

---

tempjuly   Average July temperature

```
                type:   numeric (float)

               range:   [58.099998,93.599998]         units:  0
       unique values:   196                    missing .:  0/956
      unique mv codes:  1                     missing .*:  2/956

                mean:   75.0538
            std. dev:   5.49504

         percentiles:        10%       25%       50%       75%       90%
                            68.8      71.8     74.25      78.7      82.3

     missing values:       tempjan==mv <-> tempjuly==mv
```

codebook reports that if `tempjan` is missing, `tempjuly` is also missing, and vice versa. In the output for the `cooldd` variable, codebook also reports that the pattern of missing values is the same for `cooldd` and `heatdd`. In both cases, the correspondence is indicated with "<->".

For `cooldd`, codebook also states that "`tempjan==mv --> cooldd==mv`". The one-way arrow means that a missing `tempjan` value implies a missing `cooldd` value but that a missing `cooldd` value does not necessarily imply a missing `tempjan` value.   ◁

Another feature of codebook—this one for numeric variables—is that it can determine the units of the variable. For instance, in the example above, `tempjan` and `tempjuly` both have units of .1, meaning that temperature is recorded to tenths of a degree. codebook handles precision considerations in making this determination (`tempjan` and `tempjuly` are floats; see [U] **13.10 Precision and problems therein**). If we had a variable in our dataset recorded in 100s (e.g., 21,500 or 36,800), codebook would have reported the units as 100. If we had a variable that took on only values divisible by 5 (5, 10, 15, etc.), codebook would have reported the units as 5.

▷ Example 3

We can use the `label language` command (see [D] **label language**) and the `label` command (see [D] **label**) to create German value labels for our auto dataset. These labels are reported by codebook:

```
. use http://www.stata-press.com/data/r10/auto
(1978 Automobile Data)
. label language en, rename
(language default renamed en)
. label language de, new
(language de now current language)
. label data "1978 Automobile Daten"
. label variable foreign "Art Auto"
. label values foreign origin_de
. label define origin_de 0 "Innen" 1 "Ausländish"
. codebook foreign
```

```
foreign                                                              Art Auto
-------------------------------------------------------------------------------

                 type:  numeric (byte)
                label:  origin_de

                range:  [0,1]                         units:  1
        unique values:  2                       missing .:  0/74

           tabulation:  Freq.   Numeric  Label
                           52         0  Innen
                           22         1  Ausländish
. codebook foreign, languages(en de)
-------------------------------------------------------------------------------

foreign         in en:  Car type
                in de:  Art Auto
-------------------------------------------------------------------------------

                 type:  numeric (byte)
          label in en:  origin
          label in de:  origin_de

                range:  [0,1]                         units:  1
        unique values:  2                       missing .:  0/74

           tabulation:  Freq. Numeric   origin       origin_de
                           52        0  Domestic     Innen
                           22        1  Foreign      Ausländish
```

With the `languages()` option, the value labels are shown in the specified active and available languages.

◁

▷ Example 4

`codebook, compact` summarizes the variables in your dataset, including variable labels. It is an alternative to the `summarize` command.

```
. use http://www.stata-press.com/data/r10/auto
(1978 Automobile Data)

. codebook, compact
```

Variable	Obs	Unique	Mean	Min	Max	Label
make	74	74	.	.	.	Make and Model
price	74	74	6165.257	3291	15906	Price
mpg	74	21	21.2973	12	41	Mileage (mpg)
rep78	69	5	3.405797	1	5	Repair Record 1978
headroom	74	8	2.993243	1.5	5	Headroom (in.)
trunk	74	18	13.75676	5	23	Trunk space (cu. ft.)
weight	74	64	3019.459	1760	4840	Weight (lbs.)
length	74	47	187.9324	142	233	Length (in.)
turn	74	18	39.64865	31	51	Turn Circle (ft.)
displacement	74	31	197.2973	79	425	Displacement (cu. in.)
gear_ratio	74	36	3.014865	2.19	3.89	Gear Ratio
foreign	74	2	.2972973	0	1	Car type

```
. summarize
```

Variable	Obs	Mean	Std. Dev.	Min	Max
make	0				
price	74	6165.257	2949.496	3291	15906
mpg	74	21.2973	5.785503	12	41
rep78	69	3.405797	.9899323	1	5
headroom	74	2.993243	.8459948	1.5	5
trunk	74	13.75676	4.277404	5	23
weight	74	3019.459	777.1936	1760	4840
length	74	187.9324	22.26634	142	233
turn	74	39.64865	4.399354	31	51
displacement	74	197.2973	91.83722	79	425
gear_ratio	74	3.014865	.4562871	2.19	3.89
foreign	74	.2972973	.4601885	0	1

◁

▷ Example 5

When `codebook` determines that neither a tabulation nor a listing of summary statistics is appropriate, for instance, for a string variable or for a numeric variable taking on many labeled values, it reports a few examples instead.

```
. use http://www.stata-press.com/data/r10/funnyvar

. codebook name
```

```
name                                                          (unlabeled)
```

```
              type:  string (str5), but longest is str3
      unique values:  10                    missing "":  0/10
           examples:  "1 0"
                      "3"
                      "5"
                      "7"
            warning:  variable has embedded blanks
```

codebook is also on the lookout for common problems that might cause you to make errors when dealing with the data. For string variables, this includes leading, embedded, and trailing blanks. In the output above, codebook informed us that name includes embedded blanks. If name had leading or trailing blanks, it would have mentioned that, too.

When variables are value labeled, codebook performs two checks. First, if a value label *labname* is associated with a variable, codebook checks whether *labname* is actually defined. Second, it checks whether all values are value labeled. Partial labeling of a variable may mean that the label was defined incorrectly (for instance, the variable has values 0 and 1, but the value label maps 1 to "male" and 2 to "female") or that the variable was defined incorrectly (e.g., a variable gender with three values). codebook checks whether date variables are integer valued.

If the option problems is specified, codebook does not provide detailed descriptions of each variable but reports only the potential problems in the data.

```
. codebook, problems
   Potential problems in dataset    http://www.stata-press.com/data/r10/funnyvar.dta
         potential problem    variables

    constant (or all missing) vars    human planet
        vars with nonexisting label    educ
         incompletely labeled vars    gender
    strvars that may be compressed    name address city country planet
     string vars with leading blanks    city country
    string vars with trailing blanks    planet
    string vars with embedded blanks    name address
        noninteger-valued date vars    birthdate
```

◁

In the example above, codebook, problems reported various potential problems with the dataset. These problems include

Constant variables, including variables that are always missing

Variables that are constant, taking the same value in all observations, or that are always missing, are often superfluous. Such variables, however, may also indicate problems. For instance, variables that are always missing may occur when importing data with an incorrect input specification. Such variables may also occur if you generate a new variable for a subset of the data, selected with an expression that is false for all observations.

Advice: carefully check the origin of constant variables. If you are saving a constant variable, be sure to compress the variable to use minimal storage.

Variables with nonexisting value labels

Stata treats value labels as separate objects that can be attached to one or more variables. A problem may arise if variables are linked to value labels that are not yet defined or an incorrect value label name was used.

Advice: attach the correct value label, or label define the value label; see [D] **label**.

Incompletely labeled variables

A variable is called "incompletely value labeled" if the variable is value labeled but no mapping is provided for some values of the variable. An example is a variable with values 0, 1, and 2 and value labels for 1, 2, and 3. This situation usually indicates an error, either in the data or in the value label.

Advice: change either the data or the value label.

String variables that may be compressed

The storage space used by a string variable is determined by its data type; see [D] **data types**. For instance, the storage type str20 indicates that 20 bytes are used per observation. If the declared storage type exceeds your requirements, memory and disk space is wasted.

Advice: use compress to store the data as compactly as possible.

String variables with leading or trailing blanks

In most applications, leading and trailing spaces do not affect the meaning of variables but are probably side effects from importing the data or from data manipulation. Spurious leading and trailing spaces force Stata to use more memory than required. Also, manipulating strings with leading and trailing spaces is harder.

Advice: remove leading and trailing blanks from a string variable s by

```
replace s = trim(s)
```

See [D] **functions**.

String variables with embedded blanks

String variables with embedded blanks are often appropriate; however, sometimes they indicate problems importing the data.

Advice: verify that blanks are meaningful in the variables.

Noninteger-valued date variables

Stata's date formats were designed for use with integer values but will work with noninteger values.

Advice: carefully inspect the nature of the noninteger values. In case noninteger values in a variable are the consequence of roundoff error, you may want to round the variable to the nearest integer.

```
replace time = round(time)
```

Of course, more problems not reported by codebook are possible. These might include

Numerical data stored as strings

After importing data into Stata, you may discover that some string variables can actually be interpreted as numbers. Stata can do much more with numerical data than with string data. Moreover, string representation usually makes less efficient use of computer resources. destring will convert string variables to numeric.

A string variable may contain a "field" with numeric information. An example is an address variable that contains the street name followed by the house number. The Stata string functions can extract the relevant substring.

Categorical variables stored as strings

Most statistical commands do not allow string variables. Moreover, string variables that take only a limited number of distinct values are an inefficient storage method. Use value-labeled numeric values instead. These are easily created with encode.

Duplicate observations

See [D] **duplicates**.

Observations that are always missing

Drop observations that are missing for all variables in *varlist* using the rownonmiss() egen function:

    egen nobs = rownonmiss(varlist)

    drop if nobs==0

Specify _all for *varlist* if only observations that are always missing should be dropped.

Saved Results

codebook saves the following lists of variables with potential problems in r():

Macros
r(cons)	constant (or missing)
r(labelnotfound)	undefined value labeled
r(notlabeled)	value labeled but with unlabeled categories
r(str_type)	compressible
r(str_leading)	leading blanks
r(str_trailing)	trailing blanks
r(str_embedded)	embedded blanks
r(realdate)	noninteger dates

Methods and Formulas

codebook is implemented as an ado-file.

Also See

[D] **describe** — Describe data in memory or in file

[D] **inspect** — Display simple summary of data's attributes

[D] **labelbook** — Label utilities

[D] **notes** — Place notes in data

[D] **split** — Split string variables into parts

[U] **15 Printing and preserving output**

Title

> **collapse** — Make dataset of summary statistics

Syntax

> collapse *clist* [*if*] [*in*] [*weight*] [, *options*]

where *clist* is either

$$\left[(stat) \right] \ varlist \left[\ \left[(stat) \right] \ \dots \ \right]$$

$$\left[(stat) \right] \ target\_var=varname \left[target\_var=varname \ \dots \right] \left[\ \left[(stat) \right] \ \dots \ \right]$$

or any combination of the *varlist* and *target_var* forms, and *stat* is one of

mean	means (default)	sum	sums
median	medians	rawsum	sums, ignoring optionally specified weight
p1	1st percentile	count	number of nonmissing observations
p2	2nd percentile	max	maximums
...	3rd–49th percentiles	min	minimums
p50	50th percentile (same as median)	iqr	interquartile range
...	51st–97th percentiles	first	first value
p98	98th percentile	last	last value
p99	99th percentile	firstnm	first nonmissing value
sd	standard deviations	lastnm	last nonmissing value

If *stat* is not specified, **mean** is assumed.

options	description
Options	
by(*varlist*)	groups over which *stat* is to be calculated
cw	casewise deletion instead of all possible observations
† fast	do not restore the original dataset should the user press *Break*; programmer's command

---

† fast is not shown in the dialog box.

varlist and *varname* in *clist* may contain time-series operators; see [U] **11.4.3 Time-series varlists**.

aweights, fweights, iweights, and pweights are allowed; see [U] **11.1.6 weight**, and see *Weights* below. pweights may not be used with statistic sd.

Examples:

```
. collapse age educ income, by(state)
. collapse (mean) age educ (median) income, by(state)
. collapse (mean) age educ income (median) medinc=income, by(state)
. collapse (p25) gpa [fw=number], by(year)
```

44

Description

collapse converts the dataset in memory into a dataset of means, sums, medians, etc. *clist* must refer to numeric variables exclusively.

Note: see [D] **contract** if you want to collapse to a dataset of frequencies.

Options

> Options

by(*varlist*) specifies the groups over which the means, etc., are to be calculated. If this option is not specified, the resulting dataset will contain 1 observation. If it is specified, *varlist* may refer to either string or numeric variables.

cw specifies casewise deletion. If cw is not specified, all possible observations are used for each calculated statistic.

The following option is available with collapse but is not shown in the dialog box:

fast specifies that collapse not restore the original dataset should the user press *Break*. fast is intended for use by programmers.

Remarks

collapse takes the dataset in memory and creates a new dataset containing summary statistics of the original data. collapse adds meaningful variable labels to the variables in this new dataset. Since the syntax diagram for collapse makes using it appear more complicated than it is, collapse is best explained with examples.

Remarks are presented under the following headings:

> *Introductory examples*
> *Variable-wise or casewise deletion*
> *Weights*
> *A final example*

Introductory examples

> ## Example 1

Consider the following artificial data on the grade-point average (gpa) of college students:

(*Continued on next page*)

```
. use http://www.stata-press.com/data/r10/college
. describe
Contains data from http://www.stata-press.com/data/r10/college.dta
  obs:           12
  vars:           4                        3 Jan 2007 12:05
  size:         168 (99.9% of memory free)
```

variable name	storage type	display format	value label	variable label
gpa	float	%9.0g		gpa for this year
hour	int	%9.0g		Total academic hours
year	int	%9.0g		1 = freshman, 2 = sophomore, 3 = junior, 4 = senior
number	int	%9.0g		number of students

```
Sorted by: year
. list, sep(4)
```

	gpa	hour	year	number
1.	3.2	30	1	3
2.	3.5	34	1	2
3.	2.8	28	1	9
4.	2.1	30	1	4
5.	3.8	29	2	3
6.	2.5	30	2	4
7.	2.9	35	2	5
8.	3.7	30	3	4
9.	2.2	35	3	2
10.	3.3	33	3	3
11.	3.4	32	4	5
12.	2.9	31	4	2

To obtain a dataset containing the 25th percentile of grade-point averages for each year, we type

```
. collapse (p25) gpa [fw=number], by(year)
```

We used frequency weights.

Next we want to create a dataset containing the mean of grade-point average and hour for each year. We do not have to type (mean) to specify that we want the mean because the mean is reported by default.

```
. use http://www.stata-press.com/data/r10/college, clear
. collapse gpa hour [fw=number], by(year)
. list
```

	year	gpa	hour
1.	1	2.788889	29.44444
2.	2	2.991667	31.83333
3.	3	3.233333	32.11111
4.	4	3.257143	31.71428

Now we want to create a dataset containing the mean and median of gpa and hour, and we want the median of gpa and hour to be stored as variables medgpa and medhour, respectively.

```
. use http://www.stata-press.com/data/r10/college, clear
. collapse (mean) gpa hour (median) medgpa=gpa medhour=hour [fw=num], by(year)
. list
```

	year	gpa	hour	medgpa	medhour
1.	1	2.788889	29.44444	2.8	29
2.	2	2.991667	31.83333	2.9	30
3.	3	3.233333	32.11111	3.3	33
4.	4	3.257143	31.71428	3.4	32

Here we want to create a dataset containing a count of gpa and hour and the minimums of gpa and hour. The minimums of gpa and hour will be stored as variables mingpa and minhour, respectively.

```
. use http://www.stata-press.com/data/r10/college, clear
. collapse (count) gpa hour (min) mingpa=gpa minhour=hour [fw=num], by(year)
. list
```

	year	gpa	hour	mingpa	minhour
1.	1	18	18	2.1	28
2.	2	12	12	2.5	29
3.	3	9	9	2.2	30
4.	4	7	7	2.9	31

Now we replace the values of gpa in 3 of the observations with missing values.

```
. use http://www.stata-press.com/data/r10/college, clear
. replace gpa = . in 2/4
(3 real changes made, 3 to missing)
. list, sep(4)
```

	gpa	hour	year	number
1.	3.2	30	1	3
2.	.	34	1	2
3.	.	28	1	9
4.	.	30	1	4
5.	3.8	29	2	3
6.	2.5	30	2	4
7.	2.9	35	2	5
8.	3.7	30	3	4
9.	2.2	35	3	2
10.	3.3	33	3	3
11.	3.4	32	4	5
12.	2.9	31	4	2

If we now want to list the data containing the mean of gpa and hour for each year, collapse uses all observations on hour for year $= 1$, even though gpa is missing for observations 1–3.

```
. collapse gpa hour [fw=num], by(year)
. list
```

	year	gpa	hour
1.	1	3.2	29.44444
2.	2	2.991667	31.83333
3.	3	3.233333	32.11111
4.	4	3.257143	31.71428

If we repeat this process but specify the `cw` option, `collapse` ignores all observations that have missing values.

```
. use http://www.stata-press.com/data/r10/college, clear
. replace gpa = . in 2/4
(3 real changes made, 3 to missing)
. collapse (mean) gpa hour [fw=num], by(year) cw
. list
```

	year	gpa	hour
1.	1	3.2	30
2.	2	2.991667	31.83333
3.	3	3.233333	32.11111
4.	4	3.257143	31.71428

◁

▷ Example 2

We have individual-level data from a census in which each observation is a person. Among other variables, the dataset contains the numeric variables `age`, `educ`, and `income` and the string variable `state`. We want to create a 50-observation dataset containing the means of age, education, and income for each state.

```
. collapse age educ income, by(state)
```

The resulting dataset contains means because `collapse` assumes that we want means if we do not specify otherwise. To make this explicit, we could have typed

```
. collapse (mean) age educ income, by(state)
```

Had we wanted the mean for `age` and `educ` and the median for `income`, we could have typed

```
. collapse (mean) age educ (median) income, by(state)
```

or, if we had wanted the mean for `age` and `educ` and both the mean and the median for `income`, we could have typed

```
. collapse (mean) age educ income (median) medinc=income, by(state)
```

This last dataset will contain three variables containing means—age, educ, and income—and one variable containing the median of income—medinc. Because we typed (median) medinc=income, Stata knew to find the median for income and to store those in a variable named medinc. This renaming convention is necessary in this example because a variable named income containing the mean is also being created.

◁

Variable-wise or casewise deletion

▷ Example 3

Let us assume that in our census data, we have 25,000 persons for whom age is recorded but only 15,000 for whom income is recorded; that is, income is missing for 10,000 observations. If we want summary statistics for age and income, collapse will, by default, use all 25,000 observations when calculating the summary statistics for age. If we prefer that collapse use only the 15,000 observations for which income is not missing, we can specify the cw (casewise) option:

```
. collapse (mean) age income (median) medinc=income, by(state) cw
```

◁

Weights

collapse allows all four weight types; the default is aweights. Weight normalization affects only the sum, count, and sd statistics.

Here are the definitions for count and sum with weights:

count:
    unweighted: _N, the number of physical observations
    aweight: _N, the number of physical observations
    fweight, iweight, pweight: $W = \sum w_j$, the sum of the user-specified weights
sum:
    unweighted: $\sum x_j$, the sum of the variable
    aweight: $\sum v_j x_j$; $v_j = (w_j$ normalized to sum to _N$)$
    fweight, iweight, pweight: $\sum w_j x_j$

The sd statistic with weights returns the bias-corrected standard deviation, which is based on the factor $\sqrt{N/(N-1)}$, where N is the number of observations. sd is not allowed with pweighted data. Otherwise, sd is changed by the weights through the computation of the count (N), as outlined above.

For instance, consider a case in which there are 25 physical observations in the dataset and a weighting variable that sums to 57. In the unweighted case, the weight is not specified, and $N = 25$. In the analytically weighted case, N is still 25; the scale of the weight is irrelevant. In the frequency-weighted case, however, $N = 57$, the sum of the weights.

The rawsum statistic with aweights ignores the weight, with one exception: observations with zero weight will not be included in the sum.

▷ Example 4

Using our same census data, suppose that instead of starting with individual-level data and aggregating to state level, we started with state-level data and wanted to aggregate to the region level. Also assume that our dataset contains pop, the population of each state.

To obtain unweighted means and medians of age and income, by region, along with the total population, we could type

```
. collapse (mean) age income (median) medage=age medinc=income (sum) pop,
>     by(region)
```

To obtain weighted means and medians of age and income, by region, along with the total population and using frequency weights, we could type

```
. collapse (mean) age income (median) medage=age medinc=income (count) pop
>     [fweight=pop], by(region)
```

Note: specifying (sum) pop would not have worked because that would have yielded the pop-weighted sum of pop. Specifying (count) age would have worked as well as (count) pop because count merely counts the number of nonmissing observations. The counts here, however, are frequency-weighted and equal the sum of pop.

To obtain the same mean and medians as above, but using analytic weights, we could type

```
. collapse (mean) age income (median) medage=age medinc=income (rawsum) pop
>     [aweight=pop], by(region)
```

Note: specifying (count) pop would not have worked because, with analytic weights, count would count numbers of physical observations. Specifying (sum) pop would not have worked because sum would calculate weighted sums (with a normalized weight). The rawsum function, however, ignores the weights and sums only the specified variable, with one exception: observations with zero weight will not be included in the sum. rawsum would have worked as the solution to all three cases.

◁

A final example

▷ Example 5

We have census data containing information on each state's median age, marriage rate, and divorce rate. We want to form a new dataset containing various summary statistics, by region, of the variables:

```
. use http://www.stata-press.com/data/r10/census5, clear
(1980 Census data by state)

. describe
Contains data from http://www.stata-press.com/data/r10/census5.dta
  obs:            50                          1980 Census data by state
  vars:            6                          3 Jan 2007 15:45
  size:         1,800 (98.6% of memory free)
```

variable name	storage type	display format	value label	variable label
state	str14	%14s		State
region	int	%8.0g	cenreg	Census region
pop	long	%10.0g		Population
median_age	float	%9.2f		Median age
marriage_rate	float	%9.0g		
divorce_rate	float	%9.0g		

```
Sorted by:  region

. collapse (median) median_age marriage divorce (mean) avgmrate=marriage
> avgdrate=divorce [aw=pop], by(region)
```

```
. list
```

	region	median~e	marria~e	divorc~e	avgmrate	avgdrate
1.	NE	31.90	.0080657	.0035295	.0081472	.0035359
2.	N Cntrl	29.90	.0093821	.0048636	.0096701	.004961
3.	South	29.60	.0112609	.0065792	.0117082	.0059439
4.	West	29.90	.0089093	.0056423	.0125199	.0063464

```
. describe
Contains data
  obs:            4                          1980 Census data by state
  vars:           6
  size:         104 (98.8% of memory free)
```

	storage	display	value	
variable name	type	format	label	variable label
region	int	%8.0g	cenreg	Census region
median_age	float	%9.2f		(p 50) median_age
marriage_rate	float	%9.0g		(p 50) marriage_rate
divorce_rate	float	%9.0g		(p 50) divorce_rate
avgmrate	float	%9.0g		(mean) marriage_rate
avgdrate	float	%9.0g		(mean) divorce_rate

```
Sorted by:  region
    Note:  dataset has changed since last saved
```

◁

Methods and Formulas

collapse is implemented as an ado-file.

Also See

[D] **contract** — Make dataset of frequencies and percentages

[D] **egen** — Extensions to generate

[D] **statsby** — Collect statistics for a command across a by list

[R] **summarize** — Summary statistics

Title

> **compare** — Compare two variables

Syntax

> compare *varname$_1$* *varname$_2$* $\left[\,if\,\right]$ $\left[\,in\,\right]$

by is allowed; see [D] **by**.

Description

> compare reports the differences and similarities between *varname$_1$* and *varname$_2$*.

Remarks

▷ Example 1

One of the more useful accountings made by compare is the pattern of missing values:

```
. use http://www.stata-press.com/data/r10/fullauto
(Automobile Models)
. compare rep77 rep78
```

	count	minimum	difference average	maximum
rep77<rep78	16	-3	-1.3125	-1
rep77=rep78	43			
rep77>rep78	7	1	1	1
jointly defined	66	-3	-.2121212	1
rep77 missing only	3			
jointly missing	5			
total	74			

We see that both rep77 and rep78 are missing in 5 observations and that rep77 is also missing in 3 more observations.

◁

❑ Technical Note

compare may be used with numeric variables, string variables, or both. When used with string variables, the summary of the differences (minimum, average, maximum) is not reported. When used with string and numeric variables, the breakdown by <, =, and > is also suppressed.

Stata does not normally attach any special meaning to the string ".", but some Stata users use the string "." to mean missing value.

❑

Methods and Formulas

compare is implemented as an ado-file.

Also See

[D] **cf** — Compare two datasets

[D] **codebook** — Describe data contents

[D] **inspect** — Display simple summary of data's attributes

Title

compress — Compress data in memory

Syntax

compress [*varlist*]

Description

compress attempts to reduce the amount of memory used by your data.

Remarks

compress reduces the size of your dataset by considering demoting

doubles	to	longs, ints, or bytes
floats	to	ints or bytes
longs	to	ints or bytes
ints	to	bytes
strings	to	shorter strings

compress leaves your data logically unchanged but (probably) appreciably smaller. compress never makes a mistake, results in loss of precision, or hacks off strings.

▷ Example 1

If you do not specify a *varlist*, compress considers demoting all the variables in your dataset, so typing compress by itself is usually enough:

```
. use http://www.stata-press.com/data/r10/compxmpl
. compress
mpg was float now byte
price was long now int
yenprice was double now long
weight was double now int
make was str26 now str17

. _
```

If there are no compression possibilities, compress does nothing. For instance, typing compress again results in

```
. compress

. _
```

◁

Also See

[D] **recast** — Change storage type of variable

54

Title

> **contract** — Make dataset of frequencies and percentages

Syntax

contract *varlist* [*if*] [*in*] [*weight*] [, *options*]

options	description
Options	
<u>f</u>req(*newvar*)	name of frequency variable; default is _freq
<u>cf</u>req(*newvar*)	create cumulative frequency variable
<u>p</u>ercent(*newvar*)	create percentage variable
<u>cp</u>ercent(*newvar*)	create cumulative percentage variable
float	generate percentage variables as type float
<u>f</u>ormat(*format*)	display format for new percentage variables; default is format(%8.2f)
<u>z</u>ero	include combinations with frequency zero
nomiss	drop observations with missing values

fweights are allowed; see [U] **11.1.6 weight**.

Description

contract replaces the dataset in memory with a new dataset consisting of all combinations of *varlist* that exist in the data and a new variable that contains the frequency of each combination.

Options

> Options

freq(*newvar*) specifies a name for the frequency variable. If no name is specified, _freq is used.

cfreq(*newvar*) specifies a name for the cumulative frequency variable. If no name is specified, no cumulative frequency variable is created.

percent(*newvar*) specifies a name for the percentage variable. If not specified, no percentage variable is created.

cpercent(*newvar*) specifies a name for the cumulative percentage variable. If not specified, no cumulative percentage variable is created.

float specifies that the percentage variables specified by percent() and cpercent() will be generated as variables of type float. If float is not specified, these variables will be generated as variables of type double. All generated variables are compressed to the smallest storage type possible without loss of precision; see [D] **compress**.

format(*format*) specifies a display format for the generated percentage variables specified by percent() and cpercent(). If format() is not specified, these variables will have the display format %8.2f.

zero specifies that combinations with frequency zero be included.

nomiss specifies that observations with missing values on any variable in *varlist* be dropped. If nomiss is not specified, all observations possible are used.

Remarks

contract takes the dataset in memory and creates a new dataset containing all combinations of *varlist* that exist in the data and a new variable that contains the frequency of each combination.

Sometimes you may want to collapse a dataset into frequency form. Several observations that have identical values on one or more variables will be replaced by one such observation, together with the frequency of the corresponding set of values. For example, in certain generalized linear models, the frequency of some combination of values is the response variable, so you would need to produce that response variable. The set of covariate values associated with each frequency is sometimes called a covariate class or covariate pattern. Such collapsing is reversible for the variables concerned, as the original dataset can be reconstituted by using expand (see [D] **expand**) with the variable containing the frequencies of each covariate class.

▷ Example 1

Suppose that we wish to collapse the auto dataset to a set of frequencies of the variables rep78, which takes values 1, 2, 3, 4, and 5, and foreign, which takes values labeled 'Domestic' and 'Foreign'.

```
. use http://www.stata-press.com/data/r10/auto
(1978 Automobile Data)
. contract rep78 foreign
. list
```

	rep78	foreign	_freq
1.	1	Domestic	2
2.	2	Domestic	8
3.	3	Domestic	27
4.	3	Foreign	3
5.	4	Domestic	9
6.	4	Foreign	9
7.	5	Domestic	2
8.	5	Foreign	9
9.	.	Domestic	4
10.	.	Foreign	1

By default, contract uses the variable name _freq for the new variable that contains the frequencies. If _freq is in use, you will be reminded to specify a new variable name via the freq() option.

Specifying the zero option requests that combinations with frequency zero also be listed.

```
. use http://www.stata-press.com/data/r10/auto, clear
(1978 Automobile Data)
. contract rep78 foreign, zero
```

. list

	rep78	foreign	_freq
1.	1	Domestic	2
2.	1	Foreign	0
3.	2	Domestic	8
4.	2	Foreign	0
5.	3	Domestic	27
6.	3	Foreign	3
7.	4	Domestic	9
8.	4	Foreign	9
9.	5	Domestic	2
10.	5	Foreign	9
11.	.	Domestic	4
12.	.	Foreign	1

◁

Methods and Formulas

contract is implemented as an ado-file.

Acknowledgments

contract was written by Nicholas J. Cox of Durham University (Cox 1998). The options cfreq(), percent(), cpercent(), float, and format() were written by Roger Newson of Imperial College London.

Reference

Cox, N. J. 1998. dm59: Collapsing datasets to frequencies. *Stata Technical Bulletin* 44: 2–3. Reprinted in *Stata Technical Bulletin Reprints*, vol. 8, pp. 20–21.

Also See

[D] **expand** — Duplicate observations

[D] **collapse** — Make dataset of summary statistics

[D] **duplicates** — Report, tag, or drop duplicate observations

Title

> **copy** — Copy file from disk or URL

Syntax

> copy *filename*$_1$ *filename*$_2$ $\big[$, *options* $\big]$

filename$_1$ may be a filename or a URL. *filename*$_2$ may *not* be a URL.

Note: double quotes may be used to enclose the filenames, and the quotes must be used if the filename contains embedded blanks.

options	description
public	make *filename*$_2$ readable by all
text	interpret *filename*$_1$ as text file and translate to native text format
† replace	may overwrite *filename*$_2$

† replace is not shown in the dialog box.

Description

> copy copies *filename*$_1$ to *filename*$_2$.

Options

public specifies that *filename*$_2$ be readable by everyone; otherwise, the file will be created according to the default permissions of your operating system.

text specifies that *filename*$_1$ be interpreted as a text file and be translated to the native form of text files on your computer. Computers differ on how end-of-line is recorded: Unix systems record one line feed character, Windows computers record a carriage-return/line feed combination, and Macintosh computers record just a carriage return. text specifies that *filename*$_1$ be examined to determine how it has end of line recorded and that the line-end characters be switched to whatever is appropriate for your computer when the copy is made.

There is no reason to specify text when copying a file already on your computer to a different location because the file would already be in your computer's format.

Do not specify text unless you know that the file is a text file; if the file is binary and you specify text, the copy will be useless. Most word processors produce binary files, not text files. The term *text*, as it is used here, specifies a particular ASCII way of recording textual information.

When other parts of Stata read text files, they do not care how lines are terminated, so there is no reason to translate end-of-line characters on that score. You specify text because you may want to look at the file with other software.

The following option is available with copy but is not shown in the dialog box:

replace specifies that *filename*$_2$ be replaced if it already exists.

Remarks

Examples:

Windows:

```
. copy orig.dta newcopy.dta
. copy "my document" "copy of document"
. copy ..\mydir\doc.txt document\doc.tex
. copy http://www.stata.com/examples/simple.dta simple.dta
. copy http://www.stata.com/examples/simple.txt simple.txt, text
```

Unix:

```
. copy orig.dta newcopy.dta
. copy ../mydir/doc.txt document/doc.tex
. copy http://www.stata.com/examples/simple.dta simple.dta
. copy http://www.stata.com/examples/simple.txt simple.txt, text
```

Macintosh:

```
. copy orig.dta newcopy.dta
. copy "my document" "copy of document"
. copy ../mydir/doc.txt document/doc.tex
. copy http://www.stata.com/examples/simple.dta simple.dta
. copy http://www.stata.com/examples/simple.txt simple.txt, text
```

Also See

[D] **cd** — Change directory

[D] **dir** — Display filenames

[D] **erase** — Erase a disk file

[D] **mkdir** — Create directory

[D] **rmdir** — Remove directory

[D] **shell** — Temporarily invoke operating system

[D] **type** — Display contents of a file

[U] **11.6 File-naming conventions**

Title

> **corr2data** — Create dataset with specified correlation structure

Syntax

> corr2data *newvarlist* [, *options*]

options	description
Main	
clear	replace the current dataset
double	generate variable type as double; default is float
n(#)	# of observations to be generated; default is current number
sds(*vector*)	standard deviations of generated variables
corr(*matrix | vector*)	correlation matrix
cov(*matrix | vector*)	covariance matrix
cstorage(full)	correlation/covariance structure is stored as a symmetric $k \times k$ matrix
cstorage(lower)	correlation/covariance structure is stored as a lower triangular matrix
cstorage(upper)	correlation/covariance structure is stored as an upper triangular matrix
forcepsd	force the covariance/correlation matrix to be positive semidefinite
means(*vector*)	means of generated variables; default is means(0)
Options	
seed(#)	seed for random-number generator

Description

corr2data adds new variables with specified covariance (correlation) structure to the existing dataset or creates a new dataset with a specified covariance (correlation) structure. Singular covariance (correlation) structures are permitted. The purpose of this is to allow you to perform analyses from summary statistics (correlations/covariances and maybe the means) when these summary statistics are all you know and summary statistics are sufficient to obtain results. For example, these summary statistics are sufficient for performing analysis of t tests, variance, principal components, regression, and factor analysis. The recommended process is

```
. clear                        (clear memory)
. corr2data ..., n(#) cov(...) ...   (create artificial data)
. regress ...                  (use artificial data appropriately)
```

However, for factor analyses and principal components, the commands factormat and pcamat allow you to skip the step of using corr2data; see [MV] **factor** and [MV] **pca**.

The data created by corr2data are artificial; they are not the original data, and it is not a sample from an underlying population with the summary statistics specified. See [D] **drawnorm** if you want to generate a random sample. In a sample, the summary statistics will differ from the population values and will differ from one sample to the next.

The dataset `corr2data` creates is suitable for one purpose only: performing analyses when all that is known are summary statistics, and those summary statistics are sufficient for the analysis at hand. The artificial data tricks the analysis command into producing the desired result. The analysis command, being by assumption only a function of the summary statistics, extracts from the artificial data the summary statistics, which are the same summary statistics you specified, and then makes its calculation based on those statistics.

In case you doubt whether the analysis depends only on the specified summary statistics, you can generate different artificial datasets by using different seeds of the random number generator (see option `seed()` below) and compare the results, which should be the same within rounding error.

Options

__Main__

`clear` specifies that it is okay to replace the dataset in memory, even though the current dataset has not been saved on disk.

`double` specifies that the new variables be stored as Stata `doubles`, meaning 8-byte reals. If `double` is not specified, variables are stored as `floats`, meaning 4-byte reals. See [D] **data types**.

`n(#)` specifies the number of observations to be generated; the default is the current number of observations. If `n(#)` is not specified or is the same as the current number of observations, `corr2data` adds the new variables to the existing dataset; otherwise, `corr2data` replaces the dataset in memory.

`sds(`*vector*`)` specifies the standard deviations of the generated variables. `sds()` may not be specified with `cov()`.

`corr(`*matrix* | *vector*`)` specifies the correlation matrix. If neither `corr()` nor `cov()` is specified, the default is orthogonal data.

`cov(`*matrix* | *vector*`)` specifies the covariance matrix. If neither `corr()` nor `cov()` is specified, the default is orthogonal data.

`cstorage(full` | `lower` | `upper)` specifies the storage mode for the correlation or covariance structure in `corr()` or `cov()`. The following storage modes are supported:

    `full` specifies that the correlation or covariance structure is stored (recorded) as a symmetric $k \times k$ matrix.

    `lower` specifies that the correlation or covariance structure is recorded as a lower triangular matrix. With k variables, the matrix should have $k(k+1)/2$ elements in the following order:

$$C_{11}\ C_{21}\ C_{22}\ C_{31}\ C_{32}\ C_{33}\ \ldots\ C_{k1}\ C_{k2}\ \ldots\ C_{kk}$$

    `upper` specifies that the correlation or covariance structure is recorded as an upper triangular matrix. With k variables, the matrix should have $k(k+1)/2$ elements in the following order:

$$C_{11}\ C_{12}\ C_{13}\ \ldots\ C_{1k}\ C_{22}\ C_{23}\ \ldots C_{2k}\ \ldots\ C_{(k-1k-1)}\ C_{(k-1k)}\ C_{kk}$$

    Specifying `cstorage(full)` is optional if the matrix is square. `cstorage(lower)` or `cstorage(upper)` is required for the vectorized storage methods. See *Storage modes for correlation and covariance matrices* in [D] **drawnorm** for examples.

forcepsd modifies the matrix C to be positive semidefinite (psd) and to thus be a proper covariance matrix. If C is not positive semidefinite, it will have negative eigenvalues. By setting the negative eigenvalues to 0 and reconstructing, we obtain the least-squares positive-semidefinite approximation to C. This approximation is a singular covariance matrix.

means(*vector*) specifies the means of the generated variables. The default is means(0).

⌐ Options ⌐

seed(*#*) specifies the seed of the random-number generator used to generate data. *#* defaults to 0. The random numbers generated inside corr2data do not affect the seed of the standard random-number generator.

Remarks

corr2data is designed to enable analyses of correlation (covariance) matrices by commands that expect variables rather than a correlation (covariance) matrix. corr2data creates variables with exactly the correlation (covariance) that you want to analyze. Apart from means and covariances, all aspects of the data are meaningless. Only analyses that depend on the correlations (covariances) and means produce meaningful results. Thus you may perform a paired t test ([R] **ttest**) or an ordinary regression analysis ([R] **regress**), etc.

If you are not sure that a statistical result depends only on the specified summary statistics and not on other aspects of the data, you can generate different datasets, each having the same summary statistics but other different aspects, by specifying the seed() option. If the statistical results differ beyond what is attributable to roundoff error, then using corr2data is inappropriate.

▷ Example 1

We first run a regression using the auto dataset.

```
. use http://www.stata-press.com/data/r10/auto
(1978 Automobile Data)

. regress weight length trunk
```

Source	SS	df	MS		Number of obs =	74
					F(2, 71) =	303.95
Model	39482774.4	2	19741387.2		Prob > F =	0.0000
Residual	4611403.95	71	64949.3513		R-squared =	0.8954
					Adj R-squared =	0.8925
Total	44094178.4	73	604029.841		Root MSE =	254.85

weight	Coef.	Std. Err.	t	P>|t|	[95% Conf. Interval]	
length	33.83435	1.949751	17.35	0.000	29.94666	37.72204
trunk	-5.83515	10.14957	-0.57	0.567	-26.07282	14.40252
_cons	-3258.84	283.3547	-11.50	0.000	-3823.833	-2693.846

Suppose that, for some reason, we no longer have the auto dataset. Instead we know the means and covariance matrices of weight, length, and trunk, and we want to do the same regression again. The matrix of means is

```
. mat list M

M[1,3]
           weight      length       trunk
_cons   3019.4595   187.93243   13.756757
```

and the covariance matrix is

```
. mat list V

symmetric V[3,3]
           weight      length       trunk
weight   604029.84
length   16370.922   495.78989
 trunk   2234.6612   69.202518   18.296187
```

To do the regression analysis in Stata, we need to create a dataset that has the specified correlation structure.

```
. corr2data x y z, n(74) cov(V) means(M)

. regress x y z
```

Source	SS	df	MS		Number of obs	=	74
					F(2, 71)	=	303.95
Model	39482773.3	2	19741386.6		Prob > F	=	0.0000
Residual	4611402.75	71	64949.3345		R-squared	=	0.8954
					Adj R-squared	=	0.8925
Total	44094176	73	604029.809		Root MSE	=	254.85

x	Coef.	Std. Err.	t	P>|t|	[95% Conf. Interval]	
y	33.83435	1.949751	17.35	0.000	29.94666	37.72204
z	-5.835155	10.14957	-0.57	0.567	-26.07282	14.40251
_cons	-3258.84	283.3546	-11.50	0.000	-3823.833	-2693.847

The results from the regression based on the generated data are the same as those based on the real data.

◁

Methods and Formulas

corr2data is implemented as an ado-file.

Two steps are involved in generating the desired dataset. The first step is to generate a zero-mean, zero-correlated dataset. The second step is to apply the desired correlation structure and the means to the zero-mean, zero-correlated dataset. In both steps, we take into account that, given any matrix $\mathbf{A}$ and any vector of variables $\mathbf{X}$, $\mathrm{Var}(\mathbf{A}'\mathbf{X}) = \mathbf{A}'\mathrm{Var}(\mathbf{X})\mathbf{A}$.

Reference

Cappellari, L., and S. P. Jenkins. 2006. Calculation of multivariate normal probabilities by simulation, with applications to maximum simulated likelihood estimation. *Stata Journal* 6: 156–189.

Also See

[D] **drawnorm** — Draw sample from multivariate normal distribution

[D] **data types** — Quick reference for data types

Title

> **count** — Count observations satisfying specified conditions

Syntax

<u>cou</u>nt $\left[\,if\,\right]$ $\left[\,in\,\right]$

by is allowed; see [D] **by**.

Description

count counts the number of observations that satisfy the specified conditions. If no conditions are specified, count displays the number of observations in the data.

Remarks

count may strike you as an almost useless command, but it can be one of Stata's handiest.

> ▷ Example 1

How many times have you obtained a statistical result and then asked yourself how it was possible? You think a moment and then mutter aloud, "Wait a minute. Is income ever *negative* in these data?" or "Is sex ever equal to *3*?" count can quickly answer those questions:

```
. use http://www.stata-press.com/data/r10/countxmpl
(1980 Census data by state)
. count
  641
. count if income<0
    0
. count if sex==3
    1
. by division: count if sex==3

-> division = New England
    0

-> division = Mountain
    0

-> division = Pacific
    1
```

We have 641 observations. income is never negative. sex, however, takes on the value 3 once. When we decompose the count by division, we see that it takes on that odd value in the Pacific division.

◁

Saved Results

count saves the following in r():

Scalars
    r(N)   number of observations

Also See

[R] **tabulate oneway** — One-way tables of frequencies

Title

> **cross** — Form every pairwise combination of two datasets

Syntax

cross using *filename*

Description

cross forms every pairwise combination of the data in memory with the data in *filename*. If *filename* is specified without a suffix, .dta is assumed.

Remarks

This command is rarely used; also see [D] **joinby**, [D] **merge**, and [D] **append**.

Crossing refers to merging two datasets in every way possible. That is, the first observation of the data in memory is merged with every observation of *filename*, followed by the second, and so on. Thus the result will have $N_1 N_2$ observations, where N_1 and N_2 are the number of observations in memory and in *filename*, respectively.

Typically, the datasets will have no common variables. If they do, such variables will take on only the values of the data in memory.

▷ Example 1

We wish to form a dataset containing all combinations of three age categories and two sexes to serve as a stub. The three age categories are 20, 30, and 40. The two sexes are male and female:

```
. input str6 sex
              sex
  1. male
  2. female
  3. end
. save sex
file sex.dta saved
. drop _all
. input agecat
         agecat
  1. 20
  2. 30
  3. 40
  4. end
. cross using sex
```

```
. list
```

	agecat	sex
1.	20	male
2.	30	male
3.	40	male
4.	20	female
5.	30	female
6.	40	female

◁

Methods and Formulas

cross is implemented as an ado-file.

Reference

Franklin, C. H. 2006. Stata tip 29: For all times and all places. *Stata Journal* 6: 147–148.

Also See

[D] **save** — Save datasets

[D] **append** — Append datasets

[D] **fillin** — Rectangularize dataset

[D] **joinby** — Form all pairwise combinations within groups

[D] **merge** — Merge datasets

Title

> **data types** — Quick reference for data types

Description

This entry provides a quick reference for data types allowed by Stata. See [U] **12 Data** for details.

Remarks

Storage type	Minimum	Maximum	Closest to 0 without being 0	Bytes
byte	-127	100	± 1	1
int	$-32,767$	32,740	± 1	2
long	$-2,147,483,647$	2,147,483,620	± 1	4
float	$-1.70141173319 \times 10^{38}$	$1.70141173319 \times 10^{38}$	$\pm 10^{-38}$	4
double	$-8.9884656743 \times 10^{307}$	$8.9884656743 \times 10^{307}$	$\pm 10^{-323}$	8

Precision for float is 3.795×10^{-8}.

Precision for double is 1.414×10^{-16}.

String storage type	Maximum length	Bytes
str1	1	1
str2	2	2
...	.	.
...	.	.
...	.	.
str244	244	244

Also See

[D] **compress** — Compress data in memory

[D] **destring** — Convert string variables to numeric variables and vice versa

[D] **encode** — Encode string into numeric and vice versa

[D] **format** — Set variables' output format

[D] **recast** — Change storage type of variable

[U] **12.2.2 Numeric storage types**

[U] **12.4.4 String storage types**

[U] **12.5 Formats: controlling how data are displayed**

[U] **13.10 Precision and problems therein**

Title

> **datasignature** — Determine whether data have changed

Syntax

```
datasignature

datasignature set [, reset]

datasignature confirm [, strict]

datasignature report

datasignature set, saving(filename[, replace]) [reset]

datasignature confirm using filename [, strict]

datasignature report using filename

datasignature clear
```

Note: datasignature was introduced during the Stata 9 release. This is not that command. The new datasignature command is easier to use and has new capabilities.

The original is now named _datasignature and is documented in [P] _datasignature. Under version control, datasignature becomes _datasignature.

Programmers will still be interested in [P] _datasignature. datasignature is implemented in terms of _datasignature.

Description

These commands calculate, display, save, and verify checksums of the data, which taken together form what is called a *signature*. An example signature is 162:11(12321):2725060400:4007406597. That signature is a function of the values of the variables and their names, and thus the signature can be used later to determine whether a dataset has changed.

datasignature without arguments calculates and displays the signature of the data in memory.

datasignature set does the same, and it stores the signature as a characteristic in the dataset. You should save the dataset afterward so that the signature becomes a permanent part of the dataset.

datasignature confirm verifies that, were the signature recalculated this instant, it would match the one previously set. datasignature confirm displays an error message and returns a nonzero return code if the signatures do not match.

`datasignature report` displays a full report comparing the previously set signature to the current one.

In the above, the signature is stored in the dataset and accessed from it. The signature can also be stored in a separate, small file.

`datasignature set, saving(`*filename*`)` calculates and displays the signature and, in addition to storing it as a characteristic in the dataset, also saves the signature in *filename*.

`datasignature confirm using` *filename* verifies that the current signature matches the one stored in *filename*.

`datasignature report using` *filename* displays a full report comparing the current signature with the one stored in *filename*.

In all the above, if *filename* is specified without an extension, `.dtasig` is assumed.

`datasignature clear` clears the signature, if any, stored in the characteristics of the dataset in memory.

Options

`reset` is used with `datasignature set`. It specifies that even though you have previously set a signature, you want to erase the old signature and replace it with the current one.

`saving(`*filename*`[, replace ])` is used with `datasignature set`. It specifies that, in addition to storing the signature in the dataset, you want a copy of the signature saved in a separate file. If *filename* is specified without a suffix, `.dtasig` is assumed. Suboption `replace` allows *filename* to be replaced if it already exists.

`strict` is for use with `datasignature confirm`. It specifies that, in addition to requiring that the signatures match, you also wish to require that the variables be in the same order and that no new variables have been added to the dataset. (If any variables were dropped, the signatures would not match.)

Remarks

Remarks are presented under the following headings:

> *Using datasignature interactively*
>> *Example 1: Verification at a distance*
>> *Example 2: Protecting yourself from yourself*
>> *Example 3: Working with assistants*
>> *Example 4: Working with shared data*
> *Using datasignature in do-files*
> *Interpreting data signatures*
> *The logic of data signatures*

Using datasignature interactively

`datasignature` is useful in the following cases:

1. You and a coworker, separated by distance, have both received what is claimed to be the same dataset. You wish to verify that is so.

2. You work interactively and realize the you could mistakenly modify your data. You wish to guard against that.

3. You want to give your dataset to an assistant to improve the labels and the like. You wish to verify that the data returned to you are the same data.

4. You work with an important dataset served on a network drive, and you wish to verify that others have not changed it.

Example 1: Verification at a distance

You load the data and type

```
. datasignature
74:12(71728):3831085005:1395876116
```

Your coworker does the same with his or her copy. You compare the two signatures.

Example 2: Protecting yourself from yourself

You load the data and type

```
. datasignature set
74:12(71728):3831085005:1395876116    (data signature set)
. save, replace
```

From then on, you periodically type

```
. datasignature confirm
(data unchanged since 19feb2007 14:24)
```

One day, however, you check and see the message:

```
. datasignature confirm
(data unchanged since 19feb2007 14:24, except 2 variables have been added)
```

You can find out more by typing

```
. datasignature report
(data signature set on Monday 19feb2007 14:24)
```

Data signature summary

```
1. Previous data signature      74:12(71728):3831085005:1395876116
2. Same data signature today    (same as 1)
3. Full data signature today    74:14(113906):1142538197:2410350265
```

Comparison of current data with previously set data signature

variables	number	notes
original # of variables	12	(values unchanged)
added variables	2	(1)
dropped variables	0	
resulting # of variables	14	

(1) Added variables are agesquared logincome.

(Continued on next page)

You could now either drop the added variables or decide to incorporate them:

```
. datasignature set
data signature already set -- specify option -reset-
r(198)
. datasignature set, reset
  74:14(113906):1142538197:2410350265        (data signature reset)
```

Concerning the detailed report, three data signatures are reported: (1) the stored signature, (2) the signature that would be calculated today on the basis of the same variables in their original order, and (3) the signature that would be calculated today on the basis of all the variables and in their current order.

datasignature confirm knew that new variables had been added because (1) was equal to (2). If some variables had been dropped, however, datasignature confirm would not be able to determine whether the remaining variables had changed.

Example 3: Working with assistants

You give your dataset to an assistant to have variable labels and the like added. You wish to verify that the returned data are the same data.

Saving the signature with the dataset is inadequate here. Your assistant, having your dataset, could change both your data and the signature and might even do that in a desire to be helpful. The solution is to save the signature in a separate file that you do not give to your assistant:

```
. datasignature set, saving(mycopy)
  74:12(71728):3831085005:1395876116        (data signature set)
  (file mycopy.dtasig saved)
```

You keep file mycopy.dtasig. When your assistant returns the dataset to you, you use it and compare the current signature to that you have stored in mycopy.dtasig:

```
. datasignature confirm using mycopy
  (data unchanged since 19feb2007 15:05)
```

By the way, the signature is a function of the following:

1. The number of observations and number of variables in the data

2. The values of the variables

3. The names of the variables

4. The order in which the variables occur in the dataset

5. The storage types of the individual variables

The signature is not a function of variable labels, value labels, notes, and the like.

Example 4: Working with shared data

You work on a dataset served on a network drive, which means that others could change the data. You wish to know whether this occurs.

The solution here is the same as working with an assistant: you save the signature in a separate, private file on your computer,

```
. datasignature set, saving(private)
  74:12(71728):3831085005:1395876116        (data signature set)
  (file private.dtasig saved)
```

and then you periodically check the signature by typing

```
. datasignature confirm using private
(data unchanged since 15mar2007 11:22)
```

Using datasignature in do-files

datasignature confirm aborts with error if the signatures do not match:

```
. datasignature confirm
data have changed since 19feb2007 15:05
r(9);
```

This means that, if you use datasignature confirm in a do-file, execution of the do-file will be stopped if the data have changed.

You may want to specify the strict option. strict adds two more requirements: that the variables be in the same order and that no new variables have been added. Without strict, these are not considered errors:

```
. datasignature confirm
(data unchanged since 19feb2007 15:22)
. datasignature confirm, strict
(data unchanged since 19feb2007 15:05, but order of variables has changed)
r(9);
```

and

```
. datasignature confirm
(data unchanged since 19feb2007 15:22, except 1 variable has been added)
. datasignature confirm, strict
(data unchanged since 19feb2007 15:22, except 1 variable has been added)
r(9);
```

If you keep logs of your analyses, issuing datasignature or datasignature confirm immediately after loading each dataset is a good idea. This way, you have a permanent record that you can use for comparison.

Interpreting data signatures

An example signature is 74:12(71728):3831085005:1395876116. The components are

1. 74, the number of observations;

2. 12, the number of variables;

3. 71728, a checksum function of the variable names and the order in which they occur; and

4. 3831085005 and 1395876116, checksum functions of the values of the variables, calculated two different ways.

Two signatures are equal only if all their components are equal.

Two different datasets will probably not have the same signature, and it is even more unlikely that datasets containing similar values will have equal signatures. There are two data checksums, but do not read too much into that. If either data checksum changes, even just a little, the data have changed. Whether the change in the checksum is large or small—or in one, the other, or both—signifies nothing.

The logic of data signatures

The components of a data signature are known as checksums. The checksums are many-to-one mappings of the data onto the integers. Let's consider the checksums of `auto.dta` carefully.

The data portion of `auto.dta` contains 38,184 bytes. There are 256^{38184} such datasets, or equivalently, 2^{305472}. The first checksum has 2^{48} possible values, and it can be proven that those values are equally distributed over the 2^{305472} datasets. Thus there are $2^{305472}/2^{48} - 1 = 2^{305424} - 1$ datasets that have the same first checksum value as `auto.dta`. The same can be said for the second checksum. It would be difficult to prove, but we believe that the two checksums are conditionally independent, being based on different bit shifts and bit shuffles of the same data. Of the $2^{305424} - 1$ datasets that have the same first checksum as `auto.dta`, the second checksum should be equally distributed over them. Thus there are about $2^{305376} - 1$ datasets with the same first and second checksums as `auto.dta`.

Now let's consider those $2^{305376} - 1$ other datasets. Most of them look nothing like `auto.dta`. The checksum formulas guarantee that a change of one variable in 1 observation will lead to a change in the calculated result if the value changed is stored in 4 or fewer bytes, and they nearly guarantee it in other cases. When it is not guaranteed, the change cannot be subtle—"Chevrolet" will have to change to binary junk, or a double-precision 1 to $-6.476678983751e+301$, and so on. The change will be easily detected if you `summarize` your data and just glance at the minimums and maximums. If the data look at all like `auto.dta`, which is unlikely, they will look like a corrupted version.

More interesting are offsetting changes across observations. For instance, can you change one variable in 1 observation and make an offsetting change in another observation so that, taken together, they will go undetected? You can fool one of the checksums, but fooling both of them simultaneously will prove difficult. The basic rule is that the more changes you make, the easier it is to create a dataset with the same checksums as `auto.dta`, but by the time you've done that, the data will look nothing like `auto.dta`.

Saved Results

`datasignature` without arguments and `datasignature set` save the following in `r()`:

Macros
    r(datasignature)   the signature

`datasignature confirm` saves the following in `r()`:

Scalars
    r(added)   number of variables added
Macros
    r(datasignature)   the signature

`datasignature confirm` aborts execution if the signatures do not match and so then returns nothing except a return code of 9.

`datasignature report` saves the following in `r()`:

Scalars

r(datetime)	%tc date–time when set
r(changed)	. if r(k_dropped) $\neq$ 0, otherwise
	0 if data have not changed, 1 if data have changed
r(reordered)	1 if variables reordered, 0 if not reordered,
	. if r(k_added) $\neq$ 0 $\mid$ r(k_dropped) $\neq$ 0
r(k_original)	number of original variables
r(k_added)	number of added variables
r(k_dropped)	number of dropped variables

Macros

r(origdatasignature)	original signature
r(curdatasignature)	current signature on same variables if it can be calculated
r(fulldatasignature)	current full-data signature
r(varsadded)	variable names added
r(varsdropped)	variable names dropped

`datasignature clear` saves nothing in `r()` but does clear it.

`datasignature set` stores the signature in the following characteristics:

Characteristic

_dta[datasignature_si]	signature
_dta[datasignature_dt]	%tc date–time when set in %21x format
_dta[datasignature_vl1]	part 1, original variables
_dta[datasignature_vl2]	part 2, original variables, if necessary
etc.	

To access the original variables stored in _dta[datasignature_vl1], etc., from an ado-file, code

```
mata: ado_fromlchar("vars", _dta, "datasignature_vl")
```

Thereafter, the original variable list would be found in 'vars'.

Methods and Formulas

`datasignature` is implemented using `_datasignature`; see [P] **_datasignature**.

Reference

Gould, W. 2006. Stata tip 35: Detecting whether data have changed. *Stata Journal* 6: 428–429.

Also See

[P] **_datasignature** — Determine whether data have changed

[P] **signestimationsample** — Determine whether the estimation sample has changed

Title

> **dates and times** — Date and time (%t) values and variables

Syntax

Syntax is presented under the following headings:

How Stata records dates and times
Inputting date and time data
Recommended storage types for %t variables
Typing dates and times
Constructing date and time values from numerical components
Converting date and time values
Extracting date and time components
Obtaining and working with durations
Formatting date and time values

How Stata records dates and times

Dates and times are called %t values. %t values are numerical and integral. The integral value records the number of time units that have passed from an agreed-upon base, which for Stata is 1960.

Coding and interpretation of date and time (%t) values are as follows:

Format	Meaning	—— Numerical value and interpretation ——		
		Value = −1	**Value = 0**	Value = 1
%tc	clock	31dec1959 23:59:59.999	**01jan1960 00:00:00.000**	01jan1960 00:00:00.001
%td	days	31dec1959	**01jan1960**	02jan1960
%tw	weeks	1959w52	**1960w1**	1960w2
%tm	months	1959m12	**1960m1**	1960m2
%tq	quarters	1959q4	**1960q1**	1960q2
%th	half years	1959h2	**1960h1**	1960h2
%tg	generic	−1	**0**	1

Explanation: the middle, bolded column shows the **base value**. For a %td value, 0 means 01jan1960. The table also shows that −1 means 31dec1959 and 1 means 02jan1960. A %td value records the number of days from 01jan1960. A %tc value records the number of milliseconds from the start of 01jan1960, a %tw value records the number of weeks from the first week of 1960, and so on.

That is,

For a %tc value, a 1-unit change represents 1 ms.

Integer 394,839,482,000 represents 05jul1972 21:38:02.000 because that date occurred 394,839,482,000 ms after 01jan1960 00:00:00.000.

Integer −394,839,482,000 represents 28jun1947 02:21:58.000 because that date occurred 394,839,482,000 ms before 01jan1960 00:00:00.000.

For a %td value, a 1-unit change represents 1 day.

Integer 4,569 represents 05jul1972 because that date occurred 4,569 days after 01jan1960.

Integer −4,569 represents 29jun1947 because that date occurred 4,569 days before 01jan1960.

For a %tw value, a 1-unit change represents 1 week.

Integer 650 represents 1972w27 because that date occurred 650 weeks after 1960w1.

Integer −650 represents 1947w27 because that date occurred 650 weeks before 1960w1.

For a %tm value, a 1-unit change represents 1 month.

Integer 150 represents 1972m7 because that date occurred 150 calendar months after 1960m1.

Integer −150 represents 1947m7 because that date occurred 150 calendar months before 1960m1.

For a %tq value, a 1-unit change represents one quarter (3 calendar months).

Integer 50 represents 1972q3 because that date occurred 50 quarters after 1960q1.

Integer −50 represents 1947q3 because that date occurred 50 quarters before 1960q1.

For a %th value, a 1-unit change represents one half-year, or 6 months.

Integer 25 represents 1972h2 because that date occurred 25 half years after 1960h1.

Integer −25 represents 1947h2 because that date occurred 25 half years before 1960h1.

For a %tg value, a 1-unit change represents whatever you wish.

Integer 100 might represent 100 workdays, or 100 lunar months, or anything else, after some agreed-upon event, such as 01jan1960, or the date you were born, or anything else.

Negative values would represent times before the event.

In addition to the above, there is %ty:

| Format | Meaning | —— Numerical value and interpretation —— | | |
		1959	1960	1961
%ty	year	1959	1960	1961

A %ty value is like the other %t values except that, rather than the base being 1960, the base is 0 AD. (Years 0100 through 9999 are valid.)

In addition to the above, there is %tC:

Format	Meaning	—— Numerical value and interpretation ——		
		−1	0	1
%tC	clock	31dec1959 23:59:59.999	01jan1960 00:00:00.000	01jan1960 00:00:00.001

%tC is similar to %tc, except that %tC accounts for leap seconds:

> Remember that %tc integer 394,839,482,000 represents 05jul1972 21:38:02.000.

> That integer in %tC represents 05jul1972 21:38:01.000. For those who wish their clock based on astronomical observation, 1 leap second was inserted. (The first leap second was on 30jun1972, the second on 31dec1972, and others have been inserted since then.) See *Advice on using %tc and %tC* below.

Jargon: a %td value is sometimes called an elapsed date.

Historical note: a %td value is sometimes referred to as a %d value. The t is omitted because, in Stata's history, %d values predated the other %t values. Dropping the t is still allowed but is now considered an anachronism.

Inputting date and time data

Date and time variables are best read as strings. Then use one of the string-to-numeric conversion functions to convert the string representation to the appropriate %t value:

Format	String-to-numeric conversion function
%tc	clock(*string*, *mask*)
%tC	Clock(*string*, *mask*)
%td	date(*string*, *mask*)
%tw	weekly(*string*, *mask*)
%tm	monthly(*string*, *mask*)
%tq	quarterly(*string*, *mask*)
%th	halfyearly(*string*, *mask*)
%ty	yearly(*string*, *mask*)
%tg	no function necessary; read as numeric

In the above functions, *string* is the variable or value containing the string representation to be converted and *mask* specifies the order in which the components occur:

- For %td function date(), *string* might be "August 21, 2005" or "8-21-2005" and *mask* might be "MDY", meaning that the elements occur in the order month, day, and year.

- For %tc function clock(), *string* might be "21aug2005 15:21:22" and *mask* be "DMYhms", meaning that the elements occur in the order day, month, year, hours, minutes, and seconds.

Thus one might code

```
. generate datehired = date(datehiredstr, "MDY")
. generate double timeadmitted = clock(timeadmitstr, "DMYhms")
```

See *String-to-numeric translation functions* below for details.

Recommended storage types for %t variables

In the example above, we stored %tc variable `timeadmitted` as a `double`. Doing so is important if precision is to be maintained.

The recommended storage types for %t variables are

Format	Recommended storage type
%tc	double
%tC	double
%td	float or long
%tw	float or int
%tm	float or int
%tq	float or int
%th	float or int
%ty	float or int
%tg	float or int

Storing a %tc (%tC) variable as a `double` is important if precision is to be maintained. %tc variables are integers, but being the number of milliseconds from the start of 1960, they are large integers.

What happens if you store a %tc value as a `float`:
    The largest integer that can be stored precisely in a `float` is 16,777,216, corresponding to 01jan1960 04:39:37.216. Times after that will be subject to rounding; the rounding as of recent times can be as much as 2 minutes, 11 seconds.

What happens if you store a %tc value as a `long`:
    The largest integer that can be stored in a `long` is 2,147,483,620, corresponding to 25jan1960 20:31:23.620. Times after that cannot be stored in a `long`.

What happens if you store a %tc value as a `double`:
    The largest integer that can be stored precisely in a `double` is 9,007,199,254,740,992, corresponding to a date in year 285,422,880. Stata cuts off dates at year 9999, but for other reasons.

(In the above, we use an idiosyncratic definition of "precisely": positive value x is stored precisely if x MINUS 1 is not equal to x, where MINUS is the computer's operation of subtraction. For `float` and `double`, there are larger values that are stored exactly, but not precisely. For example, both `float` and `double` can exactly store the integer 2^{100}, a value approximately equal to 1.3e+30, but 2^{100} MINUS 1 is still 2^{100} because of loss of precision.)

> **DO NOT FORGET**
>
> %tc and %tC values **MUST BE** stored as doubles.
> Doing so is your responsibility, not Stata's.

Typing dates and times

Remember, date and time values are just integers, so in an expression, you could type the appropriate integer:

```
. gen before = cond(hiredon < 16237, 1, 0) if hiredon < .
. drop if admittedon < 1402920000000
```

Easier to type is

```
. gen before = cond(hiredon < td(15jun2004), 1, 0) if hiredon < .
. drop if admittedon < tc(15jun2004 12:00:00)
```

td() and tc() are called pseudofunctions because they translate what you type into their integer equivalents. Pseudofunctions require only that you specify the date/time components in the expected order, so rather than 15jun2004 above, we could have specified 15 June 2004, 15-6-2004, or 15/6/2004.

The date and time pseudofunctions and their expected component order are

Format	Pseudofunction
%tc	tc($\left[\mathit{day\text{-}month\text{-}year}\right]$ hh:$\mathit{mm}$$\left[:\mathit{ss}\left[.\mathit{sss}\right]\right]$)
%tC	tC($\left[\mathit{day\text{-}month\text{-}year}\right]$ hh:$\mathit{mm}$$\left[:\mathit{ss}\left[.\mathit{sss}\right]\right]$)
%td	td($\mathit{day\text{-}month\text{-}year}$)
%tw	tw($\mathit{year\text{-}week}$)
%tm	tm($\mathit{year\text{-}month}$)
%tq	tq($\mathit{year\text{-}quarter}$)
%th	th($\mathit{year\text{-}half}$)
%ty	none necessary; just type year
%tg	none necessary

The *day-month-year* in tc() and tC() are optional. If you omit them, 01jan1960 is assumed. Doing so produces time as an offset, which can be useful in, for example,

```
. gen six_hrs_later = eventtime + tc(6:00)
```

Also see *Extracting date and time components* below.

Historical note: pseudofunctions td(), tw(), tm(), tq(), and th() used to be called d(), w(), m(), q(), and h(). Those names still work but are considered anachronisms.

Constructing date and time values from numerical components

If you had numeric variables M, D, and Y containing month number, day of month, and year (in the first observation, the variables might contain 12, 15, and 2006), you could code

. generate mydate = mdy(M, D, Y)

to obtain new %td variable containing the date (which would be 15dec2006 in the first observation).

The date-from-numerical-components functions are

Format	Function
%tc	$\mathrm{mdyhms}(M, D, Y, h, m, s)$
%tc	$\mathrm{dhms}(td, h, m, s)$
%tc	$\mathrm{hms}(h, m, s)$
%tC	$\mathrm{Cmdyhms}(M, D, Y, h, m, s)$
%tC	$\mathrm{Cdhms}(td, h, m, s)$
%tC	$\mathrm{Chms}(h, m, s)$
%td	$\mathrm{mdy}(M, D, Y)$
%tw	$\mathrm{yw}(Y, W)$
%tm	$\mathrm{ym}(Y, M)$
%tq	$\mathrm{yq}(Y, Q)$
%th	$\mathrm{yh}(Y, H)$
%ty	Y

where

td is a %td value,

M, D, and Y are month, day, and year values,
$$1 \leq M \leq 12$$
$$1 \leq D \leq 31$$
$$0100 \leq Y \leq 9999$$

h, m, and s are hour, minute, and second values,
$$0 \leq h \leq 23$$
$$0 \leq m \leq 59$$
$$0.000 \leq s \leq 59.999 \quad \text{(see note below)}$$

W is a week number, $1 \leq W \leq 52$

Q is a quarter number, $1 \leq Q \leq 4$

H is a half number, $1 \leq H \leq 2$

Note concerning s: functions Cmdyhms() and Cdhms() allow $0.000 \leq s \leq 60.999$ when the 60th second is a leap second. For instance, according to the authorities, 31dec1972 23:59:60 is an official leap second, but 31dec1971 23:59:60 is not. Cmdyhms(12,31,1971,23,59,60) therefore evaluates to missing (.), whereas Cmdyhms(12,31,1972,23,59,60) evaluates to 410,313,601,000, a nonmissing value. (The expanded range of s does not apply to Chms() because it is a pure time based on 01jan1960 and there were no leap seconds on that date. Functions hms() and Chms() are in fact identical.)

Functions `mdyhms()` and `dhms()` are related by

$$\text{mdyhms}(M, D, Y, h, m, s) = \text{dhms}(\text{mdy}(M,D,Y), h, m, s)$$

and similarly,

$$\text{Cmdyhms}(M, D, Y, h, m, s) = \text{Cdhms}(\text{mdy}(M,D,Y), h, m, s)$$

With `mdyhms()`, you have six variables, such as $M = 7$, $D = 5$, $Y = 1972$, $h = 21$, $m = 38$, and $s = 2$, and `mdyhms()` returns 05jul1972 21:38:02. With `dhms()` you have four variables, the first specifying the %td value of 05jul1972, and h, m, and s being the same, and `dhms()` returns the date + time, 05jul1972 21:38:02.

Converting date and time values

One type of %t value can be converted into another. The functions are

From	To ... %tc	%tC	%td	%tw	%tm	%tq	%th	%ty
%tc		Cofc()	dofc()					
%tC	cofC()		dofC()					
%td	cofd()	Cofd()		wofd()	mofd()	qofd()	hofd()	yofd()
%tw			dofw()					
%tm			dofm()					
%tq			dofq()					
%th			dofh()					
%ty			dofy()					

For instance, to convert %td to a %tc value,

```
. generate double datetimevalue = cofd(datevalue)
```

%td is the mother of all date and time values, and to convert a %tq value to a %tc value, you must first convert to a %td value:

```
. generate double datetimevalue = cofd(dofq(quartervalue))
```

Extracting date and time components

Let d be a %td variable or value. The following functions will extract components of d:

Function	Returns	Result if $d = $ td(05jul1972) (i.e., $d = 4{,}569$)
year(d)	calendar year	1972
month(d)	calendar month	7
day(d)	day within month	5
doy(d)	day of year	187
halfyear(d)	half of year	2
quarter(d)	quarter	3
week(d)	week within year	27
dow(d)	day of week (0 = Sunday)	3 (means Wednesday)

Remember, any %t value can be converted to a %td value by using the appropriate conversion function; see *Converting date and time values* above. If variable date_time_admitted is %tc and you want to obtain the day of week,

    . gen day = dow(dofc(date_time_admitted))

Let t be a %tc variable. The following functions will extract components of t:

Function	Returns	Result if $t = $ tc(05jul1972 21:38:02) (i.e., $t = 394{,}839{,}482{,}000$)
hh(t)	time of day, hours	21
mm(t)	time of day, minutes	38
ss(t)	time of day, seconds	2.000

Other components can be extracted by calculating dofc(t) and then extracting components from the %td value.

Let T be a %tC variable. The following functions will extract components of T:

Function	Returns	Result if $T = $ tC(05jul1972 21:38:01) (i.e., $T = 394{,}839{,}482{,}000$)
hhC(T)	time of day, hours	21
mmC(T)	time of day, minutes	38
ssC(T)	time of day, seconds	1.000

By convention, leap seconds came after 23:59:59 and are labeled 23:59:60. Thus ssC(T) can return 60. Other components can be extracted by calculating dofC(T) and then extracting components from the %td value.

Obtaining and working with durations

Remember that %t variables are simply durations from 1960:

Format	Units
%tC	milliseconds
%tc	milliseconds
%td	days
%tw	weeks
%tm	months
%tq	quarters
%th	half years

Thus, to obtain the duration between %t variables, subtract them:

```
. gen days_employed = curdate - hiredate
```

```
. gen qtrs_to_15jan = curqtr - qofd(td(15jan2005))
```

To add a duration to a date, add the two values:

```
. gen lastdate = hiredate + days_employed
```

```
. format lastdate %td
```

```
. gen qtr_of_merger = curqtr + quarters_to_merger
```

```
. format qtr_of_merger %tq
```

When creating new date and time variables, remember to format them so that they will be readable should you print them.

The above applies equally to %tc and %tC variables:

```
. gen double millisecs_employed = lasttime - hiretime
```

and

```
. gen double lasttime = hiretime + millisecs_employed
. format lasttime %tc
```

Note our use of double. Times are recorded in milliseconds and must be stored as doubles if precision is to be maintained.

There are 1,000 ms in a second, $60 \times 1{,}000$ in a minute, and $60 \times 60 \times 1{,}000$ in an hour. It is easy to mistype these constants when converting to more readable units, and therefore the following functions are provided:

Function	Purpose
hours(*ms*)	convert milliseconds to hours returns $ms/(60 \times 60 \times 1000)$
minutes(*ms*)	convert milliseconds to minutes returns $ms/(60 \times 1000)$
seconds(*ms*)	convert milliseconds to seconds returns $ms/1000$
msofhours(*h*)	convert hours to milliseconds returns $h \times 60 \times 60 \times 1000$
msofminutes(*m*)	convert minutes to milliseconds returns $m \times 60 \times 1000$
msofseconds(*s*)	convert seconds to milliseconds returns $s \times 1000$

Thus you can code

```
. gen double days_employed = 24*hours(lasttime-hiretime)
```

and

```
. gen double lasttime = hiretime + msofhours(24*days_employed)
```

If precision is to be preserved, the use of these functions does not alleviate the necessity of using doubles.

Also days_employed in the above will include fraction of a day. If a rounded integer result is desired, then round explicitly:

```
. gen approx_days_employed = round(24*hours(lasttime-hiretime))
```

Formatting date and time values

A variable's values are formatted to indicate (1) the units used and (2) how the variable is to be displayed:

```
. generate mydate = date(datestr, "DMY")
. list mydate in 1
```

```
. format mydate %td
. list mydate in 1
```

	mydate
1.	22oct2006

```
. generate double mytime = clock(timestr, "DMY hm")
. list mytime in 1
```

	mytime
1.	1.477e+12

```
. format mytime %tc
. list mytime in 1
```

	mytime
1.	22oct2006 13:02:00

The %t formats result in the following output:

Format	Example of output
%tC	05jul1972 21:38:01
%tc	05jul1972 21:38:02
%td	05jul1972
%tw	1972w27
%tm	1972m7
%tq	1972q3
%th	1972h2
%ty	1972
%tg	(actual integer shown)

Formats %tC and %tc do not show the milliseconds by default.

You can specify how date and times are to be formatted. Rather than 05jul1972, you could have July 5, 1972, or rather than 05jul1972 21:38:02, you could have 7-5-72 9:38 p.m. This reformatting is done by adding codes to the end of %tC, %tc, %td, etc. In fact, the default %tC, %tc, %td, . . . , formats actually mean

Format	Implied (fully specified) format
%tC	%tCDDmonCCYY_HH:MM:SS
%tc	%tcDDmonCCYY_HH:MM:SS
%td	%tdDDmonCCYY
%tw	%twCCYY!www
%tm	%tmCCYY!mnn
%th	%thCCYY!hh
%ty	%tyCCYY

Typing

```
    . format mytimevar %tc
```

has the same effect as typing

```
    . format mytimevar %tcDDmonCCYY_HH:MM:SS
```

Format %tcDDmonCCYY_HH:MM:SS is interpreted as

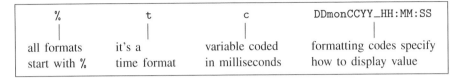

%	t	c	DDmonCCYY_HH:MM:SS
all formats start with %	it's a time format	variable coded in milliseconds	formatting codes specify how to display value

The formatting codes are

Code	Meaning	Output
CC	century − 1	01–99
cc	century − 1	1–99
YY	2-digit year	00–99
yy	2-digit year	0–99
JJJ	day within year	001–366
jjj	day within year	1–366
Mon	month	Jan, Feb, . . . , Dec
Month	month	January, February, . . . , December
mon	month	jan, feb, . . . , dec
month	month	january, february, . . . , december
NN	month	01–12
nn	month	1–12
DD	day within month	01–31
dd	day within month	1–31
DAYNAME	day of week	Sunday, Monday, . . . (aligned)
Dayname	day of week	Sunday, Monday, . . . (unaligned)
Day	day of week	Sun, Mon, . . .
Da	day of week	Su, Mo, . . .
day	day of week	sun, mon, . . .
da	day of week	su, mo, . . .
h	half	1–2
q	quarter	1–4
WW	week	01–52
ww	week	1–52
HH	hour	00–23
Hh	hour	00–12
hH	hour	0–23
hh	hour	0–12
MM	minute	00–59
mm	minute	0–59
SS	second	00–60 (sic, due to leap seconds)
ss	second	0–60 (sic, due to leap seconds)
.s	tenths	.0–.9
.ss	hundredths	.00–.99
.sss	thousandths	.000–.999

am	show am or pm	am or pm
a.m.	show a.m. or p.m.	a.m. or p.m.
AM	show AM or PM	AM or PM
A.M.	show A.M. or P.M.	A.M. or P.M.
.	display period	.
,	display comma	,
:	display colon	:
-	display hyphen	-
_	display space	
/	display slash	/
\	display backslash	\
!c	display character	c
+	separator (see note)	

Note: + displays nothing; it may be used to separate one code from the next to make the format more readable. + is never necessary. For instance, %tchh:MM+am and %tchh:MMam have the same meaning, as does %tc+hh+:+MM+am.

Thus, if you had a %td variable and wanted to display the dates as, for example, January 9, 2002, you could specify the format %tdMonth_dd,_CCYY.

If you had a %tc variable and wanted to display the time as

    Fri Aug 18 12:01:35 CDT 2006

you could specify %tcDay_Mon_DD_HH:MM:SS_!C!D!T_CCYY.

The maximum length of a format specifier is 48 characters; the example shown above is 34 characters.

Description

Complete documentation of Stata's treatment of date and time values is provided. Every feature and function is documented here, either in *Syntax* above or in *Remarks* below.

Remarks

Remarks are presented under the following headings:

> *Experimenting with the date and time functions*
> *String-to-numeric translation functions*
> > *The clock() function*
> > > *How clock() interprets the mask*
> > > *Working with two-digit years*
> > > *Working with incomplete dates and times*
> > *The Clock() function*
> > *The date() function*
> > *Translating run-together dates such as 20060125*
> > *The other translation functions*
> *Valid times*
> *When leap seconds occurred*
> *Truncated times*
> *Advice on using %tc and %tC*
> > *Summary*
> > *Explanation*

Experimenting with the date and time functions

The best way to become familiar with Stata's date and time functions is to experiment with the display command.

```
. display date("5-12-1998", "MDY")
14011
. display %td date("5-12-1998", "MDY")
12may1998
. display clock("5-12-1998 11:15", "MDY hm")
1.211e+12
. display %20.0gc clock("5-12-1998 11:15", "MDY hm")
1,210,590,900,000
. display %tc clock("5-12-1998 11:15", "MDY hm")
12may1998 11:15:00
```

Remember, when you work with display, you can specify a format in front of the expression to specify how the result is to be formatted.

String-to-numeric translation functions

The string-to-numeric date and time translation functions are

Format	String-to-numeric conversion function
%tc	clock(*string*, *mask* [, *baseyear*])
%tC	Clock(*string*, *mask* [, *baseyear*])
%td	date(*string*, *mask* [, *baseyear*])
%tw	weekly(*string*, *mask* [, *baseyear*])
%tm	monthly(*string*, *mask* [, *baseyear*])
%tq	quarterly(*string*, *mask* [, *baseyear*])
%th	halfyearly(*string*, *mask* [, *baseyear*])
%ty	yearly(*string*, *mask* [, *baseyear*])

string is the value to be translated.
mask specifies the order of the components.
baseyear is described in *Working with two-digit years* below.

These functions are typically used after reading date and/or time data. The data contain values such as "08/12/06", "12-8-2006", "12 Aug 06", "12aug2006 14:23", and "12 aug06 2:23 pm". You read the data into a string variable and then use one of the translation functions to translate the string into a %t variable.

The translation functions are used in expressions, such as

```
. generate double timeadmitted = clock(timeadmitstr, "DMYhms")
. format timeadmitted %tc
. generate datehired = date(datehiredstr, "MDY")
. format datehired %td
```

All functions require two arguments, the string to be translated and a second string specifying the order in which the date and time components occur.

The most useful of these functions are clock(), Clock(), and date(). The other functions are rarely used.

The clock() function

clock() returns a %tc value. The syntax of clock() is

clock(*string*, *mask* [, *baseyear*])

Ignore optional argument *baseyear*; we will discuss that below. Second argument *mask* is a string specifying the order of the components in *string* and consists of the following codes:

Code	Meaning
M	month
D	day within month
Y	4-digit year
19Y	2-digit year to be interpreted as 19*xx*
20Y	2-digit year to be interpreted as 20*xx*
h	hour of day
m	minutes within hour
s	seconds within minute
#	ignore one element

Examples of date strings and the mask required to translate them include

String to translate	Corresponding mask
01dec2006 14:22	"DMYhm"
01-12-2006 14.22	"DMYhm"
1dec2006 14:22	"DMYhm"
1-12-2006 14:22	"DMYhm"
01dec06 14:22	"DM20Yhm"
01-12-06 14.22	"DM20Yhm"
December 1, 2006 14:22	"MDYhm"
2006 Dec 01 14:22	"YMDhm"
2006-01-12 14:22	"YMDhm"
2006-01-12 14:22:43	"YMDhms"
2006-01-12 14:22:43.2	"YMDhms"
2006-01-12 14:22:43.21	"YMDhms"
2006-01-12 14:22:43.213	"YMDhms"
2006-01-12 2:22:43.213 pm	"YMDhms"
2006-01-12 2:22:43.213 pm.	"YMDhms"
2006-01-12 2:22:43.213 p.m.	"YMDhms"
2006-01-12 2:22:43.213 P.M.	"YMDhms"
20060112 1422	"YMDhm"
14:22	"hm" (see note)
2006-12-01	"YMD" (see note)
Wed Dec 01 14:22:43 CST 2006	"#MDhms#Y"

Note: a subset of components may be specified.
    clock("14:22", "hm") produces 01jan1960 14:22:00.
    clock("2006-12-01", "YMD") produces 01dec2006 00:00:00.

Also there is nothing special included in *mask* to process a.m. and p.m. markers.
    When you include code h, clock() automatically watches for the meridian markers.

mask may include spaces so that it is more readable; they have no meaning. Thus we can code

    . generate double admit = clock(admitstr, "#MDhms#Y")

or code

    . generate double admit = clock(admitstr, "# MD hms # Y")

and which we code makes no difference.

How clock() interprets the mask

To specify the appropriate mask, it helps to understand the rules that clock() applies. They are

1. For each string to be translated, remove all punctuation except for the period separating seconds from tenths, hundredths, and thousandths of seconds. Replace the punctuation with a space.

2. Insert a space in the string everywhere that a letter is next to a number or vice versa.

3. Interpret the resulting elements according to *mask*.

For instance, consider the string

```
01dec2006 14:22
```

Under rule (1), the string becomes

```
01dec2006 14 22
```

Under rule (2), the string becomes

```
01 dec 2006 14 22
```

Now clock() applies rule (3). If the mask is "DMYhm", then clock() interprets "01" as the day, "dec" as the month, and so on.

Or consider the string

```
Wed Dec 01 14:22:43 CST 2006
```

Under rule (1), the string becomes

```
Wed Dec 01 14 22 43 CST 2006
```

Applying rule (2) does not change the string. Now clock() applies rule (3). If the mask is "#MDhms#Y", clock() skips "Wed", interprets "Dec" as the month, and so on.

The # code serves a second purpose. If it appears at the end of the mask, it specifies that the rest of *string* is to be ignored. Consider translating

```
Wed Dec 01 14 22 43 CST 2006 patient 42
```

The mask code that previously worked when "patient 42" was not part of the string, "#MDhms#Y", will result in missing value. clock() is careful in the translation and, if all the string is not used, returns missing. If you end mask in #, however, clock() ignores the rest of the string. Changing mask from "#MDhms#Y" to "#MDhms#Y#" will produce the desired result.

Working with two-digit years

Consider translating the string 01-12-06 14:22, which is to be interpreted as 01dec2006 14:22:00. clock() provides two ways of doing this.

The first is to specify the assumed prefix in the mask. 01-12-06 14:22 can be read by specifying mask "DM20Yhm". If we instead wanted to interpret the year as 1906, we would specify mask "DM19Yhm". We could even interpret the year as 1806 by specifying "DM18Yhm".

But what if our data include 01-12-06 14:22 and include 06-15-98 11:01? We want to interpret the first as being in 2006 and the second as being in 1998. That is the purpose of optional argument *baseyear*:

clock(*string*, *mask* [, *baseyear*])

When you specify *baseyear*, you are stating that when years in *string* are two digits, the full year is to be obtained by finding the year closest to *baseyear*. Thus you could code,

```
. generate double timestamp = clock(timestr, "DMYhm", 2000)
```

Two-digit year 06 would be interpreted as 2006 because 2006 is closer to 2000 than 1906, or 2106, or 1806, etc. Two-digit 98 would be interpreted as 1998 because 1998 is closer to 2000 than 2098, 1898, etc.

Working with incomplete dates and times

Function `clock()` does not require every component of the date and time be specified.

Translating 2006-12-01 with mask `"YMD"` results in 01dec2006 00:00:00.

Translating 14:22 with mask `"hm"` results in 01jan1960 14:22:00.

Translating 11-2006 with mask `"MY"` results in 01nov2006 00:00:00.

The default for a component, if not specified in the mask, is

Code	Default if not specified
M	01
D	01
Y	1960
h	00
m	00
s	00

This feature is useful. You may have data recording "14:22", meaning a duration of 14 hours and 22 minutes, or the time 14:22 each day. See *Obtaining and working with durations* above.

The Clock() function

The syntax of function `Clock()` is

Clock(*string*, *mask* [, *baseyear*])

Function `Clock()` is identical to `clock()` except that, rather than returning a %tc value, it returns %tC.

Note: `Clock()` is almost identical to `Cofc(clock())`. The difference is that `Clock()` understands leap seconds, such as 30jun1997 23:59:60.

The date() function

The syntax of function `date()` is

date(*string*, *mask* [, *baseyear*])

Function `date()` is identical to `clock()` except that it returns a %td value rather than a %tc value. Function `date()` is the same as `dofc(clock())`.

Historical note: Stata 10's `date()` function is much improved over that of previous versions, and the mask is specified a little differently. In previous versions, the codes for year, month, and date were y, m, and d rather than Y, M, and D. Under version control, the old codes are allowed and, in fact, the original `date()` function is used.

The big advantage of Stata 10's `date()` is that it will translate run-together dates such as 20061201 (no special action by you required) and translate more complicated date strings such as Wed Dec 01 14:22:43 CST 2006 (special action required in how *mask* is specified, something that the old `date()` would not have understood).

Translating run-together dates such as 20060125

Functions clock(), Clock(), and date() will translate dates and times that are run together such as 20060125, 060125, and 20060125110215 (which is 25jan2006 11:02:15). There is nothing special that you have to do:

```
. display %td date("20060125", "YMD")
25jan2006

. display %td date("060125", "20YMD")
25jan2006

. display %tc clock("20060125110215", "YMDhms")
25jan2006 11:02:15
```

In a data context, you could type

```
. gen startdate = date(startdatestr, "YMD")

. gen double starttime = clock(starttimestr, "YMDhms")
```

Remember to read the original data into a string. If you read the data as numeric, the best advice is to read the data again. Numbers such as 20060125 and 20060125110215 will be rounded unless they are stored as doubles.

If you did read them into a double, or you have verified that rounding did not occur, you can convert the variable from numeric to string. The numeric-to-string conversion function is string(), which comes in one- and two-argument forms. You will need the two-argument form:

```
. gen str startdatestr = string(startdatedouble, "%10.0g")

. gen str starttimestr = string(starttimedouble, "%16.0g")
```

If you omitted the format, string() would produce 2.01e+07 for 20060125 and 2.01e+13 for 20060125110215. The format we used had a width 2 larger than the length of the integer number, although using a too-wide format would not hurt.

The other translation functions

The other translation functions are

Format	String-to-numeric conversion function
%tw	weekly(*string*, *mask* [, *baseyear*])
%tm	monthly(*string*, *mask* [, *baseyear*])
%tq	quarterly(*string*, *mask* [, *baseyear*])
%th	halfyearly(*string*, *mask* [, *baseyear*])

string is the value to be translated.
mask specifies the order of the components.
baseyear is described in *Working with two-digit years* above.

These functions are rarely used because data seldom arrive in these formats.

All the functions translate a pair of numbers: weekly() translates a year and a week number (1–52), monthly() translates a year and a month number (1–12), quarterly() translates a year and a quarter number (1–4), and halfyearly() translates a year and a half number (1–2).

The masks allowed are far more limited than for clock(), Clock(), and date():

Code	Meaning	
Y	4-digit year	
19Y	2-digit year to be interpreted as 19*xx*	
20Y	2-digit year to be interpreted as 20*xx*	
W	week number	(weekly() only)
M	month number	(monthly() only)
Q	quarter number	(quarterly() only)
H	half number	(halfyearly() only)

The pair of numbers to be translated must be separated by space or punctuation. No extra characters are allowed.

Historical note: Before Stata 10, the mask codes were lowercase letters. Under version control, lowercase letters are still allowed.

Valid times

27:62:90 is an invalid time. If you try to convert 27:62:90 to a %tc or %tC value, you will get missing value or an error message.

24:00:00 is also invalid. Correct is 00:00:00 of the next day.

In *hh:mm:ss*, the requirements are $0 \le hh < 24$, $0 \le mm < 60$, and $0 \le ss < 60$, although sometimes 60 is allowed.

31dec2005 23:59:60 is an invalid %tc time but a valid %tC one. 31dec2005 23:59:60 was an inserted leap second.

30dec2005 23:59:60 is an invalid time in both %tc and %tC formats. 30dec2005 23:59:60 was not an inserted leap second. Correct is 31dec2005 00:00:00.

When leap seconds occurred

Stata system file leapseconds.maint lists the dates on which leap seconds occurred. The file is updated periodically (see [R] **update**; the file is updated when you update ado-files) and Stata's %tC functions access the file to know when leap seconds occurred.

You can access it, too. To view the file, type

```
. viewsource leapseconds.maint
```

Truncated times

Consider the time 11:32:59.999. Other, less precise ways of writing that time are

```
11:32:59.99
11:32:59.9
11:32:59
11:32
```

That is, when you suppress the display of more detailed components of the time, the parts that are displayed are not rounded. Stata displays time like a digital watch; the time is 11:32 right up until the instant that it is 11:33.

Advice on using %tc and %tC

Summary

Stata provides two time formats,

%tC, also known as UTC, which accounts for leap seconds, and

%tc, which ignores them (it assumes 86,400 seconds/day).

Systems vary in how they treat time variables. SAS ignores leap seconds. Oracle includes them. Stata handles either. Our advice:

1. If you obtain data from a system that accounts for leap seconds, import using Stata's %tC.

   a. If you later need to export data to a system that does not account for leap seconds, use Stata's cofC() function to translate time values before exporting.

   b. If you intend to tsset the time variable and the analysis will be at the second level or finer, just tsset the %tC variable, specifying the appropriate delta() if necessary, e.g., delta(1000) for seconds.

   c. If you intend to tsset the time variable and the analysis will be at coarser than the second level (minute, hour, etc.), create a %tc variable from the %tC variable (generate double *tctime* = cofC(*tCtime*)) and tsset that, specifying the appropriate delta() if necessary. You must do that because, in a %tC variable, there are not necessarily 60 seconds in a minute; some minutes have 61 seconds.

2. If you obtain data from a system that ignores leap seconds, use Stata's %tc.

   a. If you later need to export data to a system that does account for leap seconds, use Stata's Cofc() function to translate time values.

   b. If you intend to tsset the time variable, just tsset it, specifying the appropriate delta().

Some users prefer to always use Stata's %tc because %tc values are a little easier to work with. You can do that if

1. you do not mind having up to 1 second of error and

2. you do not import or export numerical values (clock ticks) from other systems that are using leap seconds, because then there could be nearly 30 seconds of error.

There are two things to remember if you use %tC variables:

1. The number of seconds between two dates is a function of when the dates occurred. Five days from one date is not simply a matter of adding $5 \times 24 \times 60 \times 60 \times 1,000$ ms. You might need to add another 1,000 ms. Three hundred and sixty-five days from now might require adding 1,000 or 2,000 ms. The longer the span, the more you might have to add. (The best way to add durations to %tC variables is to extract the components, add to them, and then reconstruct from the numerical components.)

2. You cannot accurately predict date/times into the future. We do not know what the %tC value will be of 25dec2026 00:00:00 because, along the way, the authorities may (and probably will) announce leap seconds.

Explanation

Stata's %tc encoding assumes that there are $24 \times 60 \times 60 \times 1,000$ ms per day, just as an atomic clock, counting oscillations between the nucleus of an atom and its electrons, would define it.

Since 1972, leap seconds have been added once or twice a year to keep time measured that way in synchronization with the earth's rotation. Unlike leap years, however, there is no formula to predict when leap seconds will occur. The earth is on average slowing down, but there is a relatively large random component, and so leap seconds are determined by fiat and announced 6 months before they are inserted. Leap seconds are added, if necessary, on the end of the day on June 30 and December 31 and are designated 23:59:60.

You may have heard various terms such as GMT and UTC.

GMT is the old Greenwich Mean Time and is based on astronomical observation.

UTC stands for coordinated universal time and is measured by atomic clocks, occasionally corrected for leap seconds.

UT1 is the mean solar time, with which UTC is kept in sync by the occasional addition of a leap second.

TAI is atomic time on which UTC is based. TAI was set to GMT plus 10 seconds in 1958 and has been running since then.

UNK is our term for the time standard most people use. UNK stands for unknown, or unknowing. UNK is based on a recent time observation, probably UTC, and then most people just assume that there are 86,400 seconds per day after that.

The UNK standard is usually adequate, and you will want to use %tc rather than the leap second–adjusted %tC encoding. If you are using computer-timestamped data, however, you may need to find out whether the timestamping system used leap second adjustment. Problems can arise even if you do not care about losing or gaining a second here and there.

For instance, you may import timestamp values from other systems as integers, recorded in the number of milliseconds, or export them. You may do this, but as of 18aug2006, if you choose the wrong encoding scheme (choose %tc when you should choose %tC, or vice versa), your recent times will be off by 23 seconds.

To avoid such problems, you may decide to import and export data by using printable forms, such as "Fri Aug 18 14:05:36 CDT 2006". This method has advantages, but for %tC encoding, times such as 23:59:60 are possible. Some systems will refuse to decode such times.

Stata refuses to decode 23:59:60 in the %tc encoding (function clock()) and accepts it with %tC (function Clock()). (When the %tC function Clock() sees a time with a 60th second, the function verifies that the time is one of the official leap seconds.) Thus, when translating from printable forms, you can assume %tc and check for missing values. If there are none, then you can use %tc. You will never be off by more than 1 second. If there are leap seconds in your data, use Clock() to translate them and then, if you still want to work in %tc units, use function cofC() to translate %tC values into %tc. Again, you will have no more than 1 second of inaccuracy.

If precision matters, the best way to process %tC data is simply to treat them that way. The inconvenience is that you cannot assume that there are 86,400 seconds per day. To obtain the duration between dates, you must subtract the two time values involved. The other difficulty has to do with dealing with dates in the future. Under the %tC encoding, there is no set value for any date more than 6 months in the future.

Also See

[D] **format** — Set variables' output format

Title

> **describe** — Describe data in memory or in file

Syntax

Describe data in memory

> d̲escribe [*varlist*] [, *memory_options*]

Describe data in file

> d̲escribe [*varlist*] using *filename* [, *file_options*]

memory_options	description
s̲imple	display only variable names
s̲hort	display only general information
d̲etail	display additional details
f̲ullnames	do not abbreviate variable names
n̲umbers	display variable number along with name
† v̲arlist	save r(varlist) and r(sortlist) in addition to usual saved results; programmer's option

† varlist is not shown in the dialog box.

file_options	description
s̲hort	display only general information
s̲imple	display only variable names
† v̲arlist	save r(varlist) and r(sortlist) in addition to usual saved results; programmer's option

† varlist is not shown in the dialog box.

Description

describe produces a summary of the dataset in memory or of the data stored in a Stata-format dataset.

For a compact listing of variable names, use describe, simple.

Options to describe data in memory

simple displays only the variable names in a compact format. simple may not be combined with other options.

short suppresses the specific information for each variable. Only the general information (number of observations, number of variables, size, and sort order) is displayed.

detail includes information on the width of 1 observation, the maximum number of observations holding the number of variables constant, the maximum number of variables holding the number of observations constant, the maximum width for 1 observation, and the maximum size of the dataset.

fullnames specifies that describe display the full names of the variables. The default is to present an abbreviation when the variable name is longer than 15 characters. describe using always shows the full names of the variables, so fullnames may not be specified with describe using.

numbers specifies that describe present the variable number with the variable name. If numbers is specified, variable names are abbreviated when the name is longer than eight characters. Options numbers and fullnames may not be specified together. numbers may not be specified with describe using.

The following option is available with describe but is not shown in the dialog box:

varlist, an option for programmers, specifies that, in addition to the usual saved results, r(varlist) and r(sortlist) be saved, too. r(varlist) will contain the names of the variables in the dataset. r(sortlist) will contain the names of the variables by which the data are sorted.

Options to describe data in file

simple displays only the variable names in a compact format. simple may not be combined with other options.

short suppresses the specific information for each variable. Only the general information (number of observations, number of variables, size, and sort order) is displayed.

The following option is available with describe but is not shown in the dialog box:

varlist, an option for programmers, specifies that, in addition to the usual saved results, r(varlist) and r(sortlist) be saved, too. r(varlist) will contain the names of the variables in the dataset. r(sortlist) will contain the names of the variables by which the data are sorted.

Because Stata/MP and Stata/SE can create truly large datasets, there might be too many variables in a dataset for their names to be stored in r(varlist), given the current maximum length of macros, as determined by set maxvar. Should that occur, describe using will issue the error message "too many variables", r(103).

Remarks

If describe is typed with no operands, the contents of the dataset currently in memory are described.

The *varlist* in the describe using syntax differs from standard Stata varlists in two ways. First, you cannot abbreviate variable names; that is, you have to type displacement rather than displ. However, you can use the wildcard character (~) to indicate abbreviations, e.g., displ~. Second, you may not refer to a range of variables; specifying price-trunk is considered an error.

▷ Example 1

The basic description includes some general information on the number of variables and observations, along with a description of every variable in the dataset:

```
. use http://www.stata-press.com/data/r10/states
(State data)
```

```
. describe, numbers
Contains data from http://www.stata-press.com/data/r10/states.dta
  obs:           50                         State data
 vars:            5                         3 Jan 2007 15:17
 size:        1,300 (99.7% of memory free)  (_dta has notes)
```

variable name	storage type	display format	value label	variable label
1. state	str8	%9s		
2. region	int	%8.0g	reg	Census Region
3. median~e	float	%9.0g		Median Age
4. marria~e	long	%12.0g		Marriages per 100,000
5. divorc~e	long	%12.0g		Divorces per 100,000

```
Sorted by:  region
```

In this example, the dataset in memory comes from the file states.dta and contains 50 observations on five variables. This dataset occupies only a small portion of the available memory, leaving 99.7% of memory free. The dataset is labeled "State data" and was last modified on January 3, 2007, at 15:17 (3:17 p.m.). The "_dta has notes" message indicates that a note is attached to the dataset; see [U] **12.7 Notes attached to data**.

The first variable, state, is stored as a str8 and has a display format of %9s.

The next variable, region, is stored as an int and has a display format of %8.0g. This variable has associated with it a *value label* called reg, and the variable is labeled Census Region.

The third variable, which is abbreviated median~e, is stored as a float, has a display format of %9.0g, has no value label, and has a variable label of Median Age. The variables that are abbreviated marria~e and divorc~e are both stored as longs and have display formats of %12.0g. These last two variables are labeled Marriages per 100,000 and Divorces per 100,000, respectively.

The data are sorted by region.

Since we specified the numbers option, the variables are numbered; e.g., region is variable 2 in this dataset.

◁

▷ Example 2

To view the full variable names, we could omit the numbers option and specify the fullnames option.

```
. describe, fullnames
Contains data from http://www.stata-press.com/data/r10/states.dta
  obs:           50                         State data
 vars:            5                         3 Jan 2007 15:17
 size:        1,300 (99.7% of memory free)  (_dta has notes)
```

variable name	storage type	display format	value label	variable label
state	str8	%9s		
region	int	%8.0g	reg	Census Region
median_age	float	%9.0g		Median Age
marriage_rate	long	%12.0g		Marriages per 100,000
divorce_rate	long	%12.0g		Divorces per 100,000

```
Sorted by:  region
```

Here we did not need to specify the `fullnames` option to see the unabbreviated variable names since the longest variable name is 13 characters. Omitting the `numbers` option results in 15-character variable names being displayed.

◁

❏ Technical Note

The `describe` listing above also shows that the size of the dataset is 1,300. In case you are curious,

$$\{(8 + 2 + 4 + 4 + 4) + 4\} \times 50 = 1300$$

The numbers 8, 2, 4, 4, and 4 are the storage requirements for a `str8`, `int`, `float`, `long`, and `long`, respectively; see [U] **12.2.2 Numeric storage types**. The extra 4 is needed for pointers, etc. Fifty is the number of observations in the dataset.

❏

▷ Example 3

If we specify the `short` option, only general information about the data is presented:

```
. describe, short
Contains data from http://www.stata-press.com/data/r10/states.dta
  obs:            50                          State data
  vars:            5                          3 Jan 2007 15:17
  size:        1,300 (99.7% of memory free)
Sorted by:  region
```

◁

If we specify a *varlist*, only the variables in that *varlist* are described.

▷ Example 4

The `detail` option is useful for determining how many observations or variables we can add to our dataset:

```
. describe, detail
Contains data from http://www.stata-press.com/data/r10/states.dta
  obs:            50 (max=       34,950)      State data
  vars:            5 (max=        2,048)      3 Jan 2007 15:17
  width:          22 (max=       20,966)
  size:        1,300 (max=    1,048,568)      (_dta has notes)

              storage  display     value
variable name   type   format      label     variable label

state          str8    %9s
region         int     %8.0g       reg       Census Region
median_age     float   %9.0g                 Median Age
marriage_rate  long    %12.0g                Marriages per 100,000
divorce_rate   long    %12.0g                Divorces per 100,000

Sorted by:  region
```

If we did not increase the number of variables in this dataset, we could have a maximum of 34,950 observations. The maximum number of variables is 2,048, which is the maximum for Stata/IC. The maximum width allowed is 20,966. The maximum size for the dataset is 1,048,568. The maximum dataset size could possibly be increased, since many operating systems allow us to change the size of memory; see [U] **6 Setting the size of memory** and [D] **memory**.

◁

▷ Example 5

Let's change datasets. The describe *varlist* command is particularly useful when combined with the '*' abbreviation character. For instance, we can describe all the variables whose names start with pop by typing describe pop*:

```
. use http://www.stata-press.com/data/r10/census
(1980 Census data by state)

. describe pop*
```

variable name	storage type	display format	value label	variable label
pop	long	%12.0gc		Population
poplt5	long	%12.0gc		Pop, < 5 year
pop5_17	long	%12.0gc		Pop, 5 to 17 years
pop18p	long	%12.0gc		Pop, 18 and older
pop65p	long	%12.0gc		Pop, 65 and older
popurban	long	%12.0gc		Urban population

We can describe the variables state, region, and pop18p by specifying them:

```
. describe state region pop18p
```

variable name	storage type	display format	value label	variable label
state	str14	%-14s		State
region	int	%-8.0g	cenreg	Census region
pop18p	long	%12.0gc		Pop, 18 and older

◁

Typing describe using *filename* describes the data stored in *filename*. If an extension is not specified, .dta is assumed.

▷ Example 6

We can describe the contents of states.dta without disturbing the data that we currently have in memory by typing

```
. describe using http://www.stata-press.com/data/r10/states
Contains data                         State data
    obs:          50                  3 Jan 2007 15:17
   vars:           5
   size:       1,300
```

variable name	storage type	display format	value label	variable label
state	str8	%9s		
region	int	%8.0g	reg	Census Region
median_age	float	%9.0g		Median Age
marriage_rate	long	%12.0g		Marriages per 100,000
divorce_rate	long	%12.0g		Divorces per 100,000

```
Sorted by:  region
```

◁

Saved Results

describe saves the following in r():

Scalars

r(N)	number of observations	r(k_max)	maximum number of variables
r(k)	number of variables	r(widthmax)	maximum width of dataset
r(width)	width of dataset	r(changed)	flag indicating data have changed
r(N_max)	maximum number of observations		since last saved

Macros

r(varlist)	variables in dataset	r(sortlist)	variables by which data are sorted
	(if varlist specified)		(if varlist specified)

References

Cox, N. J. 1999. dm67: Numbers of missing and present values. *Stata Technical Bulletin* 49: 7–8. Reprinted in *Stata Technical Bulletin Reprints*, vol. 9, pp. 26–27.

——. 2000. dm78: Describing variables in memory. *Stata Technical Bulletin* 56: 2–4. Reprinted in *Stata Technical Bulletin Reprints*, vol. 10, pp. 15–17.

——. 2001a. dm67.1: Enhancements to numbers of missing and present values. *Stata Technical Bulletin* 60: 2–3. Reprinted in *Stata Technical Bulletin Reprints*, vol. 10, pp. 7–9.

——. 2001b. dm78.1: Describing variables in memory: Update to Stata 7. *Stata Technical Bulletin* 60: 3. Reprinted in *Stata Technical Bulletin Reprints*, vol. 10, p. 17.

Gleason, J. R. 1998. dm61: A tool for exploring Stata datasets (Windows and Macintosh only). *Stata Technical Bulletin* 45: 2–5. Reprinted in *Stata Technical Bulletin Reprints*, vol. 8, pp. 22–27.

——. 1999. dm61.1: Update to varxplor. *Stata Technical Bulletin* 51: 2. Reprinted in *Stata Technical Bulletin Reprints*, vol. 9, p. 15.

(*Continued on next page*)

Also See

Title

> **destring** — Convert string variables to numeric variables and vice versa

Syntax

Convert string variables to numeric variables

> destring [*varlist*], { generate(*newvarlist*) | replace } [*destring_options*]

Convert numeric variables to string variables

> tostring *varlist* , { generate(*newvarlist*) | replace } [*tostring_options*]

destring_options	description
* generate(*newvarlist*)	generate *newvar*$_1$, ..., *newvar*$_k$ for each variable in *varlist*
* replace	replace string variables in *varlist* with numeric variables
ignore("*chars*")	remove specified nonnumeric characters
force	convert nonnumeric strings to missing values
float	generate numeric variables as type float
percent	convert percent variables to fractional form

* Either generate(*newvarlist*) or replace is required.

tostring_options	description
* generate(*newvarlist*)	generate *newvar*$_1$, ..., *newvar*$_k$ for each variable in *varlist*
* replace	replace numeric variables in *varlist* with string variables
force	force conversion ignoring information loss
format(*format*)	convert using specified format
usedisplayformat	convert using display format

* Either generate(*newvarlist*) or replace is required.

Description

destring converts variables in *varlist* from string to numeric. If *varlist* is not specified, destring will attempt to convert all variables in the dataset from string to numeric. Characters listed in ignore() are removed. Variables in *varlist* that are already numeric will not be changed. destring treats both empty strings "" and "." as indicating sysmiss (.) and interprets the strings ".a", ".b", ..., ".z" as the extended missing values .a, .b, ..., .z. destring also ignores any leading or trailing spaces so that, for example, " " is equivalent to "" and " . " is equivalent to ".".

tostring converts variables in *varlist* from numeric to string. The most compact string format possible is used. Variables in *varlist* that are already string will not be converted.

Options for destring

Either `generate()` or `replace` must be specified. With either option, if any string variable contains nonnumeric values not specified with `ignore()`, then no corresponding variable will be generated. Nor will that variable be replaced, unless `force` is specified.

`generate`(*newvarlist*) specifies that a new variable be created for each variable in *varlist*. *newvarlist* must contain the same number of new variable names as there are variables in *varlist*. If *varlist* is not specified, `destring` attempts to generate a numeric variable for each variable in the dataset; *newvarlist* must then contain the same number of new variable names as there are variables in the dataset. Any variable labels or characteristics will be copied to the new variables created.

`replace` specifies that the variables in *varlist* be converted to numeric variables. If *varlist* is not specified, `destring` attempts to convert all variables from string to numeric. Any variable labels or characteristics will be retained.

`ignore`("*chars*") specifies nonnumeric characters to be removed. If any string variable contains any nonnumeric characters other than those specified with `ignore()`, no action will take place for that variable unless `force` is also specified.

`force` specifies that any string values containing nonnumeric characters, in addition to any specified with `ignore()`, be treated as indicating missing numeric values.

`float` specifies that any new numeric variables be created initially as type `float`. The default is type `double`. `destring` attempts automatically to compress each new numeric variable after creation.

`percent` removes any percent signs found in the values of a variable, and all values of that variable are divided by 100 to convert the values to fractional form. `percent` by itself implies that the percent sign "%" is an argument to `ignore()`, but the converse is not true.

Options for tostring

Either `generate()` or `replace` must be specified. If converting any numeric variable to string would result in loss of information, no variable will be produced unless `force` is specified. For more details, see `force` below.

`generate`(*newvarlist*) specifies that a new variable be created for each variable in *varlist*. *newvarlist* must contain the same number of new variable names as there are variables in *varlist*. Any variable labels or characteristics will be copied to the new variables created.

`replace` specifies that the variables in *varlist* be converted to string variables. Any variable labels or characteristics will be retained.

`force` specifies that conversions be forced even if they entail loss of information. Loss of information means one of two circumstances: (1) The result of `real(string(`*varname*`, "`*format*`"))` is not equal to *varname*; i.e., the conversion is not reversible without loss of information; (2) `replace` was specified, but a variable has associated value labels. In circumstance (1), it is usually best to specify `usedisplayformat` or `format()`. In circumstance (2), value labels will be ignored in a forced conversion. `decode` (see [D] **encode**) is the standard way to generate a string variable based on value labels.

`format`(*format*) specifies that a numeric format be used as an argument to the `string()` function, which controls the conversion of the numeric variable to string. For example, a format of `%7.2f` specifies that numbers are to be rounded to two decimal places before conversion to string. See *Remarks* below and [D] **functions** and [D] **format**. `format()` cannot be specified with `usedisplayformat`.

usedisplayformat specifies that the current display format be used for each variable. For example, this option could be useful when using U.S. social security numbers. usedisplayformat cannot be specified with format().

Remarks

Remarks are presented under the following headings:

> *destring*
> *tostring*

destring

▷ Example 1

We read in a dataset, but somehow all variables were created as strings. The variables contain no nonnumeric characters, and we want to convert them all from string to numeric data types.

```
. use http://www.stata-press.com/data/r10/destring1
. describe
Contains data from http://www.stata-press.com/data/r10/destring1.dta
  obs:            10
  vars:            5                          3 Mar 2007 10:15
  size:          240 (99.9% of memory free)
```

variable name	storage type	display format	value label	variable label
id	str3	%9s		
num	str3	%9s		
code	str4	%9s		
total	str5	%9s		
income	str5	%9s		

```
Sorted by:
. list
```

	id	num	code	total	income
1.	111	243	1234	543	23423
2.	111	123	2345	67854	12654
3.	111	234	3456	345	43658
4.	222	345	4567	57	23546
5.	333	456	5678	23	21432
6.	333	567	6789	23465	12987
7.	333	678	7890	65	9823
8.	444	789	8976	23	32980
9.	444	901	7654	23	18565
10.	555	890	6543	423	19234

```
. destring, replace
id has all characters numeric; replaced as int
num has all characters numeric; replaced as int
code has all characters numeric; replaced as int
total has all characters numeric; replaced as long
income has all characters numeric; replaced as long
```

```
. describe
Contains data from http://www.stata-press.com/data/r10/destring1.dta
  obs:           10
  vars:           5                           3 Mar 2007 10:15
  size:         180 (99.9% of memory free)
```

variable name	storage type	display format	value label	variable label
id	int	%10.0g		
num	int	%10.0g		
code	int	%10.0g		
total	long	%10.0g		
income	long	%10.0g		

```
Sorted by:
    Note:  dataset has changed since last saved
. list
```

	id	num	code	total	income
1.	111	243	1234	543	23423
2.	111	123	2345	67854	12654
3.	111	234	3456	345	43658
4.	222	345	4567	57	23546
5.	333	456	5678	23	21432
6.	333	567	6789	23465	12987
7.	333	678	7890	65	9823
8.	444	789	8976	23	32980
9.	444	901	7654	23	18565
10.	555	890	6543	423	19234

◁

▷ Example 2

Our dataset contains the variable date, which was accidentally recorded as a string because of spaces after the year and month. We want to remove the spaces. destring will convert it to numeric and remove the spaces.

```
. use http://www.stata-press.com/data/r10/destring2, clear
. describe date
```

variable name	storage type	display format	value label	variable label
date	str14	%10s		

```
. list date
```

	date
1.	1999 12 10
2.	2000 07 08
3.	1997 03 02
4.	1999 09 00
5.	1998 10 04
6.	2000 03 28
7.	2000 08 08
8.	1997 10 20
9.	1998 01 16
10.	1999 11 12

```
. destring date, replace ignore(" ")
date: characters space   removed; replaced as long

. describe date
```

variable name	storage type	display format	value label	variable label
date	long	%10.0g		

```
. list date
```

	date
1.	19991210
2.	20000708
3.	19970302
4.	19990900
5.	19981004
6.	20000328
7.	20000808
8.	19971020
9.	19980116
10.	19991112

◁

▷ Example 3

Our dataset contains the variables date, price, and percent. These variables were accidentally read into Stata as string variables because they contain spaces, dollar signs, commas, and percent signs. We want to remove all these characters and create new variables for date, price, and percent containing numeric values. After removing the percent sign, we want to convert the variable percent to decimal form.

(Continued on next page)

```
. use http://www.stata-press.com/data/r10/destring2, clear
. describe
Contains data from http://www.stata-press.com/data/r10/destring2.dta
  obs:            10
  vars:            3                          3 Mar 2007 22:50
  size:          320 (99.9% of memory free)
```

variable name	storage type	display format	value label	variable label
date	str14	%10s		
price	str11	%11s		
percent	str3	%9s		

```
Sorted by:
. list
```

	date	price	percent
1.	1999 12 10	$2,343.68	34%
2.	2000 07 08	$7,233.44	86%
3.	1997 03 02	$12,442.89	12%
4.	1999 09 00	$233,325.31	6%
5.	1998 10 04	$1,549.23	76%
6.	2000 03 28	$23,517.03	35%
7.	2000 08 08	$2.43	69%
8.	1997 10 20	$9,382.47	32%
9.	1998 01 16	$289,209.32	45%
10.	1999 11 12	$8,282.49	1%

```
. destring date price percent, generate(date2 price2 percent2) ignore("$ ,%")
> percent
date: characters space removed; date2 generated as long
price: characters $ , removed; price2 generated as double
percent: characters % removed; percent2 generated as double
. describe
Contains data from http://www.stata-press.com/data/r10/destring2.dta
  obs:            10
  vars:            6                          3 Mar 2007 22:50
  size:          520 (99.9% of memory free)
```

variable name	storage type	display format	value label	variable label
date	str14	%10s		
date2	long	%10.0g		
price	str11	%11s		
price2	double	%10.0g		
percent	str3	%9s		
percent2	double	%10.0g		

```
Sorted by:
    Note:  dataset has changed since last saved
```

```
. list
```

	date	date2	price	price2	percent	percent2
1.	1999 12 10	19991210	$2,343.68	2343.68	34%	.34
2.	2000 07 08	20000708	$7,233.44	7233.44	86%	.86
3.	1997 03 02	19970302	$12,442.89	12442.89	12%	.12
4.	1999 09 00	19990900	$233,325.31	233325.31	6%	.06
5.	1998 10 04	19981004	$1,549.23	1549.23	76%	.76
6.	2000 03 28	20000328	$23,517.03	23517.03	35%	.35
7.	2000 08 08	20000808	$2.43	2.43	69%	.69
8.	1997 10 20	19971020	$9,382.47	9382.47	32%	.32
9.	1998 01 16	19980116	$289,209.32	289209.32	45%	.45
10.	1999 11 12	19991112	$8,282.49	8282.49	1%	.01

◁

tostring

Conversion of numeric data to string equivalents can be problematic. Stata, like most software, holds numeric data to finite precision and in binary form. See the discussion in [U] **13.10 Precision and problems therein**. If no format() is specified, tostring uses the format %12.0g. This format is, in particular, sufficient to convert integers held as bytes, ints, or longs to string equivalent without loss of precision.

However, users will often need to specify a format themselves, especially when the numeric data have fractional parts and for some reason a conversion to string is required.

▷ Example 4

Our dataset contains a string month variable and a numeric year and day variable. We want to convert the three variables to a %td date.

```
. use http://www.stata-press.com/data/r10/tostring, clear
. list
```

	id	month	day	year
1.	123456789	jan	10	2001
2.	123456710	mar	20	2001
3.	123456711	may	30	2001
4.	123456712	jun	9	2001
5.	123456713	oct	17	2001
6.	123456714	nov	15	2001
7.	123456715	dec	28	2001
8.	123456716	apr	29	2001
9.	123456717	mar	11	2001
10.	123456718	jul	3	2001

```
. tostring year day, replace
year was float now str4
day was float now str2
. generate date = month + "/" + day + "/" + year
. generate edate = date(date, "MDY")
. format edate %td
```

. list

	id	month	day	year	date	edate
1.	123456789	jan	10	2001	jan/10/2001	10jan2001
2.	123456710	mar	20	2001	mar/20/2001	20mar2001
3.	123456711	may	30	2001	may/30/2001	30may2001
4.	123456712	jun	9	2001	jun/9/2001	09jun2001
5.	123456713	oct	17	2001	oct/17/2001	17oct2001
6.	123456714	nov	15	2001	nov/15/2001	15nov2001
7.	123456715	dec	28	2001	dec/28/2001	28dec2001
8.	123456716	apr	29	2001	apr/29/2001	29apr2001
9.	123456717	mar	11	2001	mar/11/2001	11mar2001
10.	123456718	jul	3	2001	jul/3/2001	03jul2001

◁

Saved characteristics

Each time the destring or tostring commands are issued, an entry is made in the characteristics list of each converted variable. You can type char list to view these characteristics.

After example 3, we could use char list to find out what characters were removed by the destring command.

```
. char list
date2[destring]:        Characters removed were:  space
price2[destring]:       Characters removed were:  $ ,
percent2[destring]:     Characters removed were:  %
```

Methods and Formulas

destring and tostring are implemented as ado-files.

Acknowledgment

destring and tostring were originally written by Nicholas J. Cox of Durham University.

References

Cox, N. J. 1999a. dm45.1: Changing string variables to numeric: Update. *Stata Technical Bulletin* 49: 2. Reprinted in *Stata Technical Bulletin Reprints*, vol. 9, p. 14.

——. 1999b. dm45.2: Changing string variables to numeric: Correction. *Stata Technical Bulletin* 52: 2. Reprinted in *Stata Technical Bulletin Reprints*, vol. 9, p. 14.

Cox, N. J., and W. W. Gould. 1997. dm45: Changing string variables to numeric. *Stata Technical Bulletin* 37: 4–6. Reprinted in *Stata Technical Bulletin Reprints*, vol. 7, pp. 34–37.

Cox, N. J., and J. B. Wernow. 2000a. dm80: Changing numeric variables to string. *Stata Technical Bulletin* 56: 8. Reprinted in *Stata Technical Bulletin Reprints*, vol. 10, pp. 24–28.

——. 2000b. dm80.1: Update to changing numeric variables to string. *Stata Technical Bulletin* 57: 2. Reprinted in *Stata Technical Bulletin Reprints*, vol. 10, pp. 28–29.

Also See

[D] **generate** — Create or change contents of variable

[D] **split** — Split string variables into parts

[D] **egen** — Extensions to generate

[D] **encode** — Encode string into numeric and vice versa

[D] **functions** — Functions

Title

> **dir** — Display filenames

Syntax

$\{$dir $|$ ls$\}$ $\left[\texttt{"}\right]\left[\textit{filespec}\right]\left[\texttt{"}\right]$ $\left[$, <u>w</u>ide $\right]$

Note: Double quotes must be used to enclose *filespec* if the name contains spaces.

Description

dir and ls—they work the same way—list the names of files in the specified directory; the names of the commands come from names popular on Windows and Unix computers. *filespec* may be any valid Windows, Unix, or Macintosh file path or file specification (see [U] **11.6 File-naming conventions**) and may include '*' to indicate any string of characters.

Option

<u>w</u>ide under Windows and Macintosh produces an effect similar to specifying /W with the DOS DIR command—it compresses the resulting listing by placing more than one filename on a line. Under Unix, it produces the same effect as typing ls -F -C. Without the wide option, ls is equivalent to typing ls -F -l.

Remarks

Windows: Other than minor differences in presentation format, there is only one difference between the Stata and DOS dir commands: the DOS /P option is unnecessary, since Stata always pauses when the screen is full.

Macintosh and Unix: The only difference between the Stata and Unix ls commands is that piping through the more(1) or pg(1) filter is unnecessary—Stata always pauses when the screen is full.

▷ Example 1

The only real difference between the Stata dir and DOS and Unix equivalent commands is that output never scrolls off the screen; Stata always pauses when the screen is full.

If you use Stata for Windows and wish to obtain a list of all your Stata-format data files, type

```
. dir *.dta
    3.9k   7/07/00 13:51   auto.dta
    0.6k   8/04/00 10:40   cancer.dta
    3.5k   7/06/98 17:06   census.dta
    3.4k   1/25/98  9:20   hsng.dta
    0.3k   1/26/98 16:54   kva.dta
    0.7k   4/27/00 11:39   sysage.dta
    0.5k   5/09/97  2:56   systolic.dta
   10.3k   7/13/98  8:37   Household Survey.dta
```

You could also include the wide option:

```
. dir *.dta, wide
    3.9k auto.dta              0.6k cancer.dta            3.5k census.dta
    3.4k hsng.dta              0.3k kva.dta               0.7k sysage.dta
    0.5k systolic.dta         10.3k Household Survey.dta
```

Unix users will find it more natural to type

```
. ls *.dta
-rw-r-----  1 roger      2868 Mar  4 15:34 highway.dta
-rw-r-----  1 roger       941 Apr  5 09:43 hoyle.dta
-rw-r-----  1 roger     19312 May 14 10:36 p1.dta
-rw-r-----  1 roger     11838 Apr 11 13:26 p2.dta
```

but they could type dir if they preferred. Macintosh users may also type either command.

```
. dir *.dta
-rw-r-----  1 roger      2868 Mar  4 15:34 highway.dta
-rw-r-----  1 roger       941 Apr  5 09:43 hoyle.dta
-rw-r-----  1 roger     19312 May 14 10:36 p1.dta
-rw-r-----  1 roger     11838 Apr 11 13:26 p2.dta
```

◁

Also See

[D] **cd** — Change directory

[D] **copy** — Copy file from disk or URL

[D] **erase** — Erase a disk file

[D] **mkdir** — Create directory

[D] **rmdir** — Remove directory

[D] **shell** — Temporarily invoke operating system

[D] **type** — Display contents of a file

[P] **macro** — Macro definition and manipulation

[U] **11.6 File-naming conventions**

Title

> **drawnorm** — Draw sample from multivariate normal distribution

Syntax

drawnorm *newvarlist* [, *options*]

options	description
Main	
clear	replace the current dataset
double	generate variable type as double; default is float
n(*#*)	# of observations to be generated; default is current number
<u>sds</u>(*vector*)	standard deviations of generated variables
corr(*matrix* | *vector*)	correlation matrix
cov(*matrix* | *vector*)	covariance matrix
<u>c</u>storage(<u>f</u>ull)	correlation/covariance structure is stored as a symmetric $k \times k$ matrix
<u>c</u>storage(<u>l</u>ower)	correlation/covariance structure is stored as a lower triangular matrix
<u>c</u>storage(<u>u</u>pper)	correlation/covariance structure is stored as an upper triangular matrix
forcepsd	force the covariance/correlation matrix to be positive semidefinite
<u>means</u>(*vector*)	means of generated variables; default is means(0)
Options	
seed(*#*)	seed for random-number generator

Description

drawnorm draws a sample from a multivariate normal distribution with desired means and covariance matrix. The default is orthogonal data with mean 0 and variance 1. The covariance matrix may be singular. The values generated are a function of the current random-number seed or the number specified with set seed(); see [D] **generate**.

Options

> Main

clear specifies that the dataset in memory be replaced, even though the current dataset has not been saved on disk.

double specifies that the new variables be stored as Stata doubles, meaning 8-byte reals. If double is not specified, variables are stored as floats, meaning 4-byte reals. See [D] **data types**.

n(*#*) specifies the number of observations to be generated. The default is the current number of observations. If n(*#*) is not specified or is the same as the current number of observations, drawnorm adds the new variables to the existing dataset; otherwise, drawnorm replaces the data in memory.

sds(*vector*) specifies the standard deviations of the generated variables. sds() may not be specified with cov().

corr(*matrix* | *vector*) specifies the correlation matrix. If neither corr() nor cov() is specified, the default is orthogonal data.

cov(*matrix* | *vector*) specifies the covariance matrix. If neither corr() nor cov() is specified, the default is orthogonal data.

cstorage(full | lower | upper) specifies the storage mode for the correlation or covariance structure in corr() or cov(). The following storage modes are supported:

full specifies that the correlation or covariance structure is stored (recorded) as a symmetric $k \times k$ matrix.

lower specifies that the correlation or covariance structure is recorded as a lower triangular matrix. With k variables, the matrix should have $k(k+1)/2$ elements in the following order:

$$C_{11} \; C_{21} \; C_{22} \; C_{31} \; C_{32} \; C_{33} \; \ldots \; C_{k1} \; C_{k2} \; \ldots \; C_{kk}$$

upper specifies that the correlation or covariance structure is recorded as an upper triangular matrix. With k variables, the matrix should have $k(k+1)/2$ elements in the following order:

$$C_{11} \; C_{12} \; C_{13} \; \ldots \; C_{1k} \; C_{22} \; C_{23} \; \ldots C_{2k} \; \ldots \; C_{(k-1k-1)} \; C_{(k-1k)} \; C_{kk}$$

Specifying cstorage(full) is optional if the matrix is square. cstorage(lower) or cstorage(upper) is required for the vectorized storage methods. See *Example 2: Storage modes for correlation and covariance matrices.*

forcepsd modifies the matrix C to be positive semidefinite (psd), and so be a proper covariance matrix. If C is not positive semidefinite, it will have negative eigenvalues. By setting negative eigenvalues to 0 and reconstructing, we obtain the least-squares positive-semidefinite approximation to C. This approximation is a singular covariance matrix.

means(*vector*) specifies the means of the generated variables. The default is means(0).

___
⌐ Options ⌐

seed(#) specifies the initial value of the random-number seed used by the uniform() function. The default is the current random-number seed. Specifying seed(#) is the same as typing set seed # before issuing the drawnorm command.

Remarks

▷ Example 1

Suppose that we want to draw a sample of 1,000 observations from a normal distribution $N(\mathbf{M}, \mathbf{V})$, where $\mathbf{M}$ is the mean matrix and $\mathbf{V}$ is the covariance matrix:

```
. matrix M = 5, -6, 0.5
. matrix V = (9, 5, 2 \ 5, 4, 1 \ 2, 1, 1)
. matrix list M
M[1,3]
      c1   c2   c3
r1     5   -6   .5
```

```
. matrix list V
symmetric V[3,3]
     c1  c2  c3
r1   9
r2   5   4
r3   2   1   1
. drawnorm x y z, n(1000) cov(V) means(M)
(obs 1000)
. summarize
    Variable │      Obs        Mean    Std. Dev.         Min          Max
─────────────┼──────────────────────────────────────────────────────────
           x │     1000    5.001715     3.00608    -4.572042     13.66046
           y │     1000   -5.980279    2.004755    -12.08166    -.0963039
           z │     1000     .5271135    1.011095    -2.636946     4.102734
. correlate, cov
(obs=1000)

             │        x          y          z
─────────────┼───────────────────────────────
           x │  9.03652
           y │  5.04462    4.01904
           z │  2.10142    1.08773    1.02231
```

◁

❏ Technical Note

The values generated by drawnorm are a function of the current random-number seed. To reproduce the same dataset each time drawnorm is run with the same setup, specify the same seed number in the seed() option.

❏

▷ Example 2: Storage modes for correlation and covariance matrices

The three storage modes for specifying the correlation or covariance matrix in corr2data and drawnorm can be illustrated with a correlation structure C of 4 variables. In full storage mode, this structure can be entered as a 4×4 Stata matrix:

```
. matrix C = ( 1.0000,   0.3232,   0.1112,   0.0066 \ ///
               0.3232,   1.0000,   0.6608,  -0.1572 \ ///
               0.1112,   0.6608,   1.0000,  -0.1480 \ ///
               0.0066,  -0.1572,  -0.1480,   1.0000 )
```

Elements within a row are separated by commas, and rows are separated by a backslash \. We use the input continuation operator /// for convenient multiline input; see [P] **comments**. In this storage mode, we probably want to set the row and column names to the variable names:

```
.   matrix rownames C = price trunk headroom rep78
.   matrix colnames C = price trunk headroom rep78
```

This correlation structure can be entered more conveniently in one of the two vectorized storage modes. In these modes, we enter the lower triangle or the upper triangle of C in rowwise order; these two storage modes differ only in the order in which the $k(k+1)/2$ matrix elements are recorded. The lower storage mode for C comprises a vector with $4(4+1)/2 = 10$ elements, i.e., a 1×10 or 10×1 Stata matrix, with one row or column,

```
.   matrix C = ( 1.0000,  ///
                 0.3232,   1.0000,  ///
                 0.1112,   0.6608,   1.0000,  ///
                 0.0066,  -0.1572,  -0.1480,   1.0000)
```

or more compactly as

```
.  matrix C = ( 1, 0.3232, 1, 0.1112, 0.6608, 1, 0.0066, -0.1572, -0.1480, 1 )
```

C may also be entered in upper storage mode as a vector with $4(4+1)/2 = 10$ elements, i.e., a 1×10 or 10×1 Stata matrix

```
.  matrix C = ( 1.0000,   0.3232,   0.1112,   0.0066, ///
                          1.0000,   0.6608,  -0.1572, ///
                                    1.0000,  -0.1480, ///
                                              1.0000 )
```

or more compactly as

```
.  matrix C = ( 1, 0.3232, 0.1112, 0.0066, 1, 0.6608, -0.1572, 1, -0.1480, 1 )
```

◁

Methods and Formulas

drawnorm is implemented as an ado-file.

Results are asymptotic. The more observations generated, the closer the correlation matrix of the dataset is to the desired correlation structure.

Let $\mathbf{V} = \mathbf{A}'\mathbf{A}$ be the desired covariance matrix and $\mathbf{M}$ be the desired mean matrix. We first generate $\mathbf{X}$, such that $\mathbf{X} \sim N(\mathbf{0}, \mathbf{I})$. Let $\mathbf{Y} = \mathbf{A}'\mathbf{X} + \mathbf{M}$, then $\mathbf{Y} \sim N(\mathbf{M}, \mathbf{V})$.

Also See

[D] **corr2data** — Create dataset with specified correlation structure

[D] **generate** — Create or change contents of variable

[D] **data types** — Quick reference for data types

Title

> **drop** — Eliminate variables or observations

Syntax

Drop variables

> drop *varlist*

Drop observations

> drop if *exp*

Drop a range of observations

> drop in *range* [if *exp*]

Keep variables

> keep *varlist*

Keep observations that satisfy specified condition

> keep if *exp*

Keep a range of observations

> keep in *range* [if *exp*]

by is allowed with the second syntax of drop and the second syntax of keep; see [D] **by**.

Description

drop eliminates variables or observations from the data in memory.

keep works the same way as drop, except that you specify the variables or observations to be kept rather than the variables or observations to be deleted.

Remarks

You can clear the entire dataset by typing drop _all without affecting value labels, macros, and programs. (Also see [U] **12.6 Dataset, variable, and value labels**, [U] **18.3 Macros**, and [P] **program**.)

▷ Example 1

We will systematically eliminate data until, at the end, no data are left in memory. We begin by describing the data:

```
. use http://www.stata-press.com/data/r10/census11
(1980 Census data by state)

. describe

Contains data from http://www.stata-press.com/data/r10/census11.dta
  obs:            50                          1980 Census data by state
  vars:           14                          6 Mar 2007 15:14
  size:        3,400 (99.5% of memory free)
```

| | storage | display | value | |
variable name	type	format	label	variable label
state	str14	%-14s		State
region	int	%-8.0g	cenreg	Census region
pop	long	%12.0gc		Population
poplt5	long	%12.0gc		Pop, < 5 year
pop5_17	long	%12.0gc		Pop, 5 to 17 years
pop18p	long	%12.0gc		Pop, 18 and older
pop65p	long	%12.0gc		Pop, 65 and older
popurban	long	%12.0gc		Urban population
medage	float	%9.2f		Median age
death	long	%12.0gc		Number of deaths
marriage	long	%12.0gc		Number of marriages
divorce	long	%12.0gc		Number of divorces
mrgrate	float	%9.0g		
dvcrate	float	%9.0g		

```
Sorted by:  region
```

We can eliminate all the variables with names that begin with pop by typing drop pop*:

```
. drop pop*

. describe

Contains data from http://www.stata-press.com/data/r10/census11.dta
  obs:            50                          1980 Census data by state
  vars:            8                          6 Mar 2007 15:14
  size:        2,200 (99.6% of memory free)
```

| | storage | display | value | |
variable name	type	format	label	variable label
state	str14	%-14s		State
region	int	%-8.0g	cenreg	Census region
medage	float	%9.2f		Median age
death	long	%12.0gc		Number of deaths
marriage	long	%12.0gc		Number of marriages
divorce	long	%12.0gc		Number of divorces
mrgrate	float	%9.0g		
dvcrate	float	%9.0g		

```
Sorted by:  region
    Note:  dataset has changed since last saved
```

(*Continued on next page*)

Let's eliminate more variables and then eliminate observations:

```
. drop marriage divorce mrgrate dvcrate
. describe
Contains data from http://www.stata-press.com/data/r10/census11.dta
  obs:            50                          1980 Census data by state
  vars:            4                          6 Mar 2007 15:14
  size:         1,400 (99.7% of memory free)
```

| |storage|display|value| |
variable name	type	format	label	variable label
state	str14	%-14s		State
region	int	%-8.0g	cenreg	Census region
medage	float	%9.2f		Median age
death	long	%12.0gc		Number of deaths

```
Sorted by:  region
     Note:  dataset has changed since last saved
```

Next, we will `drop` any observation for which `medage` is greater than 32.

```
. drop if medage>32
(3 observations deleted)
```

Let's drop the first observation in each region:

```
. by region: drop if _n==1
(4 observations deleted)
```

Now we drop all but the last observation in each region:

```
. by region: drop if _n !=_N
(39 observations deleted)
```

Let's now drop the first 2 observations in our dataset:

```
. drop in 1/2
(2 observations deleted)
```

Finally, let's get rid of everything:

```
. drop _all
. describe
Contains data
  obs:             0
  vars:            0
  size:            0 (100.0% of memory free)
Sorted by:
```

◁

Typing `keep in 10/l` is the same as typing `drop in 1/9`.

Typing `keep if x==3` is the same as typing `drop if x !=3`.

`keep` is especially useful for keeping a few variables from a large dataset. Typing `keep myvar1 myvar2` is the same as typing `drop` followed by all the variables in the dataset *except* `myvar1` and `myvar2`.

Reference

Cox, N. J. 2001. dm89: Dropping variables or observations with missing values. *Stata Technical Bulletin* 60: 7–8. Reprinted in *Stata Technical Bulletin Reprints*, vol. 10, pp. 44–46.

Also See

[D] **clear** — Clear memory

[U] **11 Language syntax**

[U] **13 Functions and expressions**

Title

> **duplicates** — Report, tag, or drop duplicate observations

Syntax

Report duplicates

    duplicates <u>r</u>eport [*varlist*] [*if*] [*in*]

List one example for each group of duplicates

    duplicates <u>e</u>xamples [*varlist*] [*if*] [*in*] [, *options*]

List all duplicates

    duplicates <u>l</u>ist [*varlist*] [*if*] [*in*] [, *options*]

View duplicates in browser

    duplicates <u>b</u>rowse [*varlist*] [*if*] [*in*] [, <u>nol</u>abel]

Tag duplicates

    duplicates <u>t</u>ag [*varlist*] [*if*] [*in*] , <u>g</u>enerate(*newvar*)

Drop duplicates

    duplicates drop [*if*] [*in*]

    duplicates drop *varlist* [*if*] [*in*] , force

options	description
Main	
<u>c</u>ompress	compress width of columns in both table and display formats
<u>noc</u>ompress	use display format of each variable
fast	synonym for nocompress; no delay in output of large datasets
<u>ab</u>breviate(#)	abbreviate variable names to # characters; default is ab(8)
<u>str</u>ing(#)	truncate string variables to # characters; default is string(10)
Options	
<u>t</u>able	force table format
<u>d</u>isplay	force display format
<u>h</u>eader	display variable header once; default is table mode
<u>noh</u>eader	suppress variable header
<u>h</u>eader(#)	display variable header every # lines
clean	force table format with no divider or separator lines
<u>div</u>ider	draw divider lines between columns
<u>sep</u>arator(#)	draw a separator line every # lines; default is separator(5)
sepby(*varlist*)	draw a separator line whenever *varlist* values change
<u>nol</u>abel	display numeric codes rather than label values
Summary	
mean[(*varlist*)]	add line reporting the mean for each of the (specified) variables
sum[(*varlist*)]	add line reporting the sum for each of the (specified) variables
N[(*varlist*)]	add line reporting the number of nonmissing values for each of the (specified) variables
<u>labv</u>ar(*varname*)	substitute Mean, Sum, or N for value of *varname* in last row of table
Advanced	
<u>c</u>onstant[(*varlist*)]	separate and list variables that are constant only once
<u>notr</u>im	suppress string trimming
<u>ab</u>solute	display overall observation numbers when using by *varlist*:
nodotz	display numerical values equal to .z as field of blanks
<u>subv</u>arname	substitute characteristic for variable name in header
<u>lin</u>esize(#)	columns per line; default is linesize(79)

Description

duplicates reports, displays, lists, browses, tags, or drops duplicate observations, depending on the subcommand specified. Duplicates are observations with identical values either on all variables if no *varlist* is specified or on a specified *varlist*.

duplicates report produces a table showing observations that occur as one or more copies and indicating how many observations are "surplus" in the sense that they are the second (third, . . .) copy of the first of each group of duplicates.

duplicates examples lists one example for each group of duplicated observations. Each example represents the first occurrence of each group in the dataset.

duplicates list lists all duplicated observations.

duplicates browse opens the Data Editor on the duplicated observations. You may not change the data while in the Editor. See browse in [D] **edit**.

duplicates tag generates a variable representing the number of duplicates for each observation. This will be 0 for all unique observations.

duplicates drop drops all but the first occurrence of each group of duplicated observations. The word drop may not be abbreviated.

Any observations that do not satisfy specified if and/or in conditions are ignored when you use report, examples, list, browse, or drop. The variable created by tag will have missing values for such observations.

Options for duplicates examples and duplicates list

_____ Main _____

compress, nocompress, fast, abbreviate(#), string(#); see [D] **list**.

_____ Options _____

table, display, header, noheader, header(#), clean, divider, separator(#), sepby(*varlist*), nolabel; see [D] **list**.

_____ Summary _____

mean$\left[(varlist)\right]$, sum$\left[(varlist)\right]$, N$\left[(varlist)\right]$, labvar(*varname*); see [D] **list**.

_____ Advanced _____

constant$\left[(varlist)\right]$, notrim, absolute, nodotz, subvarname, linesize(#); see [D] **list**.

Option for duplicates browse

nolabel displays the underlying numeric values rather than the label values in the Editor for variables with value labels. This option is seldom specified.

Option for duplicates tag

generate(*newvar*) is required and specifies the name of a new variable that will tag duplicates.

Option for duplicates drop

force specifies that observations duplicated with respect to a named *varlist* be dropped. The force option is required when such a *varlist* is given as a reminder that information may be lost by dropping observations, given that those observations may differ on any variable not included in *varlist*.

Remarks

Current data management and analysis may hinge on detecting (and sometimes dropping) duplicate observations. In Stata terms, *duplicates* are observations with identical values, either on all variables, if no *varlist* is specified, or on a specified *varlist*, that is, 2 or more observations that are identical on all specified variables form a group of duplicates. When the specified variables are a set of explanatory variables, such a group is often called a *covariate pattern* or a *covariate class*.

Linguistic purists will point out that duplicate observations are strictly only those that occur in pairs, and they might prefer a more literal term, although the most obvious replacement, "replicates", already has another statistical meaning. However, the looser term appears in practice to be much more frequently used for this purpose and to be as easy to understand.

Observations may occur as duplicates through some error; for example, the same observations might have been entered more than once into your dataset. For example, some researchers deliberately enter a dataset twice. Each entry is a check on the other, and all observations should occur as identical pairs, assuming that one or more variables identify unique records. If there is just one copy, or more than two copies, there has been an error in data entry.

Or duplicate observations may also arise simply because some observations just happen to be identical, which is especially likely with categorical variables or large datasets. In this second situation, consider whether `contract`, which automatically produces a count of each distinct set of observations, is more appropriate for your problem. See [D] **contract**.

Observations unique on all variables in *varlist* occur as single copies. Thus there are no surplus observations in the sense that no observation may be dropped without losing information about the contents of observations. (Information will inevitably be lost on the frequency of such observations. Again, if recording frequency is important to you, `contract` is the better command to use.) Observations that are duplicated twice or more occur as copies, and, in each case, all but one copy may be considered surplus.

This command helps you produce a dataset, usually smaller than the original, in which each observation is *unique* (literally, each occurs once only) and *distinct* (each differs from all the others). If you are familiar with Unix systems, or with sets of Unix utilities ported to other platforms, you will know the `uniq` command, which removes duplicate adjacent lines from a file, usually as part of a pipe.

▷ Example 1

Suppose that we are given a dataset in which some observations are unique (no other observation is identical on all variables) and other observations are duplicates (in each case, at least 1 other observation exists that is identical). Imagine dropping all but 1 observation from each group of duplicates, that is, dropping the surplus observations. Now all the observations are unique. This example helps clarify the difference between (1) identifying unique observations before dropping surplus copies and (2) identifying unique observations after dropping surplus copies (whether in truth or merely in imagination). `codebook` (see [D] **codebook**) reports the number of unique values for each variable in this second sense.

Suppose that we have typed in a dataset for 200 individuals. However, a simple `describe` or `count` shows that we have 202 observations in our dataset. We guess that we may have typed in 2 observations twice. `duplicates report` gives a quick report of the occurrence of duplicates:

```
. use http://www.stata-press.com/data/r10/dupxmpl
. duplicates report
Duplicates in terms of all variables
```

copies	observations	surplus
1	198	0
2	4	2

Our hypothesis is supported: 198 observations are unique (just 1 copy of each), whereas 4 occur as duplicates (2 copies of each; in each case, 1 may be dubbed surplus). We now wish to see which observations are duplicates, so the next step is to ask for a duplicates list.

```
. duplicates list
Duplicates in terms of all variables
```

group:	obs:	id	x	y
1	42	42	0	2
1	43	42	0	2
2	145	144	4	4
2	146	144	4	4

The records for id 42 and id 144 were evidently entered twice. Satisfied, we now issue duplicates drop.

```
. duplicates drop
Duplicates in terms of all variables
(2 observations deleted)
```

◁

The report, list, and drop subcommands of duplicates are perhaps the most useful, especially for a relatively small dataset. For a larger dataset with many duplicates, a full listing may be too long to be manageable, especially as you see repetitions of the same data. duplicates examples gives you a more compact listing in which each group of duplicates is represented by just 1 observation, the first to occur.

You may prefer duplicates browse, in which you open the Data Editor on the duplicate observations and then you can move around inspecting those observations at will. Specifying browse rather than edit means that you cannot modify data while using the Data Editor. In particular, you cannot delete any of the duplicate observations (see [D] **edit**). This is our way of encouraging you to adopt what we regard as better style. If you drop observations, you should drop them directly, so that any changes you make to the data are easier to understand and reproduce from a log of your session.

A subcommand that is occasionally useful is duplicates tag, which generates a new variable containing the number of duplicates for each observation. Thus unique observations are tagged with value 0, and all duplicate observations are tagged with values greater than 0. For checking double data entry, in which you expect just one surplus copy for each individual record, you can generate a tag variable and then look at observations with tag not equal to 1 because both unique observations and groups with two or more surplus copies need inspection.

```
. duplicates tag, gen(tag)
Duplicates in terms of all variables
```

Methods and Formulas

`duplicates` is implemented as an ado-file.

Acknowledgments

`duplicates` was written by Nicholas J. Cox, Durham University, who in turn thanks Thomas Steichen, RJRT, for ideas contributed to an earlier jointly written program (Steichen and Cox 1998).

References

Jacobs, M. 1991. dm4: A duplicate value identification program. *Stata Technical Bulletin* 4: 5. Reprinted in *Stata Technical Bulletin Reprints*, vol. 1, p. 30.

Steichen, T. J., and N. J. Cox. 1998. dm53: Detection and deletion of duplicate observations. *Stata Technical Bulletin* 41: 2–4. Reprinted in *Stata Technical Bulletin Reprints*, vol. 7, pp. 52–55.

Wang, D. 2000. dm77: Removing duplicate observations in a dataset. *Stata Technical Bulletin* 54: 16–17. Reprinted in *Stata Technical Bulletin Reprints*, vol. 9, pp. 87–88.

Also See

[D] **edit** — Edit and list data with Data Editor

[D] **list** — List values of variables

[D] **codebook** — Describe data contents

[D] **contract** — Make dataset of frequencies and percentages

[D] **isid** — Check for unique identifiers

Title

> **edit** — Edit and list data with Data Editor

Syntax

Edit using Data Editor

> e̲dit [*varlist*] [*if*] [*in*] [, no̲label]

List using Data Editor

> b̲rowse [*varlist*] [*if*] [*in*] [, no̲label]

Description

edit brings up a spreadsheet-style data editor for entering new data and editing existing data. edit is a better alternative to input; see [D] **input**.

browse is similar to edit, except that it will not allow you to change the data. browse is a convenient alternative to list; see [D] **list**.

Option

no̲label causes the underlying numeric values, rather than the label values (equivalent strings), to be displayed for variables with value labels; see [D] **label**.

Remarks

Remarks are presented under the following headings:

Modes
The current observation and current variable
Viewing variable properties
Buttons
Assigning value labels to variables
Changing values of existing cells
Adding new variables
Adding new observations
Copying and pasting
Exiting
Logging changes
Advice

A tutorial discussion of edit and browse is found in the *Getting Started with Stata* manual. This entry provides technical details.

Clicking Stata's **Data Editor** button is equivalent to typing edit by itself. Clicking Stata's **Data Browser** button is equivalent to typing browse by itself.

edit, typed by itself, opens the Data Editor with all observations on all variables displayed. If you specify *varlist*, only the specified variables are displayed in the editor. If you specify one or both of in *range* and if *exp*, only the observations specified are displayed.

Modes

We will refer to the Data Editor in the singular with `edit` and `browse` referring to two of its three modes.

Full-edit mode. This is the editor's mode that you enter when you type `edit` or type `edit` followed by a list of variables. All features of the editor are turned on.

Restricted-edit mode. This is the editor's mode that you enter when you use `edit` with or without a list of variables but include `in` *range*, `if` *exp*, or both. A few of the editor's features are turned off, most notably the ability to sort data, the ability to delete all observations, and the ability to paste data into the editor.

Browse mode. This is the editor's mode that you enter when you use `browse`. All the editing features are turned off, ensuring that the data cannot be changed. One feature that is left on may surprise you: the ability to sort data. Sorting, in Stata's mind, is not really a change to the dataset. On the other hand, if you enter using `browse` and specify `in` *range* or `if` *exp*, sorting is not allowed. You can think of this as restricted-browse mode.

Actually, the editor does not set its mode to restricted just because you specify an `in` *range* or `if` *exp*. It sets its mode to restricted if you specify `in` or `if` and if this restriction is effective; that is, if the `in` or `if` would actually cause some data to be omitted. For instance, typing `edit if x>0` would result in unrestricted full-edit mode if x were greater than zero for all observations.

The current observation and current variable

The Data Editor looks much like a spreadsheet, with rows and columns corresponding to observations and variables, respectively. At all times, one of the cells is highlighted. This is called the current cell. The observation (row) of the current cell is called the current observation. The variable (column) of the current cell is called the current variable.

You change the current cell by clicking with the mouse on another cell, using the arrow keys, or moving the scroll bars.

To help distinguish between the different types of variables in the editor, string values are displayed in red, value labels are displayed in blue, and all other values are displayed in black. You can change the colors for strings and value labels by right-clicking on the Data Editor window and selecting **Preferences...**.

Viewing variable properties

When you double-click on a cell, the Variable Properties dialog appears. In edit mode, this allows you to change the variable's name, variable label, format, and value label; see [U] **11.3 Naming conventions**, [U] **12.6.2 Variable labels**, [U] **12.5 Formats: controlling how data are displayed**, and [U] **12.6.3 Value labels**. You can also open the **Define value labels** dialog to define and add value labels.

Buttons

Seven buttons appear at the top of the window.

Preserve updates the backup copy of the data. By default, when you enter the Data Editor, a backup copy is made so that you can abort your changes. (See the technical note below for information about turning off the automatic backup feature.) If you are satisfied with your changes but still wish to continue, you can click **Preserve** so that you do not have to exit and reenter the editor. **Preserve** is disabled in browse mode.

Restore undoes your changes by restoring the backup copy of your data. Clicking **Restore** does not change the backup copy, so you can click **Restore** as many times as you need to restore your original data. **Restore** is disabled in browse mode. If you turn off the automatic backup feature (see technical note below), **Restore** is also disabled in edit mode until you click **Preserve**.

Sort reorders the observations in ascending sequence of the current variable. **Sort** is disabled in restricted-edit and restricted-browse modes.

≪ sets the current variable as the first variable in the dataset. In edit mode, the change is real. When you exit the Data Editor, the variables will remain in the order that you have specified. In browse mode, the change is cosmetic. While in the editor, it appears as if the variable has been moved, but when you exit, the variables remain in the same order as they were originally.

≫ sets the current variable as the last variable in the dataset. In edit mode, the change is real. When you exit the Data Editor, the variables will remain in the order that you have specified. In browse mode, the change is cosmetic. While in the editor, it appears as if the variable has been moved, but when you exit, the variables remain in the same order as they were originally.

Hide eliminates the variable from the Data Editor. The effect is cosmetic. The variable is not dropped from the dataset; the editor merely stops displaying it.

Delete... displays a dialog with options to (1) delete the current variable, (2) delete the current observation, or (3) delete all observations whose corresponding variable's value is equal to that of the current observation. The third option is not presented in restricted-edit mode. **Delete...** is disabled in browse mode.

❏ Technical Note

By default, when you enter the Data Editor, a backup copy of your data is made on disk. For large datasets, making this copy takes time. If you do not want the backup copy to be made automatically, select **Edit > Preferences > General Preferences...** from the Stata menu bar, choose **Data Editor**, and uncheck *Auto-Preserve*. When you enter the Data Editor, **Restore** will now be disabled until you click **Preserve**.

❏

Assigning value labels to variables

You can assign a value label to a nonstring variable by right-clicking any cell on the variable column and selecting a value label from the **Assign Value Label to Variable** '*varname*' menu. You can define a value label by right-clicking on the Data Editor window and selecting **Define/Modify Value Labels...**.

Changing values of existing cells

Make the cell you wish to change the current cell. Type the new value, and press *Enter*. When updating string variables, do not type double quotes around the string. For variables that have a value label, you can right-click on the cell to display a list of values for the value label. You can assign a new value to the cell by selecting a value from the list.

❑ Technical Note

Stata experts will wonder about storage types. Say that variable `mpg` is stored as an `int` and you want to change the fourth observation to contain 22.5. The Data Editor will change the storage type of the variable. Similarly, if the variable is a `str4` and you type `alpha`, it will be changed to `str5`.

The editor will not, however, change numeric variable types to strings (unless the numeric variable contains only missing values). This is intentional, as such a change could result in a loss of data and is probably the result of a mistake.

❑

Adding new variables

Go to the first empty column, and begin entering your data. The first entry that you make will create the variable and determine whether that variable is numeric or string. The variable will be given a name like `var1`, but you can rename it by double-clicking on any cell in the column.

❑ Technical Note

Stata experts: the storage type will be determined automatically. If you type a number, the created variable will be numeric; if you type a string, it will be a string. Thus if you want a string variable, be sure that your first entry cannot be interpreted as a number. A way to achieve this is to use surrounding quotes so that `"123"` will be taken as the string `"123"`, not the number 123. If you want a numeric variable, do not worry about whether it is `byte`, `int`, `float`, etc. If a `byte` will hold your first number but you need a `float` to hold your second number, the editor will recast the variable later.

❑

❑ Technical Note

If you do not type in the first empty column but instead type in one to the right of it, the editor will create variables for all the intervening columns.

❑

Adding new observations

Go to the first empty row, and begin entering your data. As soon as you add one cell below the last row of the dataset, an observation will be created using the data you typed in the previous row.

❑ Technical Note

If you do not enter data in the first empty row but, instead, enter data in a row below it, the Data Editor will create observations for all the intervening rows.

❑

Copying and pasting

You can copy and paste data between Stata's Data Editor and other spreadsheets.

First, select the data you wish to copy. In Stata, click on a cell, and drag the mouse across other cells to select a range of cells. If you want to select an entire column, click once on the variable name at the top of that column. If you want to select an entire row, click once on the observation number at the left of that row. You can hold down the mouse button after clicking and drag to select multiple columns or rows.

Once you have selected the data, copy the data to the Clipboard. In Stata, right-click on the selected data, and select **Copy**.

You can copy data to the Clipboard from Stata with or without the variable names at the top of each column by right-clicking on the Data Editor window, selecting **Preferences...**, and checking or unchecking *Include variable names on copy to Clipboard*.

You can choose to copy either the value labels or the underlying numeric values associated with the selected data by right-clicking on the Data Editor window, selecting **Preferences...**, and checking or unchecking *Copy value labels instead of numbers*. For more information about value labels, see [U] **12.6.3 Value labels** and [D] **label**.

After you have copied data to the Clipboard from Stata's Data Editor or another spreadsheet, you can paste the data into Stata's Data Editor. First, select the top-left cell of the area into which you wish to paste the data by clicking on it once. Then right-click on the cell and select **Paste**. Stata will paste the data from the Clipboard into the editor, overwriting any data below and to the right of the cell you selected as the top left of the paste area. If you entered Stata's Data Editor in restricted-edit or in browse mode, **Paste** will be disabled, meaning that you cannot paste into Stata's Data Editor.

❏ Technical Note

If you attempt to paste one or more string values into numeric variables, the original numeric values will be left unchanged for those cells. Stata will display a message box to let you know that this has happened: "You attempted to paste one or more string values into numeric variables. The contents of these cells, if any, are unchanged."

If you see this message, you should look carefully at the data that you pasted into Stata's Data Editor to make sure that you pasted into the area that you intended. We recommend that you click **Preserve** before pasting into Stata's Data Editor so that you can **Restore** the data in case you make a mistake.

❏

Exiting

If you are using a Macintosh, click on the editor's close box.

If you are using Windows, click on the editor's close box (the box with an **X** at the right of the editor's title bar). You can also hold down *Alt* and press *F4* to exit the editor.

Logging changes

When you use `edit` to change existing data (as opposed to entering new data), you will find output in the Stata Results window documenting the changes that you made. For example, a line of this output might be

```
- replace mpg=22.5 in 5
```

The syntax is that of the Stata command you could have typed to achieve the same result that you did using the editor. The dash in front of the command indicates that the change was done in the editor. If you are logging your results, you will have a permanent record of what you did.

Advice

1. People who care about data integrity know that editors are dangerous—it is too easy to make changes accidentally. Never use `edit` when you want to `browse`.

2. Protect yourself when you edit existing data by limiting exposure. If you need to change `mpg` and need to see `model` to know which value of `mpg` to change, do not click the **Data Editor** button. Instead, type `edit model mpg`. It is now impossible for you to change (damage) variables other than `model` and `mpg`. Furthermore, if you know that you need to change `mpg` only if it is missing, you can reduce your exposure even more by typing 'edit model mpg if mpg>=.'.

3. Stata's Data Editor is safer than most because it logs changes to the Results window. Use this feature—look at the log afterward, and verify that the changes you made are the changes you wanted to make.

References

Brady, T. 1998. dm63: Dialog box window for browsing, editing, and entering observations. *Stata Technical Bulletin* 46: 2–6. Reprinted in *Stata Technical Bulletin Reprints*, vol. 8, pp. 28–34.

——. 2000. dm63.1: A new version of winshow for Stata 6. *Stata Technical Bulletin* 53: 3–5. Reprinted in *Stata Technical Bulletin Reprints*, vol. 9, pp. 15–19.

Also See

[D] **input** — Enter data from keyboard

[D] **insheet** — Read ASCII (text) data created by a spreadsheet

[D] **list** — List values of variables

[D] **save** — Save datasets

[GSM] **8 Using the Data Editor**

[GSW] **8 Using the Data Editor**

[GSU] **8 Using the Data Editor**

Title

egen — Extensions to generate

Syntax

egen [*type*] *newvar* = *fcn* (*arguments*) [*if*] [*in*] [, *options*]

by is allowed with some of the egen functions as noted below.

where depending on the *fcn*, *arguments* refers to an expression, *varlist*, or *numlist*, and the *options* are also *fcn* dependent, and where *fcn* is

anycount(*varlist*), values(*integer numlist*)
    may not be combined with by. It returns the number of variables in *varlist* for which values are equal to any integer value in a supplied numlist. Values for any observations excluded by either [*if*] or [*in*] are set to 0 (not missing). Also see anyvalue(*varname*) and anymatch(*varlist*).

anymatch(*varlist*), values(*integer numlist*)
    may not be combined with by. It is 1 if any variable in *varlist* is equal to any integer value in a supplied *numlist* and 0 otherwise. Values for any observations excluded by either [*if*] or [*in*] are set to 0 (not missing). See also anyvalue(*varname*) and anycount(*varlist*).

anyvalue(*varname*), values(*integer numlist*)
    may not be combined with by. It takes the value of *varname* if *varname* is equal to any integer value in a supplied *numlist* and is missing otherwise. Also see anymatch(*varlist*) and anycount(*varlist*).

concat(*varlist*) [, format(%*fmt*) decode maxlength(#) punct(*pchars*)]
    may not be combined with by. It concatenates *varlist* to produce a string variable. Values of string variables are unchanged. Values of numeric variables are converted to string, as is, or are converted using a numeric format under option format(%*fmt*) or decoded under option decode, in which case maxlength() may also be used to control the maximum label length used. By default, variables are added end to end: punct(*pchars*) may be used to specify punctuation, such as a space, punct(" "), or a comma, punct(,).

count(*exp*)   (allows by *varlist*:)
    creates a constant (within *varlist*) containing the number of nonmissing observations of *exp*. Also see rownonmiss() and rowmiss().

cut(*varname*), { at(#,#,...,#) | group(#) } [icodes label]
    may not be combined with by. It creates a new categorical variable coded with the left-hand ends of the grouping intervals specified in the at() option, which expects an ascending numlist.

    at(#,#,...,#) supplies the breaks for the groups, in ascending order. The list of breakpoints may be simply a list of numbers separated by commas but can also include the syntax a(b)c, meaning from a to c in steps of size b. If no breaks are specified, the command expects the option group().

    group(#) specifies the number of equal frequency grouping intervals to be used in the absence of breaks. Specifying this option automatically invokes icodes.

    icodes requests that the codes 0, 1, 2, etc., be used in place of the left-hand ends of the intervals.

136

label requests that the integer-coded values of the grouped variable be labeled with the left-hand ends of the grouping intervals. Specifying this option automatically invokes icodes.

diff(*varlist*)

    may not be combined with by. It creates an indicator variable equal to 1 if the variables in *varlist* are not equal and 0 otherwise.

ends(*strvar*) $\big[$, punct(*pchars*) trim $\big[$head | last | tail$\big]\,\big]$

    may not be combined with by. It gives the first "word" or head (with the head option), the last "word" (with the last option), or the remainder or tail (with the tail option) from string variable *strvar*.

        head, last, and tail are determined by the occurrence of *pchars*, which is by default one space (" ").

        The head is whatever precedes the first occurrence of *pchars*, or the whole of the string if it does not occur. For example, the head of "frog toad" is "frog" and that of "frog" is "frog". With punct(,), the head of "frog,toad" is "frog".

        The last word is whatever follows the last occurrence of *pchars* or is the whole of the string if a space does not occur. The last word of "frog toad newt" is "newt" and that of "frog" is "frog". With punct(,), the last word of "frog,toad" is "toad".

        The remainder or tail is whatever follows the first occurrence of *pchars*, which will be the empty string "" if *pchars* does not occur. The tail of "frog toad newt" is "toad newt" and that of "frog" is "". With punct(,), the tail of "frog,toad" is "toad".

        The trim option trims any leading or trailing spaces.

fill(*numlist*)

    may not be combined with by. It creates a variable of ascending or descending numbers or complex repeating patterns. *numlist* must contain at least two numbers and may be specified using standard *numlist* notation; see [U] **11.1.8 numlist**. $\big[$*if*$\big]$ and $\big[$*in*$\big]$ are not allowed with fill().

group(*varlist*) $\big[$, missing label lname(*name*) truncate(*num*)$\big]$

    may not be combined with by. It creates one variable taking on values 1, 2, ... for the groups formed by *varlist*. *varlist* may contain numeric variables, string variables, or a combination of the two. The order of the groups is that of the sort order of *varlist*. missing indicates that missing values in *varlist* (either . or "") are to be treated like any other value when assigning groups, instead of as missing values being assigned to the group missing. The label option returns integers from 1 up according to the distinct groups of *varlist* in sorted order. The integers are labeled with the values of *varlist* or the value labels, if they exist. lname() specifies the name to be given to the value label created to hold the labels; lname() implies label. The truncate() option truncates the values contributed to the label from each variable in *varlist* to the length specified by the integer argument *num*. The truncate option cannot be used without specifying the label option. The truncate option does not change the groups that are formed; it changes only their labels.

iqr(*exp*)   (allows by *varlist*:)

    creates a constant (within *varlist*) containing the interquartile range of *exp*. See also pctile().

kurt(*varname*)   (allows by *varlist*:)

    returns the kurtosis (within *varlist*) of *varname*.

mad(*exp*)   (allows by *varlist*:)

    returns the median absolute deviation from the median (within *varlist*) of *exp*.

max(*exp*)   (allows by *varlist*:)
    creates a constant (within *varlist*) containing the maximum value of *exp*.

mdev(*exp*)   (allows by *varlist*:)
    returns the mean absolute deviation from the mean (within *varlist*) of *exp*.

mean(*exp*)   (allows by *varlist*:)
    creates a constant (within *varlist*) containing the mean of *exp*.

median(*exp*)   (allows by *varlist*:)
    creates a constant (within *varlist*) containing the median of *exp*. Also see pctile().

min(*exp*)   (allows by *varlist*:)
    creates a constant (within *varlist*) containing the minimum value of *exp*.

mode(*varname*) [, <u>min</u>mode <u>max</u>mode <u>num</u>mode(*integer*) <u>missing</u>]   (allows by *varlist*:)
    produces the mode (within *varlist*) for *varname*, which may be numeric or string. The mode
    is the value occurring most frequently. If two or more modes exist, the mode produced will
    be a missing value. To avoid this, the minmode, maxmode, or nummode() option may be used
    to specify choices for selecting among the multiple modes. minmode returns the lowest value,
    and maxmode returns the highest value. nummode(#) will return the #th mode, counting from
    the lowest up. Missing values are excluded from determination of the mode unless missing
    is specified. Even so, the value of the mode is recorded for observations for which the values
    of *varname* are missing unless they are explicitly excluded, that is, by if *varname* < . or if
    varname != "".

mtr(*year income*)
    may not be combined with by. It returns the U.S. marginal income tax rate for a married couple
    with taxable income *income* in year *year*, where $1930 \le year \le 2007$. *year* and *income* may
    be specified as variable names or constants; e.g., mtr(1993 faminc), mtr(surveyyr 28000),
    or mtr(surveyyr faminc). A blank or comma may be used to separate *income* from *year*.

pc(*exp*) [, prop]   (allows by *varlist*:)
    returns *exp* (within *varlist*) scaled to be a percentage of the total, between 0 and 100. The prop
    option returns *exp* scaled to be a proportion of the total, between 0 and 1.

pctile(*exp*) [, p(#)]   (allows by *varlist*:)
    creates a constant (within *varlist*) containing the #th percentile of *exp*. If p(#) is not specified,
    50 is assumed, meaning medians. Also see median().

rank(*exp*) [, [<u>fi</u>eld | <u>t</u>rack | <u>u</u>nique]]   (allows by *varlist*:)
    creates ranks (within *varlist*) of *exp*; by default, equal observations are assigned the average
    rank. The field option calculates the field rank of *exp*: the highest value is ranked 1, and there
    is no correction for ties. That is, the field rank is 1 + the number of values that are higher.
    The track option calculates the track rank of *exp*: the lowest value is ranked 1, and there is
    no correction for ties. That is, the track rank is 1 + the number of values that are lower. The
    unique option calculates the unique rank of *exp*: values are ranked 1, ..., #, and values and
    ties are broken arbitrarily. Two values that are tied for second are ranked 2 and 3.

rowfirst(*varlist*)
    may not be combined with by. It gives the first nonmissing value in *varlist* for each observation
    (row). If all values in *varlist* are missing for an observation, *newvar* is set to missing.

rowlast(*varlist*)
    may not be combined with by. It gives the last nonmissing value in *varlist* for each observation
    (row). If all values in *varlist* are missing for an observation, *newvar* is set to missing.

rowmax(*varlist*)

    may not be combined with by. It gives the maximum value (ignoring missing values) in *varlist* for each observation (row). If all values in *varlist* are missing for an observation, *newvar* is set to missing.

rowmean(*varlist*)

    may not be combined with by. It creates the (row) means of the variables in *varlist*, ignoring missing values; for example, if three variables are specified and, in some observations, one of the variables is missing, in those observations *newvar* will contain the mean of the two variables that do exist. Other observations will contain the mean of all three variables. Where none of the variables exist, *newvar* is set to missing.

rowmin(*varlist*)

    may not be combined with by. It gives the minimum value in *varlist* for each observation (row). If all values in *varlist* are missing for an observation, *newvar* is set to missing.

rowmiss(*varlist*)

    may not be combined with by. It gives the number of missing values in *varlist* for each observation (row).

rownonmiss(*varlist*) [, <u>s</u>trok]

    may not be combined with by. It gives the number of nonmissing values in *varlist* for each observation (row)—this is the value used by rowmean() for the denominator in the mean calculation.

        String variables may not be specified unless option strok is also specified. If strok is specified, string variables will be counted as containing missing values when they contain "". Numeric variables will be counted as containing missing when their value is "$\geq$.".

rowsd(*varlist*)

    may not be combined with by. It creates the (row) standard deviations of the variables in *varlist*, ignoring missing values.

rowtotal(*varlist*)

    may not be combined with by. It creates the (row) sum of the variables in *varlist*, treating missing values as 0.

sd(*exp*)   (allows by *varlist*:)

    creates a constant (within *varlist*) containing the standard deviation of *exp*. Also see mean().

seq() [, <u>f</u>rom(#) <u>t</u>o(#) <u>b</u>lock(#)]   (allows by *varlist*:)

    returns integer sequences. Values start from from() (default 1) and increase to to() (the default is the maximum number of values) in blocks (default size 1). If to() is less than the maximum number, sequences restart at from(). Numbering may also be separate within groups defined by *varlist* or decreasing if to() is less than from(). Sequences depend on the sort order of observations, following three rules: (1) observations excluded by if or in are not counted; (2) observations are sorted by *varlist*, if specified; and (3) otherwise, the order is that when called. No *arguments* are specified.

skew(*varname*)   (allows by *varlist*:)

    returns the skewness (within *varlist*) of *varname*.

std(*exp*) [, <u>m</u>ean(#) <u>s</u>td(#)]

    may not be combined with by. It creates the standardized values of *exp*. The options specify the desired mean and standard deviation. The default is mean(0) and std(1), producing a variable with mean 0, standard deviation 1.

tag(*varlist*) [, <u>missing</u>]

may not be combined with by. It tags just 1 observation in each distinct group defined by *varlist*. When all observations in a group have the same value for a summary variable calculated for the group, it will be sufficient to use just one such value for many purposes. The result will be 1 if the observation is tagged and never missing, and 0 otherwise. Values for any observations excluded by either [*if*] or [*in*] are set to 0 (not missing). Hence, if tag is the variable produced by egen tag = tag(*varlist*), the idiom if tag is always safe. missing specifies that missing values of *varlist* may be included.

total(*exp*)   (allows by *varlist*:)

creates a constant (within *varlist*) containing the sum of *exp*. Also see mean().

Description

egen creates *newvar* of the optionally specified storage type equal to *fcn*(*arguments*). Here *fcn*() is a function specifically written for egen, as documented below or as written by users. Only egen functions may be used with egen, and conversely, only egen may be used to run egen functions.

Depending on *fcn*(), *arguments*, if present, refers to an expression, *varlist*, or a *numlist*, and the *options* are similarly *fcn* dependent. Explicit subscripting (using _N and _n), which is commonly used with generate, should not be used with egen; see [U] **13.7 Explicit subscripting**.

Remarks

Remarks are presented under the following headings:

> *Summary statistics*
> *Generating patterns*
> *Marking differences among variables*
> *Ranks*
> *Standardized variables*
> *Row functions*
> *Categorical and integer variables*
> *String variables*
> *U.S. marginal income tax rate*

Summary statistics

The functions count(), iqr(), kurt(), mad(), max(), mdev(), mean(), median(), min(), mode(), pc(), pctile(), sd(), skew(), and total() create variables containing summary statistics. All functions take a by ... : prefix and, if specified, calculate the summary statistics within each by-group.

▷ Example 1: Without the by prefix

Without the by prefix, the result produced by these functions is a constant for every observation in the data. For instance, we have data on cholesterol levels (chol) and wish to have a variable that, for each patient, records the deviation from the average across all patients:

```
. use http://www.stata-press.com/data/r10/egenxmpl
. egen avg = mean(chol)
. generate deviation = chol - avg
```

◁

▷ Example 2: With the by prefix

These functions are most useful when the by prefix is specified. For instance, assume that our dataset includes dcode, a hospital–patient diagnostic code, and los, the number of days that the patient remained in the hospital. We wish to obtain the deviation in length of stay from the median for all patients having the same diagnostic code:

```
. use http://www.stata-press.com/data/r10/egenxmpl2, clear
. by dcode, sort: egen medstay = median(los)
. generate deltalos = los - medstay
```

◁

❑ Technical Note

Distinguish carefully between Stata's sum() function and egen's total() function. Stata's sum() function creates the running sum, whereas egen's total() function creates a constant equal to the overall sum; for example,

```
. clear
. set obs 5
obs was 0, now 5
. generate a = _n
. generate sum1=sum(a)
. egen sum2=total(a)
. list
```

	a	sum1	sum2
1.	1	1	15
2.	2	3	15
3.	3	6	15
4.	4	10	15
5.	5	15	15

❑

❑ Technical Note

The definitions and formulas used by these functions are the same as those used by summarize; see [R] **summarize**. For comparison with summarize, mean() and sd() correspond to the mean and standard deviation. total() is the numerator of the mean, and count() is its denominator. min() and max() correspond to the minimum and maximum. median()—or, equally well, pctile() with p(50)—is the median. pctile() with p(5) refers to the fifth percentile, and so on. iqr() is the difference between the 75th and 25th percentiles.

❑

The mode is the most common value of a dataset, whether it contains numeric or string variables. It is perhaps most useful for categorical variables (whether defined by integers or strings) or for other integer-valued values, but mode() can be applied to variables of any type. Nevertheless, the modes of continuous (or nearly continuous) variables are perhaps better estimated either from inspection of a graph of a frequency distribution or from the results of some density estimation (see [R] **kdensity**).

Missing values need special attention. It is possible that missing is the most common value in a variable (whether missing is defined by the period [.] or extended missing values [.a, .b, . . . , .z] for numeric variables or the empty string [""] for string variables). However, missing values are by default excluded from determination of modes. If you wish to include them, use the `missing` option.

In contrast, `egen mode = mode(`*varname*`)` allows the generation of nonmissing modes for observations for which *varname* is missing. This allows use of the mode as one simple means of imputing categorical variables. If you want the mode to be missing whenever *varname* is missing, you can specify `if` *varname* `< .` or `if` *varname* `!= ""` or, most generally, `if !missing(`*varname*`)`.

`mad()` and `mdev()` produce alternative measures of spread. The median absolute deviation from the median and even the mean deviation will both be more resistant than the standard deviation to heavy tails or outliers, in particular from distributions with heavier tails than the normal or Gaussian. The first measure was named the MAD by Andrews et al. in 1972 but was already known to K. F. Gauss in 1816, according to Hampel et al. (1986). For more historical and statistical details, see David (1998) and Wilcox (2003, 72–73).

Generating patterns

To create a sequence of numbers, simply "show" the `fill()` function how the sequence should look. It must be a linear progression to produce the expected results. Stata does not understand geometric progressions. To produce repeating patterns, you present `fill()` with the pattern twice in the *numlist*.

▷ Example 3: Sequences produced by fill( )

Here are some examples of ascending and descending sequences produced by `fill()`:

```
. clear
. set obs 12
obs was 0, now 12
. egen i=fill(1 2)
. egen w=fill(100 99)
. egen x=fill(22 17)
. egen y=fill(1 1 2 2)
. egen z=fill(8 8 8 7 7 7)
. list, sep(4)
```

	i	w	x	y	z
1.	1	100	22	1	8
2.	2	99	17	1	8
3.	3	98	12	2	8
4.	4	97	7	2	7
5.	5	96	2	3	7
6.	6	95	−3	3	7
7.	7	94	−8	4	6
8.	8	93	−13	4	6
9.	9	92	−18	5	6
10.	10	91	−23	5	5
11.	11	90	−28	6	5
12.	12	89	−33	6	5

◁

▷ Example 4: Patterns produced by fill( )

Here are examples of patterns produced by `fill()`:

```
. clear
. set obs 12
obs was 0, now 12
. egen a=fill(0 0 1 0 0 1)
. egen b=fill(1 3 8 1 3 8)
. egen c=fill(-3(3)6 -3(3)6)
. egen d=fill(10 20 to 50   10 20 to 50)
. list, sep(4)
```

	a	b	c	d
1.	0	1	-3	10
2.	0	3	0	20
3.	1	8	3	30
4.	0	1	6	40
5.	0	3	-3	50
6.	1	8	0	10
7.	0	1	3	20
8.	0	3	6	30
9.	1	8	-3	40
10.	0	1	0	50
11.	0	3	3	10
12.	1	8	6	20

◁

▷ Example 5: seq( )

`seq()` creates a new variable containing one or more sequences of integers. It is useful mainly for quickly creating observation identifiers or automatically numbering levels of factors or categorical variables.

```
. clear
. set obs 12
```

In the simplest case,

```
. egen a = seq()
```

is just equivalent to the common idiom

```
. generate a = _n
```

a may also be obtained from

```
. range a 1 _N
```

(the actual value of _N may also be used).

In more complicated cases, `seq()` with option calls is equivalent to calls to the versatile functions int and mod.

```
. egen b = seq(), b(2)
```

produces integers in blocks of 2, whereas

```
. egen c = seq(), t(6)
```

restarts the sequence after 6 is reached.

```
. egen d = seq(), f(10) t(12)
```

shows that sequences may start with integers other than 1, and

```
. egen e = seq(), f(3) t(1)
```

shows that they may decrease.

The results of these commands are shown by

```
. list, sep(4)
```

	a	b	c	d	e
1.	1	1	1	10	3
2.	2	1	2	11	2
3.	3	2	3	12	1
4.	4	2	4	10	3
5.	5	3	5	11	2
6.	6	3	6	12	1
7.	7	4	1	10	3
8.	8	4	2	11	2
9.	9	5	3	12	1
10.	10	5	4	10	3
11.	11	6	5	11	2
12.	12	6	6	12	1

All these sequences could have been generated in one line with `generate` and with the use of the `int` and `mod` functions. The variables b through e are obtained with

```
. gen b = 1 + int((_n - 1)/2)
. gen c = 1 + mod(_n - 1, 6)
. gen d = 10 + mod(_n - 1, 3)
. gen e = 3 - mod(_n - 1, 3)
```

Nevertheless, `seq()` may save users from puzzling out such solutions or from typing in the needed values.

In general, the sequences produced depend on the sort order of observations, following three rules:

1. observations excluded by `if` or `in` are not counted;

2. observations are sorted by *varlist*, if specified; and

3. otherwise, the order is that specified when `seq()` is called.

The result of applying `seq` was not guaranteed to be identical from application to application whenever sorting was required, even with identical data, because of the indeterminacy of sorting. That is, if we sort, say, integer values, it is sufficient that all the 1s are together and are followed by all the 2s. But there is no guarantee that the order of the 1s, as defined by any other variables, will be identical from sort to sort.

◁

The functions `fill()` and `seq()` are alternatives. In essence, `fill()` requires a minimal example that indicates the kind of sequence required, whereas `seq()` requires that the rule be specified through options. There are sequences that `fill()` can produce that `seq()` cannot, and vice versa. `fill()` cannot be combined with `if` or `in`, in contrast to `seq()`.

Marking differences among variables

▷ Example 6: diff( )

We have three measures of respondents' income obtained from different sources. We wish to create the variable `differ` equal to 1 for disagreements:

```
. use http://www.stata-press.com/data/r10/egenxmpl3, clear
. egen byte differ = diff(inc*)
. list if differ==1
```

	inc1	inc2	inc3	id	differ
10.	42,491	41,491	41,491	110	1
11.	26,075	25,075	25,075	111	1
12.	26,283	25,283	25,283	112	1
78.	41,780	41,780	41,880	178	1
100.	25,687	26,687	25,687	200	1
101.	25,359	26,359	25,359	201	1
102.	25,969	26,969	25,969	202	1
103.	25,339	26,339	25,339	203	1
104.	25,296	26,296	25,296	204	1
105.	41,800	41,000	41,000	205	1
134.	26,233	26,233	26,133	234	1

Rather than typing `diff(inc*)`, we could have typed `diff(inc1 inc2 inc3)`.

◁

Ranks

▷ Example 7: rank( )

Most applications of `rank()` will be to one variable, but the argument *exp* can be more general, namely, an expression. In particular, `rank(-varname)` reverses ranks from those obtained by `rank(varname)`.

The default ranking and those obtained by using one of the `track`, `field`, and `unique` options differ principally in their treatment of ties. The default is to assign the same rank to tied values such that the sum of the ranks is preserved. The `track` option assigns the same rank but resembles the convention in track events; thus, if one person had the lowest time and three persons tied for second-lowest time, their ranks would be 1, 2, 2, and 2, and the next person(s) would have rank 5. The `field` option acts similarly except that the highest is assigned rank 1, as in field events in which the greatest distance or height wins. The `unique` option breaks ties arbitrarily: its most obvious use is assigning ranks for a graph of ordered values. See also `group()` for another kind of "ranking".

```
. use http://www.stata-press.com/data/r10/auto, clear
(1978 Automobile Data)
. keep in 1/10
(64 observations deleted)
. egen rank = rank(mpg)
. egen rank_r = rank(-mpg)
. egen rank_f = rank(mpg), field
. egen rank_t = rank(mpg), track
```

```
. egen rank_u = rank(mpg), unique
. egen rank_ur = rank(-mpg), unique
. sort rank_u
. list mpg rank*
```

	mpg	rank	rank_r	rank_f	rank_t	rank_u	rank_ur
1.	15	1	10	10	1	1	10
2.	16	2	9	9	2	2	9
3.	17	3	8	8	3	3	8
4.	18	4	7	7	4	4	7
5.	19	5	6	6	5	5	6
6.	20	6.5	4.5	4	6	6	5
7.	20	6.5	4.5	4	6	7	4
8.	22	8.5	2.5	2	8	8	3
9.	22	8.5	2.5	2	8	9	2
10.	26	10	1	1	10	10	1

◁

Standardized variables

▷ Example 8: std( )

We have a variable called `age` recording the median age in the 50 states. We wish to create the standardized value of age and verify the calculation:

```
. use http://www.stata-press.com/data/r10/states1, clear
(State data)
. egen stdage = std(age)
. summarize age stdage
```

Variable	Obs	Mean	Std. Dev.	Min	Max
age	50	29.54	1.693445	24.2	34.7
stdage	50	6.41e-09	1	-3.153336	3.047044

```
. correlate age stdage
(obs=50)
```

	age	stdage
age	1.0000	
stdage	1.0000	1.0000

`summarize` shows that the new variable has a mean of approximately 0; 10^{-9} is the precision of a `float` and is close enough to zero for all practical purposes. If we wanted, we could have typed `egen double stdage = std(age)`, making `stdage` a double-precision variable, and the mean would have been 10^{-16}. In any case, `summarize` also shows that the standard deviation is 1. `correlate` shows that the new variable and the original variable are perfectly correlated.

We may optionally specify the mean and standard deviation for the new variable. For instance,

```
. egen newage1 = std(age), std(2)
. egen newage2 = std(age), mean(2) std(4)
. egen newage3 = std(age), mean(2)
```

```
. summarize age newage1-newage3
    Variable │      Obs        Mean    Std. Dev.        Min        Max
─────────────┼─────────────────────────────────────────────────────────
         age │       50       29.54    1.693445       24.2       34.7
     newage1 │       50     1.28e-08           2   -6.306671   6.094089
     newage2 │       50           2           4   -10.61334   14.18818
     newage3 │       50           2           1   -1.153336   5.047044
. correlate age newage1-newage3
(obs=50)
             │      age   newage1   newage2   newage3
─────────────┼────────────────────────────────────────
         age │   1.0000
     newage1 │   1.0000    1.0000
     newage2 │   1.0000    1.0000    1.0000
     newage3 │   1.0000    1.0000    1.0000    1.0000
```

◁

Row functions

▷ Example 9: rowtotal( )

generate's sum() function creates the vertical, running sum of its argument, whereas egen's total() function creates a constant equal to the overall sum. egen's rowtotal() function, however, creates the horizontal sum of its arguments. They all treat missing as zero:

```
. use http://www.stata-press.com/data/r10/egenxmpl4, clear
. egen hsum = rowtotal(a b c)
. generate vsum = sum(hsum)
. egen sum = total(hsum)
. list
```

```
     ┌──────────────────────────────────────────┐
     │    a     b     c    hsum    vsum    sum   │
     ├──────────────────────────────────────────┤
  1. │    .     2     3       5       5     63   │
  2. │    4     .     6      10      15     63   │
  3. │    7     8     .      15      30     63   │
  4. │   10    11    12      33      63     63   │
     └──────────────────────────────────────────┘
```

◁

▷ Example 10: rowmean( ), rowsd( ), and rownonmiss( )

summarize displays the mean and standard deviation of a variable across observations; program writers can access the mean in r(mean) and the standard deviation in r(sd) (see [R] **summarize**). egen's rowmean() function creates the means of observations across variables. rowsd() creates the standard deviations of observations across variables. rownonmiss() creates a count of the number of nonmissing observations, the denominator of the rowmean() calculation:

```
. use http://www.stata-press.com/data/r10/egenxmpl4, clear
. egen avg = rowmean(a b c)
. egen std = rowsd(a b c)
. egen n = rownonmiss(a b c)
```

```
. list
```

	a	b	c	avg	std	n
1.	.	2	3	2.5	.7071068	2
2.	4	.	6	5	1.414214	2
3.	7	8	.	7.5	.7071068	2
4.	10	11	12	11	1	3

◁

▷ Example 11: rowmiss()

rowmiss() returns $k - $ rownonmiss(), where k is the number of variables specified. rowmiss() can be especially useful for finding casewise-deleted observations caused by missing values.

```
. use http://www.stata-press.com/data/r10/auto3, clear
(1978 Automobile Data)
. correlate price weight mpg
(obs=70)
```

	price	weight	mpg
price	1.0000		
weight	0.5309	1.0000	
mpg	-0.4478	-0.7985	1.0000

```
. egen excluded = rowmiss(price weight mpg)
. list make price weight mpg if excluded !=0
```

	make	price	weight	mpg
5.	Buick Electra	.	4,080	15
12.	Cad. Eldorado	14,500	3,900	.
40.	Olds Starfire	4,195	.	24
51.	Pont. Phoenix	.	3,420	.

◁

▷ Example 12: rowmin(), rowmax(), rowfirst(), and rowlast()

rowmin(), rowmax(), rowfirst(), and rowlast() return the minimum, maximum, first, or last nonmissing value, respectively, for the specified variables within an observation (row).

```
. use http://www.stata-press.com/data/r10/egenxmpl5, clear
. egen min = rowmin(x y z)
(1 missing value generated)
. egen max = rowmax(x y z)
(1 missing value generated)
. egen first = rowfirst(x y z)
(1 missing value generated)
. egen last = rowlast(x y z)
(1 missing value generated)
```

```
. list, sep(4)
```

	x	y	z	min	max	first	last
1.	-1	2	3	-1	3	-1	3
2.	.	-6	.	-6	-6	-6	-6
3.	7	.	-5	-5	7	7	-5
4.	.	.	.	.	.	.	.
5.	4	.	.	4	4	4	4
6.	.	.	8	8	8	8	8
7.	.	3	7	3	7	3	7
8.	5	-1	6	-1	6	5	6

◁

Categorical and integer variables

▷ Example 13: anyvalue( ), anymatch( ), and anycount( )

anyvalue(), anymatch(), and anycount() are for categorical or other variables taking integer values. If we define a subset of values specified by an integer *numlist* (see [U] **14.1.8 numlist**), anyvalue() extracts the subset, leaving every other value missing; anymatch() defines an indicator variable (1 if in subset, 0 otherwise); and anycount() counts occurrences of the subset across a set of variables. Therefore, with just one variable, anymatch(*varname*) and anycount(*varname*) are equivalent.

With the auto dataset, we can generate a variable containing the high values of rep78 and a variable indicating whether rep78 has a high value:

```
. use http://www.stata-press.com/data/r10/auto, clear
(1978 Automobile Data)
. egen hirep = anyvalue(rep78), v(3/5)
(15 missing values generated)
. egen ishirep = anymatch(rep78), v(3/5)
```

Here it is easy to produce the same results with official Stata commands:

```
. generate hirep = rep78 if inlist(rep78,3,4,5)
. generate byte ishirep = inlist(rep78,3,4,5)
```

However, as the specification becomes more complicated or involves several variables, the egen functions may be more convenient.

◁

▷ Example 14: group( )

group() maps the distinct groups of a *varlist* to a categorical variable that takes on integer values from 1 to the total number of groups. The order of the groups is that of the sort order of *varlist*. The *varlist* may be of numeric variables, string variables, or a mixture of the two. The resulting variable can be useful for many purposes, including stepping through the distinct groups easily and systematically and cleaning up an untidy ordering. Suppose that the actual (and arbitrary) codes present in the data are 1, 2, 4, and 7, but we desire equally spaced numbers, as when the codes will be values on one axis of a graph. group( ) maps these to 1, 2, 3, and 4.

We have a variable `agegrp` that takes on the values 24, 40, 50, and 65, corresponding to age groups 18–24, 25–40, 41–50, and 51 and above. Perhaps we created this coding using the `recode()` function (see [U] **13.3 Functions** and [U] **25 Dealing with categorical variables**) from another age-in-years variable:

```
. generate agegrp=recode(age,24,40,50,65)
```

We now want to change the codes to 1, 2, 3, and 4:

```
. egen agegrp2 = group(agegrp)
```
                                                                                    ◁

▷ Example 15: group( ) with missing values

We have two categorical variables, `race` and `sex`, which may be string or numeric. We want to use `ir` (see [ST] **epitab**) to create a Mantel–Haenszel weighted estimate of the incidence rate. `ir`, however, allows only one variable to be specified in its `by()` option. We type

```
. use http://www.stata-press.com/data/r10/egenxmpl6, clear
. egen racesex = group(race sex)
(2 missing values generated)
. ir deaths smokes pyears, by(racesex)
  (output omitted )
```

The new numeric variable `racesex` will be missing wherever `race` or `sex` is missing (meaning . for numeric variables and "" for string variables), so missing values will be handled correctly. When we list some of the data, we see

```
. list race sex racesex in 1/7, sep(0)
```

	race	sex	racesex
1.	White	Female	1
2.	White	Male	2
3.	Black	Female	3
4.	Black	Male	4
5.	Black	Male	4
6.	.	Female	.
7.	Black	.	.

`group()` began by putting the data in the order of the grouping variables and then assigned the numeric codes. Observations 6 and 7 were assigned to `racesex==.` because, in one case, `race` was not known, and in the other, `sex` was not known. (These observations were not used by `ir`.)

If we wanted the unknown groups to be treated just as any other category, we could have typed

```
. egen rs2=group(race sex), missing
. list race sex rs2 in 1/7, sep(0)
```

	race	sex	rs2
1.	White	Female	1
2.	White	Male	2
3.	Black	Female	3
4.	Black	Male	4
5.	Black	Male	4
6.	.	Female	6
7.	Black	.	5

                                                                                    ◁

The resulting variable from `group` does not have value labels. Therefore, the values carry no indication of meaning. Interpretation requires comparison with the original *varlist*.

The `label` option produces a categorical variable with value labels. These value labels are either the actual values of *varname* or any value labels of *varname*, if they exist. The values of *varname* could be as long as those of one `str244` variable, but value labels may be no longer than 80 characters.

String variables

Concatenation of string variables is provided in Stata. In context, Stata understands the addition symbol + as specifying concatenation or adding strings end to end. `"soft" + "ware"` produces `"software"`, and given string variables `s1` and `s2`, `s1 + s2` indicates their concatenation.

The complications that may arise in practice include wanting (1) to concatenate the string versions of numeric variables and (2) to concatenate variables, together with some separator such as a space or a comma. Given numeric variables `n1` and `n2`,

```
. generate newstr = s1 + string(n1) + string(n2) + s2
```

shows how numeric values may be converted to their string equivalents before concatenation, and

```
. generate newstr = s1 + " " + s2 + " " + s3
```

shows how spaces may be added between variables. Stata will automatically assign the most appropriate data type for the new string variables.

▷ Example 16: concat( )

`concat()` allows us to do everything in one line concisely.

```
. egen newstr = concat(s1 n1 n2 s2)
```

carries with it an implicit instruction to convert numeric values to their string equivalents, and the appropriate string data type is worked out within `concat()` by Stata's automatic promotion. Moreover,

```
. egen newstr = concat(s1 s2 s3), p(" ")
```

specifies that spaces be used as separators. (The default is to have no separation of concatenated strings.)

As an example of punctuation other than a space, consider

```
. egen fullname = concat(surname forename), p(", ")
```

Noninteger numerical values can cause difficulties, but

```
. egen newstr = concat(n1 n2), format(%9.3f) p(" ")
```

specifies the use of format `%9.3f`. This is equivalent to

```
. generate str1 newstr = ""
. replace newstr = string(n1,"%9.3f") + " " + string(n2,"%9.3f")
```

See [D] **functions** for more about `string()`.

◁

As a final flourish, the decode option instructs concat() to use value labels. With that option, the maxlength() option may also be used. For more details about decode, see [D] **encode**. Unlike the decode command, however, concat() uses string(*varname*), not "", whenever values of *varname* are not associated with value labels, and the format() option, whenever specified, applies to this use of string().

▷ Example 17: ends()

The function ends(*strvar*) is used for subdividing strings. The approach is to find specified separators by using the strpos() string function and then to extract what is desired, which either precedes or follows the separators, using the substr() string function (see [U] **16.3.5 String functions**).

By default, substrings are considered to be separated by individual spaces, so we will give definitions in those terms and then generalize.

The head of the string is whatever precedes the first space, or the whole of the string if no space occurs. This could also be called the first "word". The tail of the string is whatever follows the first space. This could be nothing or one or more words. The last word in the string is whatever follows the last space, or the whole of the string if no space occurs.

To clarify, let's look at some examples. The quotation marks here just mark the limits of each string and are not part of the strings.

	head	tail	last
"frog"	"frog"	""	"frog"
"frog toad"	"frog"	"toad"	"toad"
"frog toad newt"	"frog"	"toad newt"	"newt"
"frog toad newt"	"frog"	" toad newt"	"newt"
"frog toad newt"	"frog"	"toad newt"	"newt"

The main subtlety is that these functions are literal, so the tail of "frog toad newt", in which two spaces follow "frog", includes the second of those spaces, and is thus " toad newt". Therefore, you may prefer to use the trim option to trim the result of any leading or trailing spaces, producing "toad newt" in this instance.

The punct(*pchars*) option may be used to specify separators other than spaces. The general definitions of the head, tail, and last options are therefore interpreted in terms of whatever separator has been specified; that is, they are relative to the first or last occurrence of the separator in the string value. Thus with punct(,) and the string "Darwin, Charles Robert", the head is "Darwin", and the tail and the last are both " Charles Robert". Note again the leading space in this example, which may be trimmed with trim. The punctuation (here the comma ,) is discarded, just as it is with one space.

pchars, the argument of punct(), will usually, but not always, be one character. If two or more characters are specified, these must occur together; for example, punct(:;) would mean that words are separated by a colon followed by a semicolon (that is, :;). It is not implied, in particular, that the colon and semicolon are alternatives. To do that, you would have to modify the programs presented here or resort to first principles by using split; see [D] **split**.

With personal names, the option head or last might be applied to extract surnames if strings were similar to "Darwin, Charles Robert" or "Charles Robert Darwin", with the surname coming first or last. What then happens with surnames like "von Neumann" or "de la Mare"? "von Neumann, John" is no problem, if the comma is specified as a separator, but the option last is not intelligent enough to handle "Walter de la Mare" properly. For that, the best advice is to use programs specially written for person-name extraction, such as extrname (Gould 1993).

◁

U.S. marginal income tax rate

mtr(*year income*) (Schmidt 1993, 1994) returns the U.S. marginal income tax rate for a married couple with taxable income *income* in year *year*, where $1930 \le year \le 2007$.

▷ Example 18: mtr( )

Schmidt (1993) examines the change in the progressivity of the U.S. tax schedule over the period from 1930 to 1990. As a measure of progressivity, he calculates the difference in the marginal tax rates at the 75th and 25th percentiles of income, using a dataset of percentiles of taxable income developed by Hakkio, Rush, and Schmidt (1993). (Certain aspects of the income distribution are imputed in these data.) A subset of the data contains the following:

```
. describe
Contains data from income1.dta
  obs:            61
  vars:            4                          12 Feb 2007 03:33
  size:        1,220 (99.8% of memory free)

              storage  display    value
variable name  type    format     label    variable label

year           float   %9.0g               Year
inc25          float   %9.0g               25th percentile
inc50          float   %9.0g               50th percentile
inc75          float   %9.0g               75th percentile

Sorted by:
. summarize
    Variable |      Obs        Mean    Std. Dev.       Min        Max

        year |       61        1960    17.75293       1930       1990
       inc25 |       61    6948.272    6891.921       819.4   27227.35
       inc50 |       61    11645.15    11550.71     1373.29   45632.43
       inc75 |       61    18166.43     18019.1     2142.33   71186.58
```

Given the series for income and the four-digit year, we can generate the marginal tax rates corresponding to the 25th and 75th percentiles of income:

```
. egen mtr25 = mtr(year inc25)
. egen mtr75 = mtr(year inc75)
. summarize mtr25 mtr75
    Variable |      Obs        Mean    Std. Dev.       Min        Max

       mtr25 |       61    .1664898    .0677949      .01125        .23
       mtr75 |       61    .2442053    .1148427      .01125    .424625
```

◁

Methods and Formulas

egen is implemented as an ado-file.

Stata users have written many extra functions for egen. Type net search egen to locate Internet sources of programs.

Acknowledgments

The mtr() *fcn* of egen was written by Timothy J. Schmidt of the Federal Reserve Bank of Kansas City.

The cut *fcn* was written by David Clayton, Cambridge Institute for Medical Research, and Michael Hills, London School of Hygiene and Tropical Medicine (retired), (1999a, 1999b, 1999c).

Many of the other egen *fcn*s were written by Nicholas J. Cox of Durham University, UK.

References

Andrews, D. F., P. J. Bickel, F. R. Hampel, P. J. Huber, W. H. Rogers, and J. W. Tukey. 1972. *Robust Estimates of Location: Survey and Advances*. Princeton: Princeton University Press.

Cappellari, L., and S. P. Jenkins. 2006. Calculation of multivariate normal probabilities by simulation, with applications to maximum simulated likelihood estimation. *Stata Journal* 6: 156–189.

Clayton, D., and M. Hills. 1999a. dm66: Recoding variables using grouped values. *Stata Technical Bulletin* 49: 6–7. Reprinted in *Stata Technical Bulletin Reprints*, vol. 9, pp. 23–25.

——. 1999b. dm66.1: Stata 6 version of recoding variables using grouped values. *Stata Technical Bulletin* 50: 3. Reprinted in *Stata Technical Bulletin Reprints*, vol. 9, p. 25.

——. 1999c. dm66.2: Update of cut to Stata 6. *Stata Technical Bulletin* 51: 2–3. Reprinted in *Stata Technical Bulletin Reprints*, vol. 9, pp. 25–26.

Cox, N. J. 1999. dm70: Extensions to generate, extended. *Stata Technical Bulletin* 50: 9–17. Reprinted in *Stata Technical Bulletin Reprints*, vol. 9, pp. 34–45.

——. 2000. dm70.1: Extensions to generate, extended: Corrections. *Stata Technical Bulletin* 57: 2. Reprinted in *Stata Technical Bulletin Reprints*, vol. 10, p. 9.

Cox, N. J., and R. Goldstein. 1999a. dm72: Alternative ranking procedures. *Stata Technical Bulletin* 51: 5–7. Reprinted in *Stata Technical Bulletin Reprints*, vol. 9, pp. 48–51.

——. 1999b. dm72.1: Alternative ranking procedures: Update. *Stata Technical Bulletin* 52: 2. Reprinted in *Stata Technical Bulletin Reprints*, vol. 9, p. 51.

David, H. A. 1998. Early sample measures of variability. *Statistical Science* 13: 368–377.

Esman, R. M. 1998. dm55: Generating sequences and patterns of numeric data: An extension to egen. *Stata Technical Bulletin* 43: 2–3. Reprinted in *Stata Technical Bulletin Reprints*, vol. 8, pp. 4–5.

Gould, W. W. 1993. dm13: Person name extraction. *Stata Technical Bulletin* 13: 6–11. Reprinted in *Stata Technical Bulletin Reprints*, vol. 3, pp. 25–31.

Hakkio, C. S., M. Rush, and T. J. Schmidt. 1993. The marginal income tax rate schedule from 1930 to 1990. Research Working Paper, Federal Reserve Bank of Kansas City.

Hampel, F. R., E. M. Ronchetti, P. J. Rousseeuw, and W. A. Stahel. 1986. *Robust Statistics: The Approach Based on Influence Functions*. New York: Wiley.

Ryan, P. 1999. dm71: Calculating the product of observations. *Stata Technical Bulletin* 51: 3–4. Reprinted in *Stata Technical Bulletin Reprints*, vol. 9, pp. 45–48.

——. 2001. dm87: Calculating the row product of observations. *Stata Technical Bulletin* 60: 3–4. Reprinted in *Stata Technical Bulletin Reprints*, vol. 10, pp. 39–41.

Schmidt, T. J. 1993. sss1: Calculating U.S. marginal income tax rates. *Stata Technical Bulletin* 15: 17–19. Reprinted in *Stata Technical Bulletin Reprints*, vol. 3, pp. 197–200.

——. 1994. sss1.1: Updated U.S. marginal income tax rate function. *Stata Technical Bulletin* 22: 29. Reprinted in *Stata Technical Bulletin Reprints*, vol. 4, p. 224.

Wilcox, R. R. 2003. *Applying Contemporary Statistical Techniques*. San Diego, CA: Academic Press.

Also See

[D] **collapse** — Make dataset of summary statistics

[D] **generate** — Create or change contents of variable

[U] **13.3 Functions**

Title

> **encode** — Encode string into numeric and vice versa

Syntax

String variable to numeric variable

> <u>en</u>code *varname* $\begin{bmatrix} if \end{bmatrix}$ $\begin{bmatrix} in \end{bmatrix}$, <u>g</u>enerate(*newvar*) $\begin{bmatrix} \underline{l}abel(name) & \underline{no}extend \end{bmatrix}$

Numeric variable to string variable

> <u>dec</u>ode *varname* $\begin{bmatrix} if \end{bmatrix}$ $\begin{bmatrix} in \end{bmatrix}$, <u>g</u>enerate(*newvar*) $\begin{bmatrix} \underline{max}length(\#) \end{bmatrix}$

Description

encode creates a new variable named *newvar* based on the string variable *varname*, creating, adding to, or just using (as necessary) the value label *newvar* or, if specified, *name*. Do not use encode if *varname* contains numbers that merely happen to be stored as strings; instead, use generate *newvar* = real(*varname*) or destring; see [U] **23.2 Categorical string variables**, [U] **13.3 Functions**, and [D] **destring**.

decode creates a new string variable named *newvar* based on the "encoded" numeric variable *varname* and its value label.

Options for encode

generate(*newvar*) is required and specifies the name of the variable to be created.

label(*name*) specifies the name of the value label to be created or used and added to if the named value label already exists. If label() is not specified, encode uses the same name for the label as it does for the new variable.

noextend specifies that if there are values contained in *varname* that are not present in label(*name*), *varname* not be encoded. By default, any values not present in label(*name*) will be added to that label.

Options for decode

generate(*newvar*) is required and specifies the name of the variable to be created.

maxlength(*#*) specifies how many characters of the value label to retain; *#* must be between 1 and 244. The default is maxlength(244).

Remarks

Remarks are presented under the following headings:

> *encode*
> *decode*

encode

encode is most useful in making string variables accessible to Stata's statistical routines, most of which can work only with numeric variables. encode is also useful in reducing the size of a dataset. If you are not familiar with value labels, read [U] **12.6.3 Value labels**.

The maximum number of associations within each value label is 65,536 (1,000 for Small Stata). Each association in a value label maps a string of up to 244 characters to a number. If your string has entries longer than that, only the first 244 characters are retained and are significant.

▷ Example 1

We have a dataset on high blood pressure, and among the variables is sex, a string variable containing either "male" or "female". We wish to run a regression of high blood pressure on race, sex, and age group. We type regress hbp race sex age_grp and get the message "no observations".

```
. use http://www.stata-press.com/data/r10/hbp2

. regress hbp sex race age_grp
no observations
r(2000);
```

Stata's statistical procedures cannot directly deal with string variables; as far as they are concerned, all observations on sex are missing. encode provides the solution:

```
. encode sex, gen(gender)

. regress hbp gender race age_grp
```

Source	SS	df	MS
Model	2.01013476	3	.67004492
Residual	49.3886164	1117	.044215413
Total	51.3987511	1120	.045891742

Number of obs =	1121
F(3, 1117) =	15.15
Prob > F =	0.0000
R-squared =	0.0391
Adj R-squared =	0.0365
Root MSE =	.21027

hbp	Coef.	Std. Err.	t	P>|t|	[95% Conf. Interval]	
gender	.0394747	.0130022	3.04	0.002	.0139633	.0649861
race	-.0409453	.0113721	-3.60	0.000	-.0632584	-.0186322
age_grp	.0241484	.00624	3.87	0.000	.0119049	.0363919
_cons	-.016815	.0389167	-0.43	0.666	-.093173	.059543

encode looks at a string variable and makes an internal table of all the values it takes on, here "male" and "female". It then alphabetizes that list and assigns numeric codes to each entry. Thus 1 becomes "female", and 2 becomes "male". It creates an int variable (gender) and substitutes a 1 where sex is "female", a 2 where sex is "male", and a *missing* (.) where sex is *null* (""). It creates a value label (also named gender) that records the mapping $1 \leftrightarrow$ female and $2 \leftrightarrow$ male. Finally, encode labels the values of the new variable with the value label.

◁

▷ Example 2

It is difficult to distinguish the result of encode from the original string variable. For instance, in our last two examples, we typed encode sex, gen(gender). Let's compare the two variables:

. list sex gender in 1/4

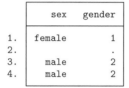

	sex	gender
1.	female	female
2.		.
3.	male	male
4.	male	male

They look almost identical, although you should notice the missing value for gender in the second observation.

The difference does show, however, if we tell list to ignore the value labels and show how the data really appear:

. list sex gender in 1/4, nolabel

	sex	gender
1.	female	1
2.		.
3.	male	2
4.	male	2

We could also ask to see the underlying value label:

```
. label list gender
gender:
            1 female
            2 male
```

gender really is a numeric variable, but since *all* Stata commands understand value labels, the variable displays as "male" and "female", just as the underlying string variable sex would.

◁

▷ Example 3

We can drastically reduce the size of our dataset by encoding strings and then discarding the underlying string variable. We have a string variable, sex, that records each person's sex as "male" and "female". Because female has six characters, the variable is stored as a str6.

We can encode the sex variable and use compress to store the variable as a byte, which takes only 1 byte. Because our dataset contains 1,130 people, the string variable takes 6,780 bytes, but the encoded variable will take only 1,130 bytes.

```
. use http://www.stata-press.com/data/r10/hbp2, clear
. describe
Contains data from http://www.stata-press.com/data/r10/hbp2.dta
  obs:          1,130
  vars:             7                          3 Mar 2007 06:47
  size:        29,380 (97.2% of memory free)
```

variable name	storage type	display format	value label	variable label
id	str10	%10s		Record identification number
city	byte	%8.0g		
year	int	%8.0g		
age_grp	byte	%8.0g	agefmt	
race	byte	%8.0g	racefmt	
hbp	byte	%8.0g	yn	high blood pressure
sex	str6	%9s		

```
Sorted by:
. encode sex, generate(gender)
. list sex gender in 1/5
```

	sex	gender
1.	female	female
2.		.
3.	male	male
4.	male	male
5.	female	female

```
. drop sex
. rename gender sex
. compress
sex was long now byte
. describe
Contains data from http://www.stata-press.com/data/r10/hbp2.dta
  obs:          1,130
  vars:             7                          3 Mar 2007 06:47
  size:        23,730 (97.7% of memory free)
```

variable name	storage type	display format	value label	variable label
id	str10	%10s		Record identification number
city	byte	%8.0g		
year	int	%8.0g		
age_grp	byte	%8.0g	agefmt	
race	byte	%8.0g	racefmt	
hbp	byte	%8.0g	yn	high blood pressure
sex	byte	%8.0g	gender	

```
Sorted by:
    Note:  dataset has changed since last saved
```

The size of our dataset has fallen from 29,380 bytes to 23,730 bytes.

◁

❑ Technical Note

In the examples given above, the value label did not exist before `encode` created it, as that is not required. If the value label does exist, `encode` uses your encoding as far as it can and adds new mappings for anything not found in your value label. For instance, if you wanted "female" to be encoded as 0 rather than 1 (possibly for use in linear regression), you could type

```
. label define gender 0 "female"
. encode sex, gen(gender)
```

can also specify the name of the value label. If you do not, the value label is assumed to have the same name as the newly created variable. For instance,

```
. label define sexlbl 0 "female"
. encode sex, gen(gender) label(sexlbl)
```

❑

decode

`decode` is used to convert numeric variables with associated value labels into true string variables.

▷ Example 4

We have a numeric variable named `female` that records the values 0 and 1. `female` is associated with a value label named `sexlbl` that says that 0 means male and 1 means female:

```
. use http://www.stata-press.com/data/r10/hbp3, clear
. describe female

              storage  display    value
variable name   type   format     label       variable label

female          byte   %8.0g      sexlbl
. label list sexlbl
sexlbl:
           0 male
           1 female
```

We see that `female` is stored as a `byte`. It is a numeric variable. Nevertheless, it has an associated value label describing what the numeric codes mean, so if we `tabulate` the variable, for instance, it appears to contain the strings "male" and "female":

```
. tabulate female
     female |     Freq.     Percent       Cum.

       male |       695       61.61       61.61
     female |       433       38.39      100.00

      Total |     1,128      100.00
```

We can create a real string variable from this numerically encoded variable by using `decode`:

```
. decode female, gen(sex)
. describe sex

              storage  display    value
variable name   type   format     label       variable label

sex             str6   %9s
```

We have a new variable called `sex`. It is a string, and Stata automatically created the shortest possible string. The word "female" has six characters, so our new variable is a `str6`. `female` and `sex` appear indistinguishable:

. list female sex in 1/4

	female	sex
1.	female	female
2.	.	
3.	male	male
4.	male	male

But when we add `nolabel`, the difference is apparent:

. list female sex in 1/4, nolabel

	female	sex
1.	1	female
2.	.	
3.	0	male
4.	0	male

◁

▷ Example 5

`decode` is most useful in instances when we wish to match-merge two datasets on a variable that has been encoded inconsistently.

For instance, we have two datasets on individual states in which one of the variables (`state`) takes on values such as "CA" and "NY". The state variable was originally a string, but along the way the variable was encoded into an integer with corresponding value label in one or both datasets.

We wish to merge these two datasets, but either (1) one of the datasets has a string variable for state and the other an encoded variable or (2) although both are numeric, we are not certain that the codings are consistent. Perhaps "CA" has been coded 5 in one dataset and 6 in another.

Since `decode` will take an encoded variable and turn it back into a string, `decode` provides the solution:

```
use first                (load the first dataset)
decode state, gen(st)    (make a string state variable)
drop state               (discard the encoded variable)
sort st                  (sort on string)
save first, replace      (save the dataset)
use second               (load the second dataset)
decode state, gen(st)    (make a string variable)
drop state               (discard the encoded variable)
sort st                  (sort on string)
merge st using first     (merge the data)
```

Of course, now we should `tabulate _merge` to make sure that the merge went as expected; see [D] **merge**.

◁

Also See

[D] **compress** — Compress data in memory

[D] **destring** — Convert string variables to numeric variables and vice versa

[D] **generate** — Create or change contents of variable

[U] **12.6.3 Value labels**

[U] **23.2 Categorical string variables**

Title

> **erase** — Erase a disk file

Syntax

$$\left\{ \, \texttt{erase} \mid \texttt{rm} \, \right\} \; \left[\,"\,\right] \textit{filename} \left[\,"\,\right]$$

Note: double quotes must be used to enclose *filename* if the name contains spaces.

Description

The `erase` command erases files stored on disk. Unix users may type `erase`, but they will probably prefer to type `rm`. Under Unix, `erase` is a synonym for `rm`.

Stata for Macintosh users: `erase` is permanent; the file is not moved to the Trash but is immediately removed from the disk.

Stata for Windows users: `erase` is permanent; the file is not moved to the Recycle Bin but is immediately removed from the disk.

Remarks

The only difference between Stata's `erase` (`rm`) and the DOS DEL or Unix rm(1) is that we may not specify groups of files. Stata requires that we erase files one at a time.

Macintosh users may prefer to discard files by dragging them to the Trash.

Windows users may prefer to discard files by dragging them to the Recycle Bin.

▷ Example 1

Stata provides seven operating system equivalent commands: `cd`, `copy`, `dir`, `erase`, `mkdir`, `rmdir`, and `type`, or, from the Unix perspective, `cd`, `copy`, `ls`, `rm`, `mkdir`, `rmdir`, and `cat`. These commands are provided for Macintosh users, too. Also, Stata users can issue any operating system command by using Stata's `shell` command, so you should never have to exit Stata to perform some housekeeping detail.

Suppose that we have the file `mydata.dta` stored on disk, and we wish to permanently eliminate it:

```
. erase mydata
file mydata not found
r(601);
. erase mydata.dta

. _
```

Our first attempt, `erase mydata`, was unsuccessful. Although Stata ordinarily supplies the file extension for you, it does not do so when you type `erase`. You must be explicit. Our second attempt eliminated the file. Unix users could have typed `rm mydata.dta` if they preferred.

◁

Also See

[D] **cd** — Change directory

[D] **copy** — Copy file from disk or URL

[D] **dir** — Display filenames

[D] **mkdir** — Create directory

[D] **rmdir** — Remove directory

[D] **shell** — Temporarily invoke operating system

[D] **type** — Display contents of a file

[U] **11.6 File-naming conventions**

Title

expand — Duplicate observations

Syntax

expand $\left[=\right] exp \left[if\right] \left[in\right]$

Description

expand replaces each observation in the dataset with *n* copies of the observation, where *n* is equal to the required expression rounded to the nearest integer. If the expression is less than 1 or equal to *missing*, it is interpreted as if it were 1, and the observation is retained but not duplicated.

Remarks

▷ Example 1

expand is, admittedly, a strange command. It can, however, be useful in tricky programs or for reformatting data for survival analysis (see examples in [ST] **epitab**). Here is a silly use of expand:

```
. use http://www.stata-press.com/data/r10/expandxmpl
. list
```

	n	x
1.	-1	1
2.	0	2
3.	1	3
4.	2	4
5.	3	5

```
. expand n
(1 negative count ignored; observation not deleted)
(1 zero count ignored; observation not deleted)
(3 observations created)
. list
```

	n	x
1.	-1	1
2.	0	2
3.	1	3
4.	2	4
5.	3	5
6.	2	4
7.	3	5
8.	3	5

The new observations are added to the end of the dataset. expand informed us that it created 3 observations. The first 3 observations were not replicated since n was less than or equal to 1. n is 2 in the fourth observation, so expand created one replication of this observation, bringing the total number of observations of this type to 2. expand created two replications of observation 5 since n is 3.

Since there were 5 observations in the original dataset, and since expand adds new observations onto the end of the dataset, we could now undo the expansion by typing drop in 6/1.

◁

Also See

[D] **contract** — Make dataset of frequencies and percentages

[D] **expandcl** — Duplicate clustered observations

[D] **fillin** — Rectangularize dataset

Title

expandcl — Duplicate clustered observations

Syntax

expandcl $\left[= \right] exp \left[if \right] \left[in \right]$, cluster(*varlist*) generate(*newvar*)

Description

expandcl duplicates clusters of observations and generates a new variable that identifies the clusters uniquely.

expandcl replaces each cluster in the dataset with *n* copies of the cluster, where *n* is equal to the required expression rounded to the nearest integer. The expression is required to be constant within cluster. If the expression is less than 1 or equal to *missing*, it is interpreted as if it were 1, and the cluster is retained but not duplicated.

Options

cluster(*varlist*) is required and specifies the variables that identify the clusters before expanding the data.

generate(*newvar*) is required and stores unique identifiers for the duplicated clusters in *newvar*. *newvar* will identify the clusters by using consecutive integers starting from 1.

Remarks

▷ Example 1

We will show how expandcl works by using a small dataset with five clusters. In this dataset, cl identifies the clusters, x contains a unique value for each observation, and n identifies how many copies we want of each cluster.

166

```
. use http://www.stata-press.com/data/r10/expclxmpl
. list, sepby(cl)
```

	cl	x	n
1.	10	1	-1
2.	10	2	-1
3.	20	3	0
4.	20	4	0
5.	30	5	1
6.	30	6	1
7.	40	7	2.7
8.	40	8	2.7
9.	50	9	3
10.	50	10	3
11.	60	11	.
12.	60	12	.

```
. expandcl n, generate(newcl) cluster(cl)
(2 missing counts ignored; observations not deleted)
(2 noninteger counts rounded to integer)
(2 negative counts ignored; observations not deleted)
(2 zero counts ignored; observations not deleted)
(8 observations created)

. sort newcl cl x
```

(Continued on next page)

```
. list, sepby(newcl)
```

	cl	x	n	newcl
1.	10	1	-1	1
2.	10	2	-1	1
3.	20	3	0	2
4.	20	4	0	2
5.	30	5	1	3
6.	30	6	1	3
7.	40	7	2.7	4
8.	40	8	2.7	4
9.	40	7	2.7	5
10.	40	8	2.7	5
11.	40	7	2.7	6
12.	40	8	2.7	6
13.	50	9	3	7
14.	50	10	3	7
15.	50	9	3	8
16.	50	10	3	8
17.	50	9	3	9
18.	50	10	3	9
19.	60	11	.	10
20.	60	12	.	10

The first three clusters were not replicated since n was less than or equal to 1. n is 2.7 in the fourth cluster, so expandcl created two replications (2.7 was rounded to 3) of this cluster, bringing the total number of clusters of this type to 3. expandcl created two replications of cluster 50 since n is 3. Finally, expandcl did not replicate the last cluster because n was missing.

◁

Methods and Formulas

expandcl is implemented as an ado-file.

Also See

[R] **bsample** — Sampling with replacement

[D] **expand** — Duplicate observations

Title

> **fdasave** — Save and use datasets in FDA (SAS XPORT) format

Syntax

Save data in memory in FDA format

> <u>fdasa</u>ve *filename* $\begin{bmatrix} if \end{bmatrix}$ $\begin{bmatrix} in \end{bmatrix}$ $\begin{bmatrix}$, *fdasave_options* $\end{bmatrix}$

> <u>fdasa</u>ve *varlist* using *filename* $\begin{bmatrix} if \end{bmatrix}$ $\begin{bmatrix} in \end{bmatrix}$ $\begin{bmatrix}$, *fdasave_options* $\end{bmatrix}$

Read SAS XPORT file into Stata

> fdause *filename* $\begin{bmatrix}$, *fdause_options* $\end{bmatrix}$

Describe contents of SAS XPORT Transport file

> <u>fdades</u>cribe *filename* $\begin{bmatrix}$, <u>m</u>ember(*mbrname*) $\end{bmatrix}$

fdasave_options	description
Main	
<u>rena</u>me	rename variables and value labels to meet SAS XPORT restrictions
replace	overwrite files if they already exist
<u>vall</u>abfile(xpf)	save value labels in `formats.xpf`
<u>vall</u>abfile(<u>sas</u>code)	save value labels in SAS command file
<u>vall</u>abfile(both)	save value labels in `formats.xpf` and in a SAS command file
<u>vall</u>abfile(none)	do not save value labels

fdause_options	description
clear	replace data in memory
<u>noval</u>labels	ignore accompanying `formats.xpf` file if it exists
<u>m</u>ember(*mbrname*)	member to use; seldom used

Description

fdasave, fdause, and fdadescribe convert datasets to and from the U.S. Food and Drug Administration (FDA) format for new drug and new device applications (NDAs)—SAS XPORT Transport format. The primary intent of these commands is to assist people making submissions to the FDA, but the commands are general enough for use in transferring data between SAS and Stata.

To save the data in memory in the FDA format, type

. fdasave *filename*

although sometimes you will want to type

. fdasave *filename*, rename

It never hurts to specify the rename option. In any case, Stata will create *filename*.xpt as an XPORT file containing the data and, if needed, will also create formats.xpf—an additional XPORT file—containing the value-label definitions. These files can be easily read into SAS.

To read a SAS XPORT Transport file into Stata, type

. fdause *filename*

Stata will read into memory the XPORT file *filename*.xpt containing the data and, if available, will also read the value-label definitions stored in formats.xpf or FORMATS.xpf.

fdadescribe describes the contents of a SAS XPORT Transport file. The display is similar to that produced by describe, as is the syntax:

. fdadescribe *filename*

If *filename* is specified without an extension, .xpt is assumed.

Options for fdasave

⌐ Main ⌐

rename specifies that fdasave may rename variables and value labels to meet the SAS XPORT restrictions, which are that names be no more than eight characters long and that there be no distinction between uppercase and lowercase letters.

We recommend specifying the rename option. If this option is specified, any name violating the restrictions is changed to a different but related name in the file. The name changes are listed. The new names are used only in the file; the names of the variables and value labels in memory remain unchanged.

If rename is not specified and one or more names violate the XPORT restrictions, an error message will be issued and no file will be saved. The alternative to the rename option is that you can rename variables yourself with the rename command:

. rename mylongvariablename myname

See [D] **rename**. Renaming value labels yourself is more difficult. The easiest way to rename value labels is to use label save, edit the resulting file to change the name, execute the file by using do, and reassign the new value label to the appropriate variables by using label values:

. label save mylongvaluelabel using myfile.do
. doedit myfile.do   *(change mylongvaluelabel to, say, mlvlab)*
. do myfile.do
. label values myvar mlvlab

See [D] **label** and [R] **do** for more information about renaming value labels.

replace permits fdasave to overwrite existing *filename*.xpt, formats.xpf, and *filename*.sas files.

`vallabfile(xpf | sascode | both | none)` specifies whether and how value labels are to be stored. SAS XPORT Transport files do not really have value labels. In preparing datasets for submission to the FDA, value-label definitions should be provided in one of two ways:

1. In an additional SAS XPORT Transport file whose data contain the value-label definitions

2. In a SAS command file that will create the value labels

`fdasave` can create either or both of these files.

`vallabfile(xpf)`, the default, specifies that value labels be written into a separate SAS XPORT Transport file named `formats.xpf`. Thus `fdasave` creates two files: *filename*`.xpt`, containing the data, and `formats.xpf`, containing the value labels. No `formats.xpf` file is created if there are no value labels.

SAS users can easily use the resulting `.xpt` and `.xpf` XPORT files.
See http://www.sas.com/govedu/fda/macro.html for SAS-provided macros for reading the XPORT files. The SAS macro `fromexp()` reads the XPORT files into SAS. The SAS macro `toexp()` creates XPORT files. When obtaining the macros, remember to save the macros at SAS's web page as a plain-text file and to remove the examples at the bottom.

If the SAS macro file is saved as `C:\project\macros.mac` and the files `mydat.xpt` `formats.xpf` created by `fdasave` are in `C:\project\`, the following SAS commands would create the corresponding SAS dataset and format library and list the data:

```
──────────── SAS commands ────────────
%include "C:\project\macros.mac" ;
%fromexp(C:\project, C:\project) ;
libname library 'C:\project' ;
data _null_ ; set library.mydat ; put _all_ ; run ;
proc print data = library.mydat ;
quit ;
```

`vallabfile(sascode)` specifies that the value labels be written into a SAS command file, *filename*`.sas`, containing SAS `proc format` and related commands. Thus `fdasave` creates two files: *filename*`.xpt`, containing the data, and *filename*`.sas`, containing the value labels. SAS users may wish to edit the resulting *filename*`.sas` file to change the "libname datapath" and "libname xptfile xport" lines at the top to correspond to the location that they desire. `fdasave` sets the location to the current working directory at the time `fdasave` was issued. No `.sas` file will be created if there are no value labels.

`vallabfile(both)` specifies that both the actions described above be taken and that three files be created: *filename*`.xpt`, containing the data; `formats.xpf`, containing the value labels in XPORT format; and *filename*`.sas`, containing the value labels in SAS command-file format.

`vallabfile(none)` specifies that value-label definitions not be saved. Only one file is created: *filename*`.xpt`, which contains the data.

Options for fdause

`clear` permits the data to be loaded, even if there is a dataset already in memory and even if that dataset has changed since the data were last saved.

`novallabels` specifies that value-label definitions stored in `formats.xpf` or `FORMATS.xpf` not be looked for or loaded. By default, if variables are labeled in *filename*`.xpt`, `fdause` looks for `formats.xpf` to obtain and load the value-label definitions. If the file is not found, Stata looks for `FORMATS.xpf`. If that file is not found, a warning message is issued.

fdause can use only a `formats.xpf` or `FORMATS.xpf` file to obtain value-label definitions. fdause cannot understand value-label definitions from a SAS command file.

member(*mbrname*) is a rarely specified option indicating which member of the `.xpt` file is to be loaded. It is not used much anymore, but the original XPORT definition allowed multiple datasets to be placed in one file. Option `member()` allows you to read these old files. You can obtain a list of member names using fdadescribe. If `member()` is not specified—and it usually is not—fdause reads the first (and usually only) member.

Option for fdadescribe

member(*mbrname*) is a rarely specified option indicating which member of the `.xpt` file is to be described. See the description of the `member()` option for fdause directly above. If `member()` is not specified, all members are described, one after the other. It is rare for an XPORT file to have more than one member.

Remarks

SAS XPORT Transport format has been adopted by the U.S. Food and Drug Administration (FDA) for datasets submitted in support of new drug and device applications. For the FDA submission guidance document, see http://www.fda.gov/cder/guidance/2867fnl.pdf.

All users, of course, may use these commands to transfer data between SAS and Stata, but there are limitations in the SAS XPORT Transport format, such as the eight-character limit on the names of variables (specifying fdasave's `rename` option works around that). For a complete listing of limitations and issues concerning the SAS XPORT Transport format, and an explanation of how fdasave and fdause work around these limitations, see the Technical Appendix below. For non-FDA applications, you may find it more convenient to use translation packages such as Stat/Transfer; see http://www.stata.com/products/transfer.html.

Remarks are presented under the following headings:

> Saving XPORT files for transferring to SAS
> Determining the contents of XPORT files received from SAS
> Using XPORT files received from SAS

Saving XPORT files for transferring to SAS

▷ Example 1

To save the current dataset in `clindata.xpt` and the value labels in `formats.xpf`, type

    . fdasave clindata

To save the data as above but automatically rename variable names and value labels that are too long or are case sensitive, type

    . fdasave clindata, rename

To allow the replacement of any preexisting files, type

    . fdasave clindata, rename replace

To save the current dataset in `clindata.xpt` and the value labels in SAS command file `clindata.sas` and to automatically rename variable and value-label names:

    . fdasave clindata, rename vallab(sas)

To save the data as above but save the value labels in both `formats.xpf` and `clindata.sas`, type

     `. fdasave clindata, rename vallab(both)`

To not save the value labels at all, thus creating only `clindata.xpt`, type

     `. fdasave clindata, rename vallab(none)`

                                                               ◁

Determining the contents of XPORT files received from SAS

▷ Example 2

To determine the contents of `drugdata.xpt`, you might type

     `. fdadescribe drugdata`

                                                               ◁

Using XPORT files received from SAS

▷ Example 3

To read data from `drugdata.xpt` and obtain value labels from `formats.xpf` (or `FORMATS.xpf`), if the file exists, you would type

     `. fdause drugdata`

To read the data as above and discard any data in memory, type

     `. fdause drugdata, clear`

                                                               ◁

Saved Results

`fdadescribe` saves the following in `r()`:

Scalars

`r(N)`	number of observations	`r(size)`	size of data
`r(k)`	number of variables	`r(n_members)`	number of members

Macros

     `r(members)`   names of members

Technical appendix

Technical details concerning the SAS XPORT Transport format and how `fdasave` and `fdause` handle issues regarding the format are presented under the following headings:

         A1. Overview of SAS XPORT *Transport format*
         A2. Implications for writing XPORT *datasets from Stata*
         A3. Implications for reading XPORT *datasets into Stata*

A1. Overview of SAS XPORT Transport format

A SAS XPORT Transport file may contain one or more separate datasets, known as members. It is rare for a SAS XPORT Transport file to contain more than one member. See http://support.sas.com/techsup/technote/ts140.html for the SAS technical document describing the layout of the SAS XPORT Transport file.

A SAS XPORT dataset (member) is subject to certain restrictions:

1. The dataset may contain only 9,999 variables.

2. The names of the variables and value labels may not be longer than eight characters and are case insensitive; e.g., `myvar`, `Myvar`, `MyVar`, and `MYVAR` are all the same name.

3. Variable labels may not be longer than 40 characters.

4. The contents of a variable may be numeric or string:

    4.1 Numeric variables may be integer or floating but may not be smaller than $5.398e-79$ or greater than $9.046e+74$, absolutely. Numeric variables may contain missing, which may be `.`, `._`, `.a`, `.b`, ..., `.z`.

    4.2 String variables may not exceed 200 characters. String variables are recorded in a "padded" format, meaning that, when variables are read, it cannot be determined whether the variable had trailing blanks.

5. Value labels are *not* written in the XPORT dataset. Suppose that you have variable `sex` in the data with values 0 and 1, and the values are labeled for gender (0=male, and 1=female). When the dataset is written in SAS XPORT Transport format, you can record that the variable label `gender` is associated with the `sex` variable, but you cannot record the association with the value labels male and female.

Value-label definitions are typically stored in a second XPORT dataset or in a text file containing SAS commands. You can use the `vallabfile()` option of `fdasave` to produce these datasets or files.

Value labels and formats are recorded in the same position in an XPORT file, meaning that names corresponding to formats used in SAS cannot be used. Thus value labels may not be named

> best, binary, comma, commax, d, date, datetime, dateampm, day, ddmmyy,
> dollar, dollarx, downame, e, eurdfdd, eurdfde, eurdfdn, eurdfdt, eu-
> rdfdwn, eurdfmn, eurdfmy, eurdfwdx, eurdfwkx, float, fract, hex, hhmm,
> hour, ib, ibr, ieee, julday, julian, percent, minguo, mmddyy, mmss, mmyy,
> monname, month, monyy, negparen, nengo, numx, octal, pd, pdjulg, pdjuli,
> pib, pibr, pk, pvalue, qtr, qtrr, rb, roman, s370ff, s370fib, s370fibu,
> s370fpd, s370fpdu, s370fpib, s370frb, s370fzd, s370fzdl, s370fzds,
> s370fzdt, s370fzdu, ssn, time, timeampm, tod, weekdate, weekdatx, week-
> day, worddate, worddatx, wordf, words, year, yen, yymm, yymmdd, yymon,
> yyq, yyqr, z, zd,

or any uppercase variation of these.

We refer to this as the "Known Reserved Word List" in this documentation. Other words may also be reserved by SAS; the technical documentation for the SAS XPORT Transport format provides no guidelines. This list was created by examining the formats defined in *SAS Language Reference: Dictionary, Version 8*. If SAS adds new formats, the list will grow.

6. A flaw in the XPORT design can make it impossible, in rare instances, to determine the exact number of observations in a dataset. This problem can occur only if (1) all variables in the dataset are string and (2) the sum of the lengths of all the string variables is less than 80. Actually, the above is the restriction, assuming that the code for reading the dataset is written well. If it is not, the flaw could occur if (1) the last variable or variables in the dataset are string and (2) the sum of the lengths of all variables is less than 80.

   To prevent stumbling over this flaw, make sure that the last variable in the dataset is not a string variable. This is always sufficient to avoid the problem.

7. There is no provision for saving the Stata concepts `notes` and `characteristics`.

A2. Implications for writing XPORT datasets from Stata

Stata datasets for the most part fit well into the SAS XPORT Transport format. With the same numbering scheme as above,

1. Stata refuses to write the dataset if it contains more than 9,999 variables.

2. Stata issues an error message if any variable or label name violates the naming restrictions, or if option `rename` is specified, Stata fixes any names that violate the restrictions.

   Whether or not `rename` is specified, names will be recorded case insensitively: you do not have to name all your variables with all lowercase or all uppercase letters. Stata verifies that ignoring case does not lead to problems, complaining or, if option `rename` is specified, fixing them.

3. Stata truncates variable labels to 40 characters to fit within the XPORT limit.

4. Stata treats variable contents as follows:

   • If a numeric variable records a value greater than 9.046e+74 in absolute value, Stata issues an error message. If a variable records a value less than 5.398e-79 in absolute value, 0 is written.

   • If you have string variables longer than 200 characters, Stata issues an error message. Also, if any string variable has trailing blanks, Stata issues an error message. To remove trailing blanks from string variable `s`, you can type

   ```
   . replace s = rtrim(s)
   ```

   To remove leading and trailing blanks, type

   ```
   . replace s = trim(s)
   ```

5. Value-label names are written in the XPORT dataset. The contents of the value label are not written in the same XPORT dataset. By default, `formats.xpf`, a second XPORT dataset, is created containing the value-label definitions.

   SAS recommends creating a `formats.xpf` file containing the value-label definitions (what SAS calls format definitions). They have provided SAS macros, making the reading of `.xpt` and `formats.xpf` files easy. See http://www.sas.com/govedu/fda/macro.html for details.

   Alternatively, a SAS command file containing the value-label definitions can be produced. The `vallabfile()` option of `fdasave` is used to indicate which, if any, of the formats to use for recording the value-label definitions.

   If a value-label name matches a name on the Known Reserved Word List, and option `rename` is not specified, Stata issues an error message.

If a variable has no value label, the following format information is recorded:

Stata format	SAS format
%td...	MMDDYY10.
%-td...	MMDDYY10.
%#s	$CHAR#.
%-#s	$CHAR#.
% #s	$CHAR#.
all other	BEST12.

6. If you have a dataset that could provoke the XPORT design flaw, a warning message is issued. Remember, the best way to avoid this flaw is to ensure that the last variable in the dataset is numeric. This is easily done. You could, for instance, type

```
. gen ignoreme = 0
. fdasave ...
```

7. Because the XPORT file format does not support notes and characteristics, Stata ignores them when it creates the XPORT file. You may wish to incorporate important notes into the documentation that you provide to the user of your XPORT file.

A3. Implications for reading XPORT datasets into Stata

Reading SAS XPORT Transport format files into Stata is easy, but sometimes there are issues to consider:

1. If the dataset is too large to fit into memory, Stata issues an error message. You can increase the amount of memory allocated to Stata with the set memory command; see [D] **memory**.

If there are too many variables, Stata issues an error message. If you are using Stata/MP or Stata/SE, you can increase the maximum number of variables with the set maxvar command; see [D] **memory**.

2. The XPORT format variable naming restrictions are more restrictive than those of Stata, so no problems should arise. However, Stata reserves the following names:

_all, _b, byte, _coef, _cons, double, float, if, in, int, long, _n, _N, _pi, _pred, _rc, _se, _skip, str, str#, using, with

If the XPORT file contains variables with any of these names, Stata issues an error. Also, the error message

```
. fdause ...
_____ already defined
r(110);
```

indicates that the XPORT file was incorrectly prepared by some other software and that two or more variables share the same name.

3. The XPORT variable-label-length limit is more restrictive than that of Stata, so no problems can arise.

4. Variable contents may cause problems:

- The range of numeric variables in an XPORT dataset is a subset of that allowed by Stata, so no problems can arise. All variables are brought back as `doubles`; we recommend that you run `compress` after loading the dataset:

  ```
  . fdause ...
  . compress
  ```

  See [D] **compress**.

  Stata has no missing-value code corresponding to . _. If any value records . _, .u is stored.

- String variables are brought back as recorded but with all trailing blanks stripped.

5. Value-label names are read directly from the XPORT dataset. Any value-label definitions are obtained from a separate XPORT dataset, if available. If a value-label name matches any in the Known Reserved Word List, no value-label name is recorded, and instead, the variable display format is set to `%9.0g`, `%10.0g`, or `%td`.

   The `%td` Stata format is used when the following SAS Formats are encountered:

   DATE, EURDFDN, JULDAY, MONTH, QTRR, YEAR, DAY, EURDFDWN, JULIAN, MONYY, WEEKDATE, YYMM, DDMMYY, EURDFMN, MINGUO, NENGO, WEEKDATX, YYMMDD, DOW-NAME, EURDFMY, MMDDYY, PDJULG, WEEKDAY, YYMON, EURDFDD, EURDFWDX, MMYY, PDJULI, WORDDATE, YYQ, EURDFDE, EURDFWKX, MONNAME, QTR, WORDDATX, YYQR

   If the XPORT file indicates that one or more variables have value labels, `fdause` looks for the value-label definitions in `formats.xpf`, another XPORT file. If it does not find this file, it looks for `FORMATS.xpf`. If this file is not found, `fdause` issues a warning unless the `novallabels` option is specified.

   Stata does not allow value-label ranges or string variables with value labels. If the `.xpt` file or `formats.xpf` file contains any of these, an error message is issued. The `novallabels` option allows you to read the data, ignoring all value labels.

6. If a dataset is read that provokes the all-strings XPORT design flaw, the dataset with the minimum number of possible observations is returned, and a warning message is issued. This duplicates the behavior of SAS.

7. SAS XPORT format does not allow notes or characteristics, so no issues can arise.

Also See

[D] **outfile** — Write ASCII-format dataset

[D] **save** — Save datasets

[D] **infile** — Overview of reading data into Stata

[D] **odbc** — Load, write, or view data from ODBC sources

[D] **describe** — Describe data in memory or in file

Title

filefilter — Convert ASCII text or binary patterns in a file

Syntax

<u>filefil</u>ter *oldfile newfile* , <u>fr</u>om(*oldpattern*) <u>to</u>(*newpattern*) [*options*]

where *oldpattern* and *newpattern* for ASCII characters are

    "*string*" or *string*

string	:=	[*char*[*char*[*char*[. . .]]]]
char	:=	*regchar* | *code*
regchar	:=	ASCII 32–91, 93–128, 161–255; excludes '\'
code	:=	\BS backslash

\r	carriage return
\n	newline
\t	tab
\M	Macintosh EOL, or \r
\W	Windows EOL, or \r\n
\U	Unix EOL, or \n
\LQ	left single quote '
\RQ	right single quote '
\Q	double quote "
\$	dollar $
\###d	3-digit [0–9] decimal ASCII
\##h	2-digit [0–9, A–F] hexadecimal ASCII

options	description
*<u>from</u>(oldpattern)	find *oldpattern* to be replaced
*<u>to</u>(newpattern)	use *newpattern* to replace occurrences of **from()**
<u>rep</u>lace	replace *newfile* if it already exists

* from(*oldpattern*) and to(*newpattern*) are required.

Description

filefilter reads an ASCII input file, searching for *oldpattern*. Whenever a matching pattern is found, it is replaced with *newpattern*. All resulting data, whether matching or nonmatching, are then written to the new ASCII file.

Because of the buffering design of **filefilter**, arbitrarily large files can be converted quickly. **filefilter** is also useful when traditional editors cannot edit a file, such as when unprintable ASCII characters are involved. In fact, converting end-of-line characters between Macintosh, Windows, and Unix is convenient with the EOL codes.

Unicode is not directly supported at this time, but you can attempt to operate on a Unicode system by breaking a 2-byte character into the corresponding two-character ASCII representation. However, this goes beyond the original design of the command and is technically unsupported. If you attempt to use `filefilter` in this manner, you might encounter problems with variable-length encoded Unicode.

Although it is not mandatory, you may want to use quotes to delimit a pattern, protecting the pattern from Stata's parsing routines. A pattern that contains blanks must be in quotes.

Options

`from(`*oldpattern*`)` is required and specifies the pattern to be found and replaced.

`to(`*newpattern*`)` is required and specifies the pattern used to replace occurrences of `from()`.

`replace` specifies that *newfile* be replaced if it already exists.

Remarks

▷ Example 1

```
. filefilter macfile.txt winfile.txt, from(\M) to(\W) replace
. filefilter auto1.csv auto2.csv, from(\LQ) to("left quote")
. filefilter auto1.csv auto2.csv, from(\60h) to("left quote")
. filefilter auto1.csv auto2.csv, from(\096d) to("left quote")
. filefilter file1.txt file2.txt, from("\6BhText\100dText") to("")
```

◁

Saved Results

`filefilter` saves the following in `r()`:

Scalars

`r(occurrences)`	number of *oldpattern* found
`r(bytes_from)`	# of bytes represented by *oldpattern*
`r(bytes_to)`	# of bytes represented by *newpattern*

Also See

[P] **file** — Read and write ASCII text and binary files

[D] **hexdump** — Display hexadecimal report on file

Title

> **fillin** — Rectangularize dataset

Syntax

`fillin` *varlist*

Description

`fillin` adds observations with missing data so that all interactions of *varlist* exist, thus making a complete rectangularization of *varlist*. `fillin` also adds the variable `_fillin` to the dataset. `_fillin` is 1 for observations created by using `fillin` and 0 for previously existing observations.

Remarks

▷ Example 1

We have data on something by sex, race, and age group. We suspect that some of the combinations of sex, race, and age do not exist, but if so, we want them to exist with whatever remaining variables there are in the dataset set to missing. That is, rather than having a missing observation for black females aged 20–24, we want to create an observation that contains missing values:

```
. use http://www.stata-press.com/data/r10/fillin1
. list
```

	sex	race	age_gr~p	x1	x2
1.	female	white	20-24	20393	14.5
2.	male	white	25-29	32750	12.7
3.	female	black	30-34	39399	14.2

```
. fillin sex race age_group
. list, sepby(sex)
```

	sex	race	age_gr~p	x1	x2	_fillin
1.	female	white	20-24	20393	14.5	0
2.	female	white	25-29	.	.	1
3.	female	white	30-34	.	.	1
4.	female	black	20-24	.	.	1
5.	female	black	25-29	.	.	1
6.	female	black	30-34	39399	14.2	0
7.	male	white	20-24	.	.	1
8.	male	white	25-29	32750	12.7	0
9.	male	white	30-34	.	.	1
10.	male	black	20-24	.	.	1
11.	male	black	25-29	.	.	1
12.	male	black	30-34	.	.	1

◁

Methods and Formulas

`fillin` is implemented as an ado-file.

Reference

Cox, N. J. 2005. Stata tip 17: Filling in the gaps. *Stata Journal* 5: 135–136.

Also See

[D] **save** — Save datasets

[D] **cross** — Form every pairwise combination of two datasets

[D] **expand** — Duplicate observations

[D] **joinby** — Form all pairwise combinations within groups

Title

format — Set variables' output format

Syntax

Set formats

> <u>form</u>at *varlist* %*fmt*

> <u>form</u>at %*fmt varlist*

Set style of decimal point

> <u>se</u>t dp { <u>comma</u> | <u>period</u> } [, <u>perm</u>anently]

Display long formats

> <u>form</u>at [*varlist*]

where %*fmt* can be numerical, date, or string format.

Numerical % *fmt*	Description	Example
right justified		
%#.#g	general	%9.0g
%#.#f	fixed	%9.2f
%#.#e	exponential	%10.7e
%21x	hexadecimal	%21x
%16H	binary, hilo	%16H
%16L	binary, lohi	%16L
%8H	binary, hilo	%8H
%8L	binary, lohi	%8L
right justified with commas		
%#.#gc	general	%9.0gc
%#.#fc	fixed	%9.2fc
right justified with leading zeros		
%0#.#f	fixed	%09.2f
left justified		
%-#.#g	general	%-9.0g
%-#.#f	fixed	%-9.2f
%-#.#e	exponential	%-10.7e
left justified with commas		
%-#.#gc	general	%-9.0gc
%-#.#fc	fixed	%-9.2fc

You may substitute a comma (,) for period (.) in any
  of the above formats to make comma the decimal point. In
  %9,2fc, 1000.03 is 1.000,03. Or you can set dp comma.

date % *fmt*	Description	Example
right justified		
%tc	date/time	%tc
%tC	date/time	%tC
%td	date	%td
%tw	week	%tw
%tm	month	%tm
%tq	quarter	%tq
%th	halfyear	%th
%ty	year	%ty
%tg	generic	%tg
left justified		
%-tc	date/time	%-tc
%-tC	date/time	%-tC
%-td	date	%-td
etc.		

There are lots of variations allowed. See *Formatting date
  and time values* in [D] **dates and times**.

string %*fmt*	Description	Example
right justified		
%#s	string	%15s
left justified		
%-#s	string	%-20s
centered		
%~#s	string	%~12s

The centered format is for use with display only.

Description

format *varlist* %*fmt* and format %*fmt* *varlist* are the same commands. They set the display format associated with the variables specified. The default formats are a function of the type of the variable:

byte	%8.0g
int	%8.0g
long	%12.0g
float	%9.0g
double	%10.0g
str#	%#s

set dp sets the symbol that Stata uses to represent the decimal point. The default is period, meaning that one and a half is displayed as 1.5.

format [*varlist*] displays the current formats associated with the variables. format by itself lists all variables that have formats too long to be listed in their entirety by describe. format *varlist* lists the formats for the specified variables regardless of their length. format * lists the formats for all the variables.

Option

permanently specifies that, in addition to making the change right now, the dp setting be remembered and become the default setting when you invoke Stata.

Remarks

Remarks are presented under the following headings:

Setting formats
Setting European formats
Details of formats
    The %f format
    The %fc format
    The %g format
    The %gc format
    The %e format
    The %21x format
    The %16H and %16L formats
    The %8H and %8L formats
    The %t format
    The %s format
Other effects of formats
Displaying current formats

Setting formats

See [U] **12.5 Formats: controlling how data are displayed** for an explanation of %*fmt*. To review: Stata's three numeric formats are denoted by a leading percent sign, %, followed by the string *w.d* (or *w,d* for European format), where *w* and *d* stand for two integers. The first integer, *w*, specifies the width of the format. The second integer, *d*, specifies the number of digits that are to follow the decimal point; *d* must be less than *w*. Finally, a character denoting the format type (e, f, or g) is appended. For example, %9.2f specifies the f format that is nine characters wide and has two digits following the decimal point. For f and g, a c may also be suffixed to indicate comma formats. Other "numeric" formats known collectively as the %t formats are used to display dates and times; see [D] **dates and times**. String formats are denoted by %*w*s, where *w* indicates the width of the format.

▷ Example 1

We have census data by region and state on median age and population in 1980.

```
. use http://www.stata-press.com/data/r10/census10
(1980 Census data by state)

. describe

Contains data from http://www.stata-press.com/data/r10/census10.dta
  obs:            50                          1980 Census data by state
  vars:            4                          11 Dec 2006 10:33
  size:         1,400 (99.8% of memory free)
```

variable name	storage type	display format	value label	variable label
state	str14	%14s		State
region	int	%8.0g	cenreg	Census region
pop	long	%11.0g		Population
medage	float	%9.0g		Median age

```
Sorted by:
. list in 1/8
```

	state	region	pop	medage
1.	Alabama	South	3893888	29.3
2.	Alaska	West	401851	26.1
3.	Arizona	West	2718215	29.2
4.	Arkansas	South	2286435	30.6
5.	California	West	23667902	29.9
6.	Colorado	West	2889964	28.6
7.	Connecticut	NE	3107576	32
8.	Delaware	South	594338	29.8

(Continued on next page)

The `state` variable has a display format of `%14s`. To left-align the state data, we type

```
. format state %-14s
. list in 1/8
```

	state	region	pop	medage
1.	Alabama	South	3893888	29.3
2.	Alaska	West	401851	26.1
3.	Arizona	West	2718215	29.2
4.	Arkansas	South	2286435	30.6
5.	California	West	23667902	29.9
6.	Colorado	West	2889964	28.6
7.	Connecticut	NE	3107576	32
8.	Delaware	South	594338	29.8

Although it seems like `region` is a string variable, it is really a numeric variable with an attached value label. You do the same thing to left-align a numeric variable as you do a string variable: insert a negative sign.

```
. format region %-8.0g
. list in 1/8
```

	state	region	pop	medage
1.	Alabama	South	3893888	29.3
2.	Alaska	West	401851	26.1
3.	Arizona	West	2718215	29.2
4.	Arkansas	South	2286435	30.6
5.	California	West	23667902	29.9
6.	Colorado	West	2889964	28.6
7.	Connecticut	NE	3107576	32
8.	Delaware	South	594338	29.8

The `pop` variable would probably be easier to read if we inserted commas by appending a 'c':

```
. format pop %11.0gc
. list in 1/8
```

	state	region	pop	medage
1.	Alabama	South	3,893,888	29.3
2.	Alaska	West	401,851	26.1
3.	Arizona	West	2,718,215	29.2
4.	Arkansas	South	2,286,435	30.6
5.	California	West	23667902	29.9
6.	Colorado	West	2,889,964	28.6
7.	Connecticut	NE	3,107,576	32
8.	Delaware	South	594,338	29.8

Look at the value of `pop` for observation 5. There are no commas. This number was too large for Stata to insert commas and still respect the current width of 11. Let's try again:

```
. format pop %12.0gc
. list in 1/8
```

	state	region	pop	medage
1.	Alabama	South	3,893,888	29.3
2.	Alaska	West	401,851	26.1
3.	Arizona	West	2,718,215	29.2
4.	Arkansas	South	2,286,435	30.6
5.	California	West	23,667,902	29.9
6.	Colorado	West	2,889,964	28.6
7.	Connecticut	NE	3,107,576	32
8.	Delaware	South	594,338	29.8

Finally, medage would look better if the decimal points were vertically aligned.

```
. format medage %8.1f
. list in 1/8
```

	state	region	pop	medage
1.	Alabama	South	3,893,888	29.3
2.	Alaska	West	401,851	26.1
3.	Arizona	West	2,718,215	29.2
4.	Arkansas	South	2,286,435	30.6
5.	California	West	23,667,902	29.9
6.	Colorado	West	2,889,964	28.6
7.	Connecticut	NE	3,107,576	32.0
8.	Delaware	South	594,338	29.8

Display formats are permanently attached to variables by the format command. If we save the data, the next time we use it, state will still be formatted as %-14s, region will still be formatted as %-8.0g, etc.

◁

▷ Example 2

Suppose that we have an employee identification variable empid and that we want to retain the leading zeros when we list our data. format has a leading-zero option that allows this.

```
. use http://www.stata-press.com/data/r10/fmtxmpl
. describe empid
```

variable name	storage type	display format	value label	variable label
empid	float	%9.0g		

. list empid in 83/87

	empid
83.	98
84.	99
85.	100
86.	101
87.	102

. format empid %05.0f

. list empid in 83/87

	empid
83.	00098
84.	00099
85.	00100
86.	00101
87.	00102

◁

❑ Technical Note

The syntax of the `format` command allows a *varlist* and not just a *varname*. Thus you can attach the `%9.2f` format to the variables `myvar`, `thisvar`, and `thatvar` by typing

. format myvar thisvar thatvar %9.2f

❑

▷ Example 3

We have employee data that includes `hiredate` and `login` and `logout` times. `hiredate` is stored as a `float`, but we were careful to store `login` and `logout` as `doubles`. We need to attach a date format to these three variables.

. use http://www.stata-press.com/data/r10/fmtxmpl2

. format hiredate login logout

variable name	display format
hiredate	%9.0g
login	%10.0g
logout	%10.0g

. format login logout %tcDDmonCCYY_HH:MM:SS.ss

. list login logout in 1/5

	login	logout
1.	08nov2006 08:16:42.30	08nov2006 05:32:23.53
2.	08nov2006 08:07:20.53	08nov2006 05:57:13.40
3.	08nov2006 08:10:29.48	08nov2006 06:17:07.51
4.	08nov2006 08:30:02.19	08nov2006 05:42:23.17
5.	08nov2006 08:29:43.25	08nov2006 05:29:39.48

```
. format hiredate %td
. list hiredate in 1/5
```

	hiredate
1.	24jan1986
2.	10mar1994
3.	29sep2006
4.	14apr2006
5.	03dec1999

We remember that the project manager requested that hire dates be presented in the same form as they were previously.

```
. format hiredate %tdDD/NN/CCYY
. list hiredate in 1/5
```

	hiredate
1.	24/01/1986
2.	10/03/1994
3.	29/09/2006
4.	14/04/2006
5.	03/12/1999

◁

Setting European formats

Do you prefer that one and one half be written as 1,5 and that one thousand one and a half be written as 1.001,5? Stata will present numbers in that format if, when you set the format, you specify ',' rather than '.' as follows:

```
. use http://www.stata-press.com/data/r10/census10
(1980 Census data by state)
. format pop %12,0gc
. format medage %9,2f
. list in 1/8
```

	state	region	pop	medage
1.	Alabama	South	3.893.888	29,30
2.	Alaska	West	401.851	26,10
3.	Arizona	West	2.718.215	29,20
4.	Arkansas	South	2.286.435	30,60
5.	California	West	23.667.902	29,90
6.	Colorado	West	2.889.964	28,60
7.	Connecticut	NE	3.107.576	32,00
8.	Delaware	South	594.338	29,80

You can also leave the formats just as they were and instead type `set dp comma`. That tells Stata to interpret all formats as if you had typed the comma instead of the period:

```
. format pop %12.0gc                    (put the formats back as they were)
. format medage %9.2f
. set dp comma                          (tell Stata to use European format)
. list in 1/8
(same output appears as above)
```

`set dp comma` affects all Stata output, so if you run a regression, display summary statistics, or make a table, commas will be used instead of periods in the output:

```
. tabulate region [fw=pop]
```

Census region	Freq.	Percent	Cum.
NE	49.135.283	21,75	21,75
N Cntrl	58.865.670	26,06	47,81
South	74.734.029	33,08	80,89
West	43.172.490	19,11	100,00
Total	225.907.472	100,00	

You can return to using periods by typing

```
. set dp period
```

Setting a variable's display format to European affects how the variable's values are displayed by `list` and in a few other places. Setting dp to `comma` affects every bit of Stata.

Also, `set dp comma` affects only how Stata displays output, not how it gets input. When you need to type one and a half, you must type `1.5` regardless of context.

❏ Technical Note

`set dp comma` makes drastic changes inside Stata, and we mention this because some older, user-written programs may not be able to deal with those changes. If you are using an older, user-written program, you might `set dp comma` only to find that the program does not work and instead presents some sort of syntax error.

If, using any program, you get an unanticipated error, try setting dp back to `period`.

Even with `set dp comma`, you might still see some output with the decimal symbol shown as a period rather than a comma. There are two places in Stata where Stata ignores `set dp comma` because the features are generally used to produce what will be treated as input, and `set dp comma` does not affect how Stata inputs numbers. First,

```
local x = sqrt(2)
```

stores the string "1.414213562373095" in x and not "1,414213562373095", so if some program were to display 'x' as a string in the output, the period would be displayed. Most programs, however, would use 'x' in subsequent calculations or, at the least, when the time came to display what was in 'x', would display it as a number. They would code

```
display ... 'x' ...
```

and not

```
display ... "'x'" ...
```

so the output would be

```
... 1,4142136 ...
```

The other place where Stata ignores `set dp comma` is the `string()` function. If you type

    . gen res = string(numvar)

new variable `res` will contain the string representation of numeric variable `numvar`, with the decimal symbol being a period, even if you have previously `set dp comma`. Of course, if you explicitly ask that `string()` use European format,

    . gen res = string(numvar,"%9,0g")

then `string()` honors your request; `string()` merely ignores the global `set dp comma`.

❏

Details of formats

The %f format

In `%w.df`, w is the total output width, including sign and decimal point, and d is the number of digits to appear to the right of the decimal point. The result is right justified.

The number 5.139 in `%12.2f` format displays as

    ----+----1--
          5.14

When $d = 0$, the decimal point is not displayed. The number 5.14 in `%12.0f` format displays as

    ----+----1--
             5

`%-w.df` works the same way, except that the output is left justified in the field. The number 5.139 in `%-12.2f` displays as

    ----+----1--
    5.14

The %fc format

`%w.dfc` works like `%w.df` except that commas are inserted to make larger numbers more readable. w records the total width of the result, including commas.

The number 5.139 in `%12.2fc` format displays as

    ----+----1--
          5.14

The number 5203.139 in `%12.2fc` format displays as

    ----+----1--
      5,203.14

As with `%f`, if $d = 0$, the decimal point is not displayed. The number 5203.139 in `%12.0fc` format displays as

    ----+----1--
         5,203

As with `%f`, a minus sign may be inserted to left justify the output. The number 5203.139 in `%-12.0fc` format displays as

    ----+----1--
    5,203

The %g format

In %$w.dg$, w is the overall width, and d is usually specified as 0, which leaves up to the format the number of digits to be displayed to the right of the decimal point. If $d \neq 0$ is specified, then not more than d digits will be displayed. As with %f, a minus sign may be inserted to left-justify results.

%g differs from %f in that (1) it decides how many digits to display to the right of the decimal point, and (2) it will switch to a %e format if the number is too large or too small.

The number 5.139 in %12.0g format displays as

```
----+----1--
       5.139
```

The number 5231371222.139 in %12.0g format displays as

```
----+----1--
  5231371222
```

The number 52313712223.139 displays as

```
----+----1--
  5.23137e+10
```

The number 0.0000029394 displays as

```
----+----1--
  2.93940e-06
```

The %gc format

%$w.dg$c is %$w.dg$, with commas. It works in the same way as the %g and %fc formats.

The %e format

%$w.de$ displays numeric values in exponential format. w records the width of the format. d records the number of digits to be shown after the decimal place. w should be greater than or equal to $d+7$ or, if 3-digit exponents are expected, $d+8$.

The number 5.139 in %12.4e format is

```
----+----1--
  5.1390e+00
```

The number 5.139×10^{220} is

```
----+----1--
  5.1390e+220
```

The %21x format

The %21x format is for those, typically programmers, who wish to analyze routines for numerical round-off error. There is no better way to look at numbers than how the computer actually records them.

The number 5.139 in %21x format is

```
----+----1----+----2-
+1.48e5604189375X+002
```

The number 5.125 is

```
----+----1----+----2-
+1.4800000000000X+002
```

Reported is a signed, base-16 number with base-16 point, the letter X, and a signed, 3-digit base-16 integer. Call the two numbers f and e. The interpretation is $f \times 2^e$.

The %16H and %16L formats

The %16H and %16L formats show the value in the IEEE floating point, double-precision form. %16H shows the value in most-significant-byte-first (hilo) form. %16L shows the number in least-significant-byte-first (lohi) form.

The number 5.139 in %16H is

```
----+----1----+-
40148e5604189375
```

The number 5.139 in %16L is

```
----+----1----+-
75931804568e1440
```

The format is sometimes used by programmers who are simultaneously studying a hexadecimal dump of a binary file.

The %8H and %8L formats

%8H and %8L are similar to %16H and %16L but show the number in IEEE single-precision form.

The number 5.139 in %8H is

```
----+---
40a472b0
```

The number 5.139 in %8L is

```
----+---
b072a440
```

The %t format

The %t format displays numerical variables as dates and times. See *Formatting date and time values* in [D] **dates and times**.

The %s format

The %*w*s format displays a string in a right-justified field of width *w*. %-*w*s displays the string left justified.

"Mary Smith" in %16s format is

```
----+----1----+-
      Mary Smith
```

"Mary Smith" in %-16s format is

```
----+----1----+-
Mary Smith
```

In addition, in some contexts, particularly display (see [P] **display**), %~*w*s is allowed, which centers the string. "Mary Smith" in %~16s format is

```
----+----1----+-
   Mary Smith
```

Other effects of formats

You have data on the age of employees, and you type `summarize age` to obtain the mean and standard deviation. By default, Stata uses its default g format to provide as much precision as possible:

```
. use http://www.stata-press.com/data/r10/fmtxmpl
```

```
. summarize age
```

Variable	Obs	Mean	Std. Dev.	Min	Max
age	204	30.18627	10.38067	18	66

If you attach a `%9.2f` format to the variable and specify the `format` option, Stata uses that specification to format the results:

```
. format age %9.2f
```

```
. summarize age, format
```

Variable	Obs	Mean	Std. Dev.	Min	Max
age	204	30.19	10.38	18.00	66.00

Displaying current formats

`format` *varlist* is not much used to display the formats associated with variables because using `describe` (see [D] **describe**) is easier and provides more information. The exceptions are date variables. Unless you use the default %tc, %tC, ... formats (and most people do), the format specifier itself can become very long, such as

```
. format admittime %tcDDmonCCYY_HH:MM:SS.sss
```

Such formats are too long for `describe` to display, so it gives up. In such cases, you can use `format` to display the format:

```
. format admittime
```

variable name	display format
admittime	%tcDDmonCCYY_HH:MM:SS.sss

Type `format *` to see the formats for all the variables.

Also See

[P] **display** — Display strings and values of scalar expressions

[D] **dates and times** — Date and time (%t) values and variables

[U] **12.5 Formats: controlling how data are displayed**

[U] **12.6 Dataset, variable, and value labels**

Title

---

functions — Functions

Description

This entry describes the functions allowed by Stata. For information on Mata functions, see [M-4] **intro**.

A quick note about missing values: Stata denotes a numeric missing value by ., .a, .b, ..., or .z. A string missing value is denoted by "" (the empty string). Here any one of these may be referred to by *missing*. If a numeric value x is missing, then $x \geq .$ is true. If a numeric value x is not missing, then $x < .$ is true.

Functions are listed under the following headings:

Mathematical functions
Probability distributions and density functions
Random-number functions
String functions
Programming functions
Date and time functions
Selecting time spans
Matrix functions returning a matrix
Matrix functions returning a scalar

Mathematical functions

abs(x)
  Domain: −8e+307 to 8e+307
  Range: 0 to 8e+307
  Description: returns the absolute value of x.

acos(x)
  Domain: −1 to 1
  Range: 0 to π
  Description: returns the radian value of the arccosine of x.

asin(x)
  Domain: −1 to 1
  Range: $-\pi/2$ to $\pi/2$
  Description: returns the radian value of the arcsine of x.

atan(x)
  Domain: −8e+307 to 8e+307
  Range: $-\pi/2$ to $\pi/2$
  Description: returns the radian value of the arctangent of x.

atan2(y, x)
  Domain y: −8e+307 to 8e+307
  Domain x: −8e+307 to 8e+307
  Range: $-\pi$ to π
  Description: returns the radian value of the arctangent of y/x where the signs of the parameters y and x are used to determine the quadrant of the answer.

`atanh(x)`

> Domain: -1 to 1
> Range: $-8e+307$ to $8e+307$
> Description: returns the arc-hyperbolic tangent of x, $\mathtt{atanh}(x) = \frac{1}{2}\{\ln(1 + x) - \ln(1 - x)\}$.

`ceil(x)`

> Domain: $-8e+307$ to $8e+307$
> Range: integers in $-8e+307$ to $8e+307$
> Description: returns the unique integer n such that $n - 1 < x \le n$.
> returns x (not ".") if x is missing, meaning that `ceil(.a)` $=$ `.a`.
>
> Also see `floor(x)`, `int(x)`, and `round(x)`.

`cloglog(x)`

> Domain: 0 to 1
> Range: $-8e+307$ to $8e+307$
> Description: returns the complementary log-log of x.
> $\mathtt{cloglog}(x) = \ln\{-\ln(1 - x)\}$

`comb(n,k)`

> Domain n: integers 1 to $1e+305$
> Domain k: integers 0 to n
> Range: 0 to $8e+307$ and *missing*
> Description: returns the combinatorial function $n!/\{k!(n - k)!\}$.

`cos(x)`

> Domain: $-1e+18$ to $1e+18$
> Range: -1 to 1
> Description: returns the cosine of x, where x is in radians.

`digamma(x)`

> Domain: $-1e+15$ to $8e+307$
> Range: $-8e+307$ to $8e+307$ and *missing*
> Description: returns the `digamma()` function, $d\ln\Gamma(x)/dx$. This is the derivative of `lngamma(x)`.
>
> The `digamma(x)` function is sometimes called the psi function, $\psi(x)$.

`exp(x)`

> Domain: $-8e+307$ to 709
> Range: 0 to $8e+307$
> Description: returns the exponential function e^x. This function is the inverse of `ln(x)`.

`floor(x)`

> Domain: $-8e+307$ to $8e+307$
> Range: integers in $-8e+307$ to $8e+307$
> Description: returns the unique integer n such that $n \le x < n + 1$.
> returns x (not ".") if x is missing, meaning that `floor(.a)` $=$ `.a`.
>
> Also see `ceil(x)`, `int(x)`, and `round(x)`.

int(x)

    Domain: –8e+307 to 8e+307

    Range: integers –8e+307 to 8e+307

    Description: returns the integer obtained by truncating x toward 0; thus,

> int(5.2) = 5
> int(-5.8) = −5

returns x (not ".") if x is missing, meaning that int(.a) = .a.

One way to obtain the closest integer to x is int(x+sign(x)/2), which simplifies to int(x+0.5) for $x \geq 0$. However, use of the round() function is preferred. Also see round(x), ceil(x), and floor(x).

invcloglog(x)

    Domain: –8e+307 to 8e+307

    Range: 0 to 1 and *missing*

    Description: returns the inverse of the complementary log-log function of x.

$$\text{invcloglog}(x) = 1 - \exp\{-\exp(x)\}$$

invlogit(x)

    Domain: –8e+307 to 8e+307

    Range: 0 to 1 and *missing*

    Description: returns the inverse of the logit function of x.

$$\text{invlogit}(x) = \exp(x)/\{1 + \exp(x)\}$$

ln(x)

    Domain: 1e–323 to 8e+307

    Range: −744 to 709

    Description: returns the natural logarithm $\ln(x)$. This function is the inverse of exp(x).

The logarithm of x in base b can be calculated via $\log_b(x) = \log_a(x)/\log_a(b)$. Hence,

$$\log_5(x) = \ln(x)/\ln(5) = \log(x)/\log(5) = \log10(x)/\log10(5)$$
$$\log_2(x) = \ln(x)/\ln(2) = \log(x)/\log(2) = \log10(x)/\log10(2)$$

You can calculate $\log_b(x)$ by using the formula that best suits your needs.

lnfactorial(n)

    Domain: integers 0 to 1e+305

    Range: 0 to 8e+307

    Description: returns the natural log of factorial = $\ln(n!)$.

To calculate $n!$, use round(exp(lnfactorial(n)),1) to ensure that the result is an integer. Logs of factorials are generally more useful than the factorials themselves because of overflow problems.

lngamma(x)

    Domain: –2,147,483,648 to 1e+305 (excluding negative integers)

    Range: –8e+307 to 8e+307

    Description: returns $\ln\{\Gamma(x)\}$. Here the gamma function $\Gamma(x)$ is defined by $\Gamma(x) = \int_0^\infty t^{x-1}e^{-t}dt$. For integer values of $x > 0$, this is $\ln((x-1)!)$.

lngamma(x) for $x < 0$ returns a number such that exp(lngamma(x)) is equal to the absolute value of the gamma function $\Gamma(x)$. That is, lngamma(x) always returns a real (not complex) result.

log(x)

> Domain: 1e–323 to 8e+307
> Range: −744 to 709
> Description: returns the natural logarithm $\ln(x)$, which is a synonym for ln(x). Also see ln(x)
>   for more information.

log10(x)

> Domain: 1e–323 to 8e+307
> Range: −323 to 308
> Description: returns the base-10 logarithm of x.

logit(x)

> Domain: 0 to 1
> Range: –8e+307 to 8e+307 and *missing*
> Description: returns the log of the odds ratio of x.
> $$\text{logit}(x) = \ln\left\{x/(1-x)\right\}.$$

max($x_1, x_2, \ldots, x_n$)

> Domain x_1: –8e+307 to 8e+307 and *missing*
> Domain x_2: –8e+307 to 8e+307 and *missing*
> . . .
> Domain x_n: –8e+307 to 8e+307 and *missing*
> Range: –8e+307 to 8e+307 and *missing*
> Description: returns the maximum value of $x_1, x_2, \ldots, x_n$. Unless all arguments are *missing*,
>   missing values are ignored.
> $$\text{max}(2,10,.,7) = 10$$
> $$\text{max}(.,.,.) = .$$

min($x_1, x_2, \ldots, x_n$)

> Domain x_1: –8e+307 to 8e+307 and *missing*
> Domain x_2: –8e+307 to 8e+307 and *missing*
> . . .
> Domain x_n: –8e+307 to 8e+307 and *missing*
> Range: –8e+307 to 8e+307 and *missing*
> Description: returns the minimum value of $x_1, x_2, \ldots, x_n$. Unless all arguments are *missing*,
>   missing values are ignored.
> $$\text{min}(2,10,.,7) = 2$$
> $$\text{min}(.,.,.) = .$$

mod(x, y)

> Domain x: –8e+307 to 8e+307
> Domain y: 0 to 8e+307
> Range: 0 to 8e+307
> Description: returns the modulus of x with respect to y.
> $$\text{mod}(x,y) = x - y\ \text{int}(x/y)$$
> $$\text{mod}(x,0) = .$$

reldif(x, y)

> Domain x: –8e+307 to 8e+307 and *missing*
> Domain y: –8e+307 to 8e+307 and *missing*
> Range: –8e+307 to 8e+307 and *missing*
> Description: returns the "relative" difference $|x - y|/(|y| + 1)$.
>   returns 0 if both arguments are the same type of extended missing value.
>   returns *missing* if only one argument is missing, or the two arguments are
>   two different types of *missing*.

round(x,y) or round(x)
   Domain x: −8e+307 to 8e+307
   Domain y: −8e+307 to 8e+307
   Range: −8e+307 to 8e+307
   Description: returns x rounded in units of y or x rounded to the nearest integer if the argument
                   y is omitted.
                   returns x (not ".") if x is missing, meaning that round(.a) = .a and
                       round(.a,y) = .a if y is not missing; if y is missing, "." is returned.

                   For $y = 1$, or with y omitted, this amounts to the closest integer to x; round(5.2,1)
                   is 5, as is round(4.8,1); round(-5.2,1) is −5, as is round(-4.8,1). The
                   rounding definition is generalized for $y \neq 1$. With $y = .01$, for instance, x is rounded
                   to two decimal places; round(sqrt(2),.01) is 1.41. y may also be larger than 1;
                   round(28,5) is 30, which is 28 rounded to the closest multiple of 5. For $y = 0$,
                   the function is defined as returning x unmodified. Also see int(x), ceil(x), and
                   floor(x).

sign(x)
   Domain: −8e+307 to 8e+307 and *missing*
   Range: −1, 0, 1 and *missing*
   Description: returns the sign of x: −1 if $x < 0$, 0 if $x = 0$, 1 if $x > 0$, and *missing*
                  if x is missing.

sin(x)
   Domain: −1e+18 to 1e+18
   Range: −1 to 1
   Description: returns the sine of x, where x is in radians.

sqrt(x)
   Domain: 0 to 8e+307
   Range: 0 to 1e+154
   Description: returns the square root of x.

sum(x)
   Domain: −8e+307 to 8e+307 and *missing*
   Range: −8e+307 to 8e+307 (and excluding *missing*)
   Description: returns the running sum of x treating missing values as zero.

                  For example, following the command generate y=sum(x), the jth observation
                  on y contains the sum of the first through jth observations on x. See [D] **egen** for
                  an alternative sum function, total(), that produces a constant equal to the overall
                  sum.

tan(x)
   Domain: −1e+18 to 1e+18
   Range: −1e+17 to 1e+17 and *missing*
   Description: returns the tangent of x, where x is in radians.

tanh(x)
   Domain: −8e+307 to 8e+307
   Range: −1 to 1 and *missing*
   Description: returns the hyperbolic tangent of x.
                  $$\texttt{tanh}(x) = \{\exp(x) - \exp(-x)\}/\{\exp(x) + \exp(-x)\}$$

trigamma(x)
> Domain: $-1\text{e}+15$ to $8\text{e}+307$
> Range: 0 to $8\text{e}+307$ and *missing*
> Description: returns the second derivative of lngamma$(x) = d^2 \ln\Gamma(x)/dx^2$. The trigamma()
> function is the derivative of digammma(x).

trunc(x) is a synonym for int(x).

❏ Technical Note

The trigonometric functions are defined in terms of *radians*. There are 2π radians in a circle. If you prefer to think in terms of *degrees*, since there are also 360 degrees in a circle, you may convert degrees into radians by using the formula $r = d\pi/180$, where d represents degrees and r represents radians. Stata includes the built-in constant _pi, equal to π to machine precision. Thus, to calculate the sine of theta, where theta is measured in degrees, you could type

> sin(theta*_pi/180)

atan() similarly returns radians, not degrees. The arccotangent can be obtained as

> acot(x) =_pi/2 - atan(x)

❏

Probability distributions and density functions

betaden(a,b,x)
> Domain a: $1\text{e}{-}323$ to $8\text{e}+307$
> Domain b: $1\text{e}{-}323$ to $8\text{e}+307$
> Domain x: $1\text{e}{-}323$ to $8\text{e}+307$
>   Interesting domain is $0 \le x \le 1$
> Range: 0 to $8\text{e}+307$
> Description: returns the probability density of the beta distribution,

$$\text{betaden}(a,b,x) = \frac{x^{a-1}(1-x)^{b-1}}{\int_0^\infty t^{a-1}(1-t)^{b-1}dt} = \frac{\Gamma(a+b)}{\Gamma(a)\Gamma(b)}x^{a-1}(1-x)^{b-1}$$

> where a and b are the shape parameters.
> returns 0 if $x < 0$ or $x > 1$.

binomial(n,k,θ)
> Domain n: 0 to $1\text{e}+17$
> Domain k: $-8\text{e}+307$ to $8\text{e}+307$
>   Interesting domain is $0 \le k \le n$
> Domain θ: 0 to 1
> Range: 0 to 1
> Description: returns the probability of observing floor(k) or fewer successes in floor(n) trials
> when the probability of a success on one trial is θ.
> returns 0 if $k < 0$.
> returns 1 if $k > n$.

`binomialtail(`n`,`k`,`θ`)`

  Domain n: 0 to 1e+17
  Domain k: −8e+307 to 8e+307
              Interesting domain is $0 \leq k \leq n$
  Domain θ: 0 to 1
  Range: 0 to 1
  Description: returns the probability of observing `floor(`k`)` or more successes in `floor(`n`)` trials
              when the probability of a success on one trial is θ.
              returns 1 if $k < 0$.
              returns 0 if $k > n$.

`binormal(`h`,`k`,`ρ`)`

  Domain h: −8e+307 to 8e+307
  Domain k: −8e+307 to 8e+307
  Domain ρ: −1 to 1
  Range: 0 to 1
  Description: returns the joint cumulative distribution $\Phi(h, k, \rho)$ of bivariate normal
              with correlation ρ; cumulative over $(-\infty, h] \times (-\infty, k]$:

$$\Phi(h, k, \rho) = \frac{1}{2\pi\sqrt{1-\rho^2}} \int_{-\infty}^{h} \int_{-\infty}^{k} \exp\left\{ -\frac{1}{2(1-\rho^2)} \left(x_1^2 - 2\rho x_1 x_2 + x_2^2\right) \right\} dx_1 \, dx_2$$

`chi2(`n`,`x`)`

  Domain n: 2e−10 to 2e+17 (may be nonintegral)
  Domain x: −8e+307 to 8e+307
              Interesting domain is $x \geq 0$
  Range: 0 to 1
  Description: returns the cumulative χ^2 distribution with n degrees of freedom.
              `chi2(`n`,`x`)` `= gammap(`$n/2, x/2$`)`
              returns 0 if $x < 0$.

`chi2tail(`n`,`x`)`

  Domain n: 2e−10 to 2e+17 (may be nonintegral)
  Domain x: −8e+307 to 8e+307
              Interesting domain is $x \geq 0$
  Range: 0 to 1
  Description: returns the reverse cumulative (upper-tail, survival) χ^2 distribution with n degrees
              of freedom. `chi2tail(`n`,`x`)` $= 1 -$ `chi2(`n`,`x`)`
              returns 1 if $x < 0$.

`dgammapda(`a`,`x`)`

  Domain a: 1e−7 to 1e+17
  Domain x: −8e+307 to 8e+307
              Interesting domain is $x \geq 0$
  Range: −16 to 0
  Description: returns $\frac{\partial P(a,x)}{\partial a}$, where $P(a, x) =$ `gammap(`a, x`)`.
              returns 0 if $x < 0$.

dgammapdada(a,x)

Domain a: 1e−7 to 1e+17

Domain x: −8e+307 to 8e+307

Interesting domain is $x \geq 0$

Range: −0.02 to 4.77e+5

Description: returns $\frac{\partial^2 P(a,x)}{\partial a^2}$, where $P(a, x) = $ gammap(a, x).
returns 0 if $x < 0$.

dgammapdadx(a,x)

Domain a: 1e−7 to 1e+17

Domain x: −8e+307 to 8e+307

Interesting domain is $x \geq 0$

Range: −0.04 to 8e+307

Description: returns $\frac{\partial^2 P(a,x)}{\partial a \partial x}$, where $P(a, x) = $ gammap(a, x).
returns 0 if $x < 0$.

dgammapdx(a,x)

Domain a: 1e−10 to 1e+17

Domain x: −8e+307 to 8e+307

Interesting domain is $x \geq 0$

Range: 0 to 8e+307

Description: returns $\frac{\partial P(a,x)}{\partial x}$, where $P(a, x) = $ gammap(a, x).
returns 0 if $x < 0$.

dgammapdxdx(a,x)

Domain a: 1e−10 to 1e+17

Domain x: −8e+307 to 8e+307

Interesting domain is $x \geq 0$

Range: 0 to 1e+40

Description: returns $\frac{\partial^2 P(a,x)}{\partial x^2}$, where $P(a, x) = $ gammap(a, x).
returns 0 if $x < 0$.

F(n_1,n_2,f)

Domain n_1: 2e−10 to 2e+17 (may be nonintegral)

Domain n_2: 2e−10 to 2e+17 (may be nonintegral)

Domain f: −8e+307 to 8e+307

Interesting domain is $f \geq 0$

Range: 0 to 1

Description: returns the cumulative F distribution with n_1 numerator and n_2 denominator
degrees of freedom: F$(n_1, n_2, f) = \int_0^f$ Fden$(n_1, n_2, t)\, dt$
returns 0 if $f < 0$.

`Fden(`n_1`,`n_2`,`f`)`

    Domain n_1:   1e–323 to 8e+307 (may be nonintegral)
    Domain n_2:   1e–323 to 8e+307 (may be nonintegral)
    Domain f: –8e+307 to 8e+307
                Interesting domain is $f \geq 0$
    Range: 0 to 8e+307
    Description: returns the probability density function of the F distribution with n_1 numerator
                     and n_2 denominator degrees of freedom:

$$\text{Fden}(n_1, n_2, f) = \frac{\Gamma(\frac{n_1+n_2}{2})}{\Gamma(\frac{n_1}{2})\Gamma(\frac{n_2}{2})} \left(\frac{n_1}{n_2}\right)^{\frac{n_1}{2}} \cdot f^{\frac{n_1}{2}-1} \left(1 + \frac{n_1}{n_2}f\right)^{-\frac{1}{2}(n_1+n_2)}$$

                 returns 0 if $f < 0$.

`Ftail(`n_1`,`n_2`,`f`)`

    Domain n_1:   2e−10 to 2e+17 (may be nonintegral)
    Domain n_2:   2e−10 to 2e+17 (may be nonintegral)
    Domain f: –8e+307 to 8e+307
                Interesting domain is $f \geq 0$
    Range: 0 to 1
    Description: returns the reverse cumulative (upper-tail, survival) F distribution with n_1 numerator
                     and n_2 denominator degrees of freedom. $\text{Ftail}(n_1, n_2, f) = 1 - \text{F}(n_1, n_2, f)$
                 returns 1 if $f < 0$.

`gammaden(`a`,`b`,`g`,`x`)`

    Domain a: 1e–323 to 8e+307
    Domain b: 1e–323 to 8e+307
    Domain g: –8e+307 to 8e+307
    Domain x: –8e+307 to 8e+307
               Interesting domain is $x \geq g$
    Range: 0 to 8e+307
    Description: returns the probability density function of the gamma distribution defined by

$$\frac{1}{\Gamma(a)b^a}(x - g)^{a-1}e^{-(x-g)/b}$$

                 where a is the shape parameter, b is the scale parameter, and g is the
                 location parameter.
                 returns 0 if $x < g$.

(Continued on next page)

gammap(a,x)

    Domain a: 1e−10 to 1e+17

    Domain x: −8e+307 to 8e+307

                 Interesting domain is $x \geq 0$

    Range: 0 to 1

    Description: returns the cumulative gamma distribution with shape parameter a defined by

$$\frac{1}{\Gamma(a)} \int_0^x e^{-t} t^{a-1} \, dt$$

returns 0 if $x < 0$.

The cumulative Poisson (the probability of observing k or fewer events if the expected is x) can be evaluated as 1-gammap(k+1,x). The reverse cumulative (the probability of observing k or more events) can be evaluated as gammap(k,x). See Press et al. (1992, 216–221) for a more complete description and for suggested uses for this function.

gammap() is also known as the incomplete gamma function (ratio).

Probabilities for the three-parameter gamma distribution (see gammaden()) can be calculated by shifting and scaling x; i.e., gammap(a,($x - g$)/b).

gammaptail(a,x)

    Domain a: 1e−10 to 1e+17

    Domain x: −8e+307 to 8e+307

                 Interesting domain is $x \geq 0$

    Range: 0 to 1

    Description: returns the reverse cumulative (upper-tail, survival) gamma distribution with shape parameter a defined by

$$\text{gammaptail}(a, x) = 1 - \text{gammap}(a, x) = \int_x^{\infty} \text{gammaden}(a, t) \, dt$$

returns 1 if $x < 0$.

gammaptail() is also known as the complement to the incomplete gamma function (ratio).

ibeta(a,b,x)

  Domain a: 1e−10 to 1e+17
  Domain b: 1e−10 to 1e+17
  Domain x: −8e+307 to 8e+307
                 Interesting domain is $0 \le x \le 1$
  Range: 0 to 1
  Description: returns the cumulative beta distribution with shape parameters a and b defined by

$$I_x(a,b) = \frac{\Gamma(a+b)}{\Gamma(a)\Gamma(b)} \int_0^x t^{a-1}(1-t)^{b-1}\,dt$$

  returns 0 if $x < 0$.
  returns 1 if $x > 1$.

  Although Stata has a cumulative binomial function (see binomial()), the probability that an event occurs k or fewer times in n trials, when the probability of one event is p, can be evaluated as cond(k==n,1,1-ibeta(k+1,n-k,p)). The reverse cumulative binomial (the probability that an event occurs k or more times) can be evaluated as cond(k==0,1,ibeta(k,n-k+1,p)). See Press et al. (1992, 226–229) for a more complete description and for suggested uses for this function.

  ibeta() is also known as the incomplete beta function (ratio).

ibetatail(a,b,x)

  Domain a: 1e−10 to 1e+17
  Domain b: 1e−10 to 1e+17
  Domain x: −8e+307 to 8e+307
                 Interesting domain is $0 \le x \le 1$
  Range: 0 to 1
  Description: returns the reverse cumulative (upper-tail, survival) beta distribution with shape parameters a and b defined by

$$\text{ibetatail}(a,b,x) = 1 - \text{ibeta}(a,b,x) = \int_x^1 \text{betaden}(a,b,t)\,dt$$

  returns 1 if $x < 0$.
  returns 0 if $x > 1$.

  ibetatail() is also known as the complement to the incomplete beta function (ratio).

invbinomial(n,k,p)

  Domain n: 1 to 1e+17
  Domain k: 0 to n−1
  Domain p: 0 to 1 (exclusive)
  Range: 0 to 1
  Description: returns the inverse of the cumulative binomial; i.e., it returns θ (θ = probability of success on one trial) such that the probability of observing floor(k) or fewer successes in floor(n) trials is p.

invbinomialtail(n,k,p)
  Domain n: 1 to 1e+17
  Domain k: 1 to n
  Domain p: 0 to 1 (exclusive)
  Range: 0 to 1
  Description: returns the inverse of right cumulative binomial; i.e., it returns θ (θ = probability of success on one trial) such that the probability of observing floor(k) or more successes in floor(n) trials is p.

invchi2(n,p)
  Domain n: 2e−10 to 2e+17 (may be nonintegral)
  Domain p: 0 to 1
  Range: 0 to 8e+307
  Description: returns the inverse of chi2(): if chi2(n,x) = p, then invchi2(n,p) = x.

invchi2tail(n,p)
  Domain n: 2e−10 to 2e+17 (may be nonintegral)
  Domain p: 0 to 1
  Range: 0 to 8e+307
  Description: returns the inverse of chi2tail(): if chi2tail(n,x) = p, then invchi2tail(n,p) = x.

invF(n_1,n_2,p)
  Domain n_1: 2e−10 to 2e+17 (may be nonintegral)
  Domain n_2: 2e−10 to 2e+17 (may be nonintegral)
  Domain p: 0 to 1
  Range: 0 to 8e+307
  Description: returns the inverse cumulative F distribution: if F(n_1,n_2,f) = p, then invF(n_1,n_2,p) = f.

invFtail(n_1,n_2,p)
  Domain n_1: 2e−10 to 2e+17 (may be nonintegral)
  Domain n_2: 2e−10 to 2e+17 (may be nonintegral)
  Domain p: 0 to 1
  Range: 0 to 8e+307
  Description: returns the inverse reverse cumulative (upper-tail, survival) F distribution: if Ftail(n_1,n_2,f) = p, then invFtail(n_1,n_2,p) = f.

invgammap(a,p)
  Domain a: 1e−10 to 1e+17
  Domain p: 0 to 1
  Range: 0 to 8e+307
  Description: returns the inverse cumulative gamma distribution: if gammap(a,x) = p, then invgammap(a,p) = x.

invgammaptail(a,p)
  Domain a: 1e−10 to 1e+17
  Domain p: 0 to 1
  Range: 0 to 8e+307
  Description: returns the inverse reverse cumulative (upper-tail, survival) gamma distribution: if gammaptail(a,x) = p, then invgammaptail(a,p) = x.

invibeta(a,b,p)
  Domain a: 1e−10 to 1e+17
  Domain b: 1e−10 to 1e+17
  Domain p: 0 to 1
  Range: 0 to 1
  Description: returns the inverse cumulative beta distribution: if ibeta(a,b,x) = p,
                then invibeta(a,b,p) = x.

invibetatail(a,b,p)
  Domain a: 1e−10 to 1e+17
  Domain b: 1e−10 to 1e+17
  Domain p: 0 to 1
  Range: 0 to 1
  Description: returns the inverse reverse cumulative (upper-tail, survival) beta distribution:
                if ibetatail(a,b,x) = p, then invibetatail(a,b,p) = x.

invnchi2(n,λ,p)
  Domain n: integers 1 to 200
  Domain λ: 0 to 1,000
  Domain p: 0 to 1
  Range: 0 to 8e+307
  Description: returns the inverse cumulative noncentral χ^2 distribution:
                if nchi2(n,λ,x) = p, then invnchi2(n,λ,p) = x; n must be an integer.

invnFtail(n_1,n_2,λ,p)
  Domain n_1: 1e–323 to 8e+307 (may be nonintegral)
  Domain n_2: 1e–323 to 8e+307 (may be nonintegral)
  Domain λ: 0 to 1,000
  Domain p: 0 to 1
  Range: 0 to 8e+307
  Description: returns the inverse reverse cumulative (upper-tail, survival) noncentral F distribution:
                if nFtail(n_1,n_2,λ,x) = p, then invnFtail(n_1, n_2, λ, p) = x.

invnibeta(a,b,λ,p)
  Domain a: 1e–323 to 8e+307
  Domain b: 1e–323 to 8e+307
  Domain λ: 0 to 1,000
  Domain p: 0 to 1
  Range: 0 to 1
  Description: returns the inverse cumulative noncentral beta distribution:
                if nibeta(a,b,λ,x) = p, then invibeta(a,b,λ,p) = x.

invnormal(p)
  Domain: 1e–323 to $1 - 2^{-53}$
  Range: −39 to 8.2095362
  Description: returns the inverse cumulative standard normal distribution:
                if normal(z) = p, then invnormal(p) = z.

invttail(n,p)
  Domain n: 2e−10 to 2e+17 (may be nonintegral)
  Domain p: 0 to 1
  Range: 0 to 1e+10
  Description: returns the inverse reverse cumulative (upper-tail, survival) Student's t distribution:
                if ttail(n,t) = p, then invttail(n,p) = t.

`lnnormal(z)`
>   Domain: −1e+99 to 8e+307
>   Range: −5e+197 to 0
>   Description: returns the natural logarithm of the cumulative standard normal distribution:

$$\texttt{lnnormal}(z) = \ln\left(\int_{-\infty}^{z} \frac{1}{\sqrt{2\pi}} e^{-x^2/2} dx\right)$$

`lnnormalden(z)`
>   Domain: −1e+154 to 1e+154
>   Range: −5e+307 to −.91893853 = lnnormalden(0)
>   Description: returns the natural logarithm of the standard normal density, $N(0, 1)$.

`lnnormalden(z,σ)`
>   Domain z: −1e+154 to 1e+154
>   Domain $σ$:   1e−323 to 8e+307
>   Range: −5e+307 to 742.82799
>   Description: returns the natural logarithm of the rescaled standard normal density, $N(0, \sigma^2)$.
>
> $$\texttt{lnnormalden}(z,1) = \texttt{lnnormalden}(z)$$
> $$\texttt{lnnormalden}(z,\sigma) = \texttt{lnnormalden}(z) / \ln(\sigma)$$

`lnnormalden(x,μ,σ)`
>   Domain x: −8e+307 to 8e+307
>   Domain $μ$:   −8e+307 to 8e+307
>   Domain $σ$:   1e−323 to 8e+307
>   Range: 1e−323 to 8e+307
>   Description: returns the natural logarithm of the normal density with mean μ and standard deviation σ, $N(\mu, \sigma^2)$: $\texttt{lnnormalden}(x,0,1) = \texttt{lnnormalden}(x)$ and $\texttt{lnnormalden}(x,\mu,\sigma) = \texttt{lnnormalden}((x - \mu)/\sigma) - \ln(\sigma)$. In general,

$$\texttt{lnnormalden}(z, \mu, \sigma) = \ln\left\{ \frac{1}{\sigma\sqrt{2\pi}} e^{-\frac{1}{2}\left(\frac{(z-\mu)}{\sigma}\right)^2} \right\}$$

`nbetaden(a,b,λ,x)`
>   Domain a: 1e−323 to 8e+307
>   Domain b: 1e−323 to 8e+307
>   Domain $λ$:   0 to 1,000
>   Domain x: −8e+307 to 8e+307
>   Interesting domain is $0 \le x \le 1$
>   Range: 0 to 8e+307
>   Description: returns the probability density function of the noncentral beta distribution,

$$\sum_{j=0}^{\infty} \frac{e^{-\lambda/2}(\lambda/2)^j}{\Gamma(j+1)} \left\{ \frac{\Gamma(a+b+j)}{\Gamma(a+j)\Gamma(b)} x^{a+j-1}(1-x)^{b-1} \right\}$$

>   where a and b are shape parameters, λ is the noncentrality parameter, and x is the value of a beta random variable.
>   returns 0 if $x < 0$ or $x > 1$.
>
> $\texttt{nbetaden}(a,b,0,x) = \texttt{betaden}(a,b,x)$, but `betaden()` is the preferred function to use for the central beta distribution. `nbetaden()` is computed using an algorithm described in Johnson, Kotz, and Balakrishnan (1995).

nchi2(n,λ,x)
  Domain n: integers 1 to 200
  Domain λ: 0 to 1,000
  Domain x: −8e+307 to 8e+307
          Interesting domain is $x \geq 0$
  Range: 0 to 1
  Description: returns the cumulative noncentral χ^2 distribution,

$$\int_0^x \frac{e^{-t/2} e^{-\lambda/2}}{2^{n/2}} \sum_{j=0}^{\infty} \frac{t^{n/2+j-1} \lambda^j}{\Gamma(n/2+j)\, 2^{2j}\, j!}\, dt$$

where n denotes the degrees of freedom, λ is the noncentrality parameter, and x is the value of χ^2.
returns 0 if $x < 0$.

nchi2($n,0,x$) = chi2(n,x), but chi2() is the preferred function to use for the central χ^2 distribution. nchi2() is computed using the algorithm of Haynam, Govindarajulu, and Leone (1970).

nFden(n_1,n_2,λ,f)
  Domain n_1: 1e–323 to 8e+307 (may be nonintegral)
  Domain n_2: 1e–323 to 8e+307 (may be nonintegral)
  Domain λ: 0 to 1,000
  Domain f: −8e+307 to 8e+307
          Interesting domain is $f \geq 0$
  Range: 0 to 8e+307
  Description: returns the probability density function of the noncentral F distribution with n_1 numerator and n_2 denominator degrees of freedom and noncentrality parameter λ.
returns 0 if $f < 0$.

nFden($n_1,n_2,0,F$) = Fden(n_1,n_2,F), but Fden() is the preferred function to use for the central F distribution.

Also, if F follows the noncentral F distribution with n_1 and n_2 degrees of freedom and noncentrality parameter λ, then

$$\frac{n_1 F}{n_2 + n_1 F}$$

follows a noncentral beta distribution with shape parameters $a = \nu_1/2$, $b = \nu_2/2$, and noncentrality parameter λ, as given in nbetaden(). nFden() is computed based on this relationship.

(*Continued on next page*)

nFtail(n_1,n_2,λ,f)

   Domain n_1:   1e–323 to 8e+307 (may be nonintegral)
   Domain n_2:   1e–323 to 8e+307 (may be nonintegral)
   Domain λ: 0 to 1,000
   Domain f: –8e+307 to 8e+307
              Interesting domain is $f \geq 0$
   Range: 0 to 1
   Description: returns the reverse cumulative (upper-tail, survival) noncentral F distribution with n_1 numerator and n_2 denominator degrees of freedom and noncentrality parameter λ.
              returns 1 if $f < 0$.

   nFtail() is computed using nibeta() based on the relationship between the noncentral beta and F distributions. See Johnson, Kotz, and Balakrishnan (1995) for more details.

nibeta(a,b,λ,x)

   Domain a: 1e–323 to 8e+307
   Domain b: 1e–323 to 8e+307
   Domain λ: 0 to 1,000
   Domain x: –8e+307 to 8e+307
              Interesting domain is $0 \leq x \leq 1$
   Range: 0 to 1
   Description: returns the cumulative noncentral beta distribution

$$I_x(a, b, \lambda) = \sum_{j=0}^{\infty} \frac{e^{-\lambda/2}(\lambda/2)^j}{\Gamma(j+1)} I_x(a+j, b)$$

              where a and b are shape parameters, λ is the noncentrality parameter, x is the value of a beta random variable, and $I_x(a, b)$ is the cumulative beta distribution, ibeta().
              returns 0 if $x < 0$.
              returns 1 if $x > 1$.

   nibeta(a,b,0,x) = ibeta(a,b,x), but ibeta() is the preferred function to use for the central beta distribution. nibeta() is computed using an algorithm described in Johnson, Kotz, and Balakrishnan (1995).

normal(z)

   Domain: –8e+307 to 8e+307
   Range: 0 to 1
   Description: returns the cumulative standard normal distribution.
              $$\texttt{normal}(z) = \int_{-\infty}^{z} \frac{1}{\sqrt{2\pi}} e^{-x^2/2} dx$$

normalden(z)

   Domain: –8e+307 to 8e+307
   Range: 0 to .39894 . . .
   Description: returns the standard normal density, $N(0,1)$.

normalden(z,σ)

    Domain z: –8e+307 to 8e+307

    Domain σ: 1e–308 to 8e+307

    Range: 0 to 8e+307

    Description: returns the rescaled standard normal density, $N(0, \sigma^2)$.

                normalden(z,1) = normalden(z)

                normalden(z,σ) = normalden(z)$/\sigma$

normalden(x,μ,σ)

    Domain x: –8e+307 to 8e+307

    Domain μ: –8e+307 to 8e+307

    Domain σ: 1e–308 to 8e+307

    Range: 0 to 8e+307

    Description: returns the normal density with mean μ and standard deviation σ, $N(\mu, \sigma^2)$:

                normalden(x,0,1) = normalden(x) and

                normalden(x,μ,σ) = normalden(($x - \mu$)$/\sigma$)$/\sigma$

                In general,

$$\text{normalden}(z, \mu, \sigma) = \frac{1}{\sigma\sqrt{2\pi}} e^{-\frac{1}{2}\left(\frac{(z-\mu)}{\sigma}\right)^2}$$

npnchi2(n,x,p)

    Domain n: integers 1 to 200

    Domain x: 0 to 8e+307

    Domain p: 1e–138 to $1 - 2^{-52}$

    Range: 0 to 1,000

    Description: returns the noncentrality parameter λ for noncentral χ^2:

                if nchi2(n,λ,x) = p, then npnchi2(n,x,p) = λ.

tden(n,t)

    Domain n: 1e–323 to 8e+307

    Domain t: –8e+307 to 8e+307

    Range: 0 to .39894 . . .

    Description: returns the probability density function of Student's t distribution:

$$\text{tden}(n, t) = \frac{\Gamma((n+1)/2)}{\sqrt{\pi n}\,\Gamma(n/2)} \cdot \left(1 + t^2/n\right)^{-(n+1)/2}$$

ttail(n,t)

    Domain n: 2e–10 to 2e+17

    Domain t: –8e+307 to 8e+307

    Range: 0 to 1

    Description: returns the reverse cumulative (upper-tail, survival) Student's t distribution; it returns the probability $T > t$:

$$\text{ttail}(n, t) = \int_t^\infty \frac{\Gamma((n+1)/2)}{\sqrt{\pi n}\,\Gamma(n/2)} \cdot \left(1 + x^2/n\right)^{-(n+1)/2} \, dx$$

Random-number functions

uniform()
>Range: 0 to nearly 1 (0 to $1 - 2^{-32}$)
>Description: returns uniform pseudorandom numbers.

>>uniform() returns uniformly distributed pseudorandom numbers on the interval $[0, 1)$. uniform() takes no arguments, but the parentheses must be typed. uniform() can be seeded with the set seed command; see the technical note at the end of this subsection.

>>To generate pseudorandom numbers over the interval $[a, b)$, use $a+(b-a)*$uniform().

>>To generate pseudorandom integers over $[a, b]$, use $a+$int$(((b-a+1)*$uniform$()))$.

>>To generate normally distributed random numbers with mean 0 and standard deviation 1, use invnormal(uniform()).

>>To generate normally distributed random numbers with mean μ and standard deviation σ, use $\mu+\sigma*$invnormal(uniform()).

invnormal(uniform())
>Range: 0 to 1
>Description: returns normally distributed random numbers with mean 0 and standard deviation 1. (uniform()) within invnormal() takes no arguments, but the parentheses must be typed.

❏ Technical Note

The uniform pseudorandom number function uniform() is based on George Marsaglia's (1994) 32-bit pseudorandom number generator KISS (keep it simple stupid). The KISS generator is composed of two 32-bit pseudorandom number generators and two 16-bit generators (combined to make one 32-bit generator). The four generators are defined by the recursions

$$x_n = 69069\, x_{n-1} + 1234567 \quad \text{mod } 2^{32} \tag{1}$$

$$y_n = y_{n-1}(I + L^{13})(I + R^{17})(I + L^5) \tag{2}$$

$$z_n = 65184\left(z_{n-1} \text{ mod } 2^{16}\right) + \text{int}\left(z_{n-1}/2^{16}\right) \tag{3}$$

$$w_n = 63663\left(w_{n-1} \text{ mod } 2^{16}\right) + \text{int}\left(w_{n-1}/2^{16}\right) \tag{4}$$

In recursion (2), the 32-bit word y_n is viewed as a 1×32 binary vector; L is the 32×32 matrix that produces a left shift of one (L has 1s on the first left subdiagonal, 0s elsewhere); and R is L transpose, affecting a right shift by one. In recursions (3) and (4), int(x) is the integer part of x.

The KISS generator produces the 32-bit random number

$$R_n = x_n + y_n + z_n + 2^{16}w_n \quad \text{mod } 2^{32}$$

uniform() takes the output from the KISS generator and divides it by 2^{32} to produce a real number on the interval $[0, 1)$.

The recursions (1)–(4) have, respectively, the periods

$$2^{32} \tag{1}$$
$$2^{32} - 1 \tag{2}$$
$$(65184 \cdot 2^{16} - 2)/2 \approx 2^{31} \tag{3}$$
$$(63663 \cdot 2^{16} - 2)/2 \approx 2^{31} \tag{4}$$

Thus the overall period for the KISS generator is

$$2^{32} \cdot (2^{32} - 1) \cdot (65184 \cdot 2^{15} - 1) \cdot (63663 \cdot 2^{15} - 1) \approx 2^{126}$$

When Stata first comes up, it initializes the four recursions in KISS by using the seeds

$$x_0 = 123456789 \tag{1}$$
$$y_0 = 521288629 \tag{2}$$
$$z_0 = 362436069 \tag{3}$$
$$w_0 = 2262615 \tag{4}$$

Successive calls to `uniform()` then produce the sequence

$$\frac{R_1}{2^{32}}, \ \frac{R_2}{2^{32}}, \ \frac{R_3}{2^{32}}, \ \cdots$$

Hence, `uniform()` gives the same sequence of random numbers in every Stata session (measured from the start of the session) unless you reinitialize the seed. The full seed is the set of four numbers (x, y, z, w), but you can reinitialize the seed by simply issuing the command

```
. set seed #
```

where # is any integer between 0 and $2^{31} - 1$, inclusive. When this command is issued, the initial value x_0 is set equal to #, and the other three recursions are restarted at the seeds y_0, z_0, and w_0 given above. The first 100 random numbers are discarded, and successive calls to `uniform()` give the sequence

$$\frac{R'_{101}}{2^{32}}, \ \frac{R'_{102}}{2^{32}}, \ \frac{R'_{103}}{2^{32}}, \ \cdots$$

However, if the command

```
. set seed 123456789
```

is given, the first 100 random numbers are not discarded, and you get the same sequence of random numbers that `uniform()` produces by default; also see [D] **generate**.

❑

❑ Technical Note

You may "capture" the current seed (x, y, z, w) by coding

```
. local curseed = "`c(seed)'"
```

and, later in your code, reestablish that seed by coding

```
. set seed `curseed'
```

When the seed is set this way, the first 100 random numbers are not discarded.

c(seed) contains a 30-plus long character string similar to

$$\text{X075bcd151f123bb5159a55e50022865746ad}$$

The string contains an encoding of the four numbers (x, y, z, w) along with checksums and redundancy to ensure that, at set seed time, it is valid.

❏

String functions

Stata includes the following *string functions*. In the display below, s indicates a string subexpression (a string literal, a string variable, or another string expression), n indicates a numeric subexpression (a number, a numeric variable, or another numeric expression), and re indicates a regular expression based on Henry Spencer's NFA algorithms and this is nearly identical to the POSIX.2 standard.

abbrev(s,n)
Domain s: strings
Domain n: 5 to 32
Range: strings
Description: returns s, abbreviated to n characters.

If any of the characters of s are a period, ".", and $n < 8$, then the value of n defaults to a value of 8. Otherwise, if $n < 5$, then n defaults to a value of 5. If n is *missing*, abbrev() will return the entire string s. abbrev() is typically used with variable names and variable names with time-series operators (the period case). abbrev("displacement",8) is displa~t.

char(n)
Domain: integers 1 to 255
Range: ASCII characters
Description: returns the character corresponding to ASCII code n.
    returns "" if n is not in the domain.

indexnot(s_1,s_2)
Domain s_1:   strings (to be searched)
Domain s_2:   strings of individual characters (to search for)
Range: integers 0 to 244
Description: returns the position in s_1 of the first character of s_1 not found in s_2, or 0 if all characters of s_1 are found in s_2.

itrim(s)
Domain: strings
Range: strings with no multiple, consecutive internal blanks
Description: returns s with multiple, consecutive internal blanks collapsed to one blank.
    itrim("hello   there") = "hello there"

length(s)
Domain: strings
Range: integers 0 to 244
Description: returns the length of s. length("ab") = 2

`lower(`*s*`)`
>   Domain: strings
>   Range: strings with lowercased characters
>   Description: returns the lowercased variant of *s*. `lower("THIS") = "this"`

`ltrim(`*s*`)`
>   Domain: strings
>   Range: strings without leading blanks
>   Description: returns *s* without leading blanks. `ltrim(" this") = "this"`

`plural(`*n*`,`*s*`)` or `plural(`*n*`,`s_1`,`s_2`)`
>   Domain *n*:   real numbers
>   Domain *s*:   strings
>   Domain s_1:   strings
>   Domain s_2:   strings
>   Range: strings
>   Description: returns the plural of *s*, or s_1 in the 3-argument case, if $n \neq \pm 1$.
>   The plural is formed by adding "s" to *s* if you called `plural(`*n*`,`*s*`)`. If
>   you called `plural(`*n*`,`s_1`,`s_2`)` and s_2 begins with the character "+", the plural
>   is formed by adding the remainder of s_2 to s_1. If s_2 begins with the character
>   "−", the plural is formed by subtracting the remainder of s_2 from s_1. If s_2
>   begins with neither "+" nor "−", then the plural is formed by returning s_2.
>   returns *s*, or s_1 in the 3-argument case, if $n = \pm 1$.

>   `plural(1, "horse") = "horse"`
>   `plural(2, "horse") = "horses"`
>   `plural(2, "glass", "+es") = "glasses"`
>   `plural(1, "mouse", "mice") = "mouse"`
>   `plural(2, "mouse", "mice") = "mice"`
>   `plural(2, "abcdefg", "-efg") = "abcd"`

`proper(`*s*`)`
>   Domain: strings
>   Range: strings
>   Description: returns a string with the first letter capitalized, and capitalizes any other letters
>   immediately following characters that are not letters; all other
>   letters converted to lowercase.
>   `proper("mR. joHn a. sMitH") = "Mr. John A. Smith"`
>   `proper("jack o'reilly") = "Jack O'Reilly"`
>   `proper("2-cent's worth") = "2-Cent'S Worth"`

`real(`*s*`)`
>   Domain: strings
>   Range: −8e+307 to 8e+307 and *missing*
>   Description: returns *s* converted to numeric, or returns *missing*.
>   `real("5.2")+1 = 6.2`
>   `real("hello") = .`

regexm(s,re)

Domain s: strings
Domain re: regular expression
Range: strings
Description: performs a match of a regular expression and evaluates to 1 if regular expression re is satisfied by the string s, otherwise returns 0. Regular expression syntax is based on Henry Spencer's NFA algorithm and this is nearly identical to the POSIX.2 standard.

regexr(s_1,re,s_2)

Domain s_1: strings
Domain re: regular expression
Domain s_2: strings
Range: strings
Description: replaces the first substring within s_1 that matches re with s_2 and returns the resulting string. If s_1 contains no substring that matches re, the unaltered s_1 is returned.

regexs(n)

Domain: 0 to 9
Range: strings
Description: returns subexpression n from a previous **regexm**() match, where $0 \leq n < 10$. Subexpression 0 is reserved for the entire string that satisfied the regular expression.

reverse(s)

Domain: strings
Range: reversed strings
Description: returns s reversed. **reverse("hello")** = **"olleh"**

rtrim(s)

Domain: strings
Range: strings without trailing blanks
Description: returns s without trailing blanks. **rtrim("this ")** = **"this"**.

string(n)

Domain: –8e+307 to 8e+307 and *missing*
Range: strings
Description: returns n converted to a string.
```
string(4)+"F" = "4F"
string(1234567) = "1234567"
string(12345678) = "1.23e+07"
string(.) = "."
```

string(n,s)

Domain n: –8e+307 to 8e+307 and *missing*
Domain s: strings containing %*fmt* numeric display format
Range: strings
Description: returns n converted to a string.
```
string(4,"%9.2f") = "4.00"
string(123456789,"%11.0g") = "123456789"
string(123456789,"%13.0gc") = "123,456,789"
string(0,"%td") = "01jan1960"
string(225,"%tq") = "2016q2"
string(225,"not a format") = ""
```

strlen(s) is a synonym for length(s).

strmatch(s_1,s_2)
    Domain s: strings
    Range: 0 or 1
    Description: returns 1 if s_1 matches the pattern s_2; otherwise, it returns 0.
                strmatch("17.4","1??4") returns 1. In s_2, "?" means that one character goes here, and "*" means that zero or more characters go here. Also see regexm(), regexr(), and regexs().

strofreal(n) is a synonym for string(n).

strofreal(n,s) is a synonym for string(n,s).

strpos(s_1,s_2)
    Domain s_1: strings (to be searched)
    Domain s_2: strings (to search for)
    Range: integers 0 to 244
    Description: returns the position in s_1 at which s_2 is first found; otherwise, it returns 0.
                strpos("this","is") = 3
                strpos("this","it") = 0

subinstr(s_1,s_2,s_3,n)
    Domain s_1: strings (to be substituted into)
    Domain s_2: strings (to be substituted from)
    Domain s_3: strings (to be substituted with)
    Domain n: integers 0 to 244 and *missing*
    Range: strings
    Description: returns s_1, where the first n occurrences in s_1 of s_2 have been replaced with s_3. If n is *missing*, all occurrences are replaced. Also see regexm(), regexr(), and regexs().

                subinstr("this is this","is","X",1) = "thX is this"
                subinstr("this is this","is","X",2) = "thX X this"
                subinstr("this is this","is","X",.) = "thX X thX"

subinword(s_1,s_2,s_3,n)
    Domain s_1: strings (to be substituted for)
    Domain s_2: strings (to be substituted from)
    Domain s_3: strings (to be substituted with)
    Domain n: integers 0 to 244 and *missing*
    Range: strings
    Description: returns s_1, where the first n occurrences in s_1 of s_2 as a word have been replaced with s_3. A word is defined as a space-separated token. A token at the beginning or end of s_1 is considered space separated. If n is *missing*, all occurrences are replaced. Also see regexm(), regexr(), and regexs().

                subinword("this is this","is","X",1) = "this X this"
                subinword("this is this","is","X",.) = "this X this"
                subinword("this is this","th","X",.) = "this is this"

substr(s,n_1,n_2)

    Domain s: strings
    Domain n_1: integers 1 to 244 and -1 to -244
    Domain n_2: integers 1 to 244 and -1 to -244
    Range: strings
    Description: returns the substring of s, starting at column n_1, for a length of n_2.
                  If $n_1 < 0$, n_1 is interpreted as distance from the end of the string;
                  if $n_2 = .$ (*missing*), the remaining portion of the string is returned.

```
substr("abcdef",2,3) = "bcd"
substr("abcdef",-3,2) = "de"
substr("abcdef",2,.) = "bcdef"
substr("abcdef",-3,.) = "def"
substr("abcdef",2,0) = ""
substr("abcdef",15,2) = ""
```

trim(s)

    Domain: strings
    Range: strings without leading or trailing blanks
    Description: returns s without leading and trailing blanks; equivalent to
                  ltrim(rtrim(s)). trim(" this ") = "this"

upper(s)

    Domain: strings
    Range: strings with uppercased characters
    Description: returns the uppercased variant of s. upper("this") = "THIS"

word(s, n)

    Domain s: strings
    Domain n: integers $\ldots, -2, -1, 0, 1, 2, \ldots$
    Range: strings
    Description: returns the nth word in s. Positive numbers count words from the beginning of s,
                  and negative numbers count words from the end of s. (1 is the first word in s,
                  and -1 is the last word in s.) Returns *missing* ("") if n is missing.

wordcount(s)

    Domain: strings
    Range: nonnegative integers 0, 1, 2, $\ldots$
    Description: returns the number of words in s. A word is a set of characters that start
                  and terminate with spaces, start with the beginning of the string,
                  or terminate with the end of the string.

Programming functions

autocode(x,n,x_0,x_1)
  Domain x: −8e+307 to 8e+307
  Domain n: integers 1 to 8e+307
  Domain x_0:   −8e+307 to 8e+307
  Domain x_1: x_0 to 8e+307
  Range: x_0 to x_1
  Description: partitions the interval from x_0 to x_1 into n equal-length intervals and
         returns the upper bound of the interval that contains x. This function is an
         automated version of recode() (see below).
         See [U] **25 Dealing with categorical variables** for an example.

         The algorithm for autocode() is
            if $(n \geq . \mid x_0 \geq . \mid x_1 \geq . \mid n \leq 0 \mid x_0 \geq x_1)$
              then return *missing*
              if $x \geq .$, then return x
            otherwise
              for $i = 1$ to $n - 1$
                $xmap = x_0 + i * (x_1 - x_0)/n$
                if $x \leq xmap$ then return *xmap*
              end
              otherwise
                return x_1

byteorder()
  Range: 1 and 2
  Description: returns 1 if your computer stores numbers by using a hilo byte order and evaluates
         to 2 if your computer stores numbers by using a lohi byte order. Consider the
         number 1 written as a 2-byte integer. On some computers (called hilo), it is
         written as "00 01", and on other computers (called lohi), it is written as
         "01 00" (with the least significant byte written first). There are similar issues
         for 4-byte integers, 4-byte floats, and 8-byte floats. Stata automatically handles
         byte order differences for Stata created files. Users need not be concerned about
         this issue. Programmers producing customary binary files can use byteorder()
         to determine the native byte ordering; see [P] **file**.

c(*name*)
  Domain: names
  Range: real values, strings, and *missing*
  Description: returns the value of the system or constant result c(*name*); see [P] **creturn**.
         Referencing c(*name*) will return an error if the result does not exist.
         returns a scalar if the result is scalar.
         returns a string of the result containing the first 244 characters.

_caller()
>   Range: 1.0 to 10.0
>   Description: returns version of the program or session that invoked the currently running program; see [P] **version**. The current version at the time of this writing is 10.0, so 10.0 is the upper end of this range. If Stata 10.1 were the current version, 10.1 would be the upper end of this range, and likewise, if Stata 11.0 were the current version, 11.0 would be the upper end of this range. This is a function for use by programmers.

chop(x, ϵ)
>   Domain x: $-8e+307$ to $8e+307$
>   Domain ϵ: $-8e+307$ to $8e+307$
>   Range: $-8e+307$ to $8e+307$
>   Description: returns round(x) if abs($x - $ round(x)) $< \epsilon$; otherwise, returns x. returns x if x is missing.

clip(x,a,b)
>   Domain x: $-8e+307$ to $8e+307$
>   Domain a: $-8e+307$ to $8e+307$
>   Domain b: $-8e+307$ to $8e+307$
>   Range: $-8e+307$ to $8e+307$
>   Description: returns x if $a < x < b$, b if $x \geq b$, a if $x \leq a$, and *missing* if x is missing or if $a > b$. If a or b is *missing*, this is interpreted as $a = -\infty$ or $b = +\infty$, respectively. returns x if x is missing.

cond(x,a,b,c) or cond(x,a,b)
>   Domain x: $-8e+307$ to $8e+307$ and *missing*; $0 \Rightarrow$ *false*, otherwise interpreted as *true*
>   Domain a: numbers and strings
>   Domain b: numbers if a is a number; strings if a is a string
>   Domain c: numbers if a is a number; strings if a is a string
>   Range: a, b, and c
>   Description: returns a if x is *true*, b if x is *false*, and c if x is *missing*. returns a if c is not specified and x evaluates to *missing*.
>
>   cond(a>2,50,70) $= 50$ if a > 2 or a $\geq$.
>   cond(a>2,"this","that") $=$ "that" if a ≤ 2
>   cond(a>2,"this","that","missing") $=$ "missing" if a $\geq$.
>   cond(a>2,"this","that","missing") $=$ "this" if a ≥ 2 and a $<$.

e(*name*)
>   Domain: names
>   Range: strings, scalars, matrices, and *missing*
>   Description: returns the value of saved result e(*name*); see [U] **18.8 Accessing results calculated by other programs**
>   e(*name*) $=$ scalar missing if the saved result does not exist
>   e(*name*) $=$ specified matrix if the saved result is a matrix
>   e(*name*) $=$ scalar numeric value if the saved result is a scalar
>   e(*name*) $=$ a string containing the first 244 characters if the saved result is a string

e(sample)
>   Range: 0 and 1
>   Description: returns 1 if the observation is in the estimation sample and 0 otherwise.

epsdouble()

Range: a double-precision number close to 0

Description: returns the machine precision of a double-precision number. If $d <$ epsdouble() and (double) $x = 1$, then $x + d =$ (double) 1. This function takes no arguments, but the parentheses must be included.

epsfloat()

Range: a floating-point number close to 0

Description: returns the machine precision of a floating-point number. If $d <$ epsfloat() and (float) $x = 1$, then $x + d =$ (float) 1. This function takes no arguments, but the parentheses must be included.

float(x)

Domain: −1e+38 to 1e+38

Range: −1e+38 to 1e+38

Description: returns the value of x rounded to float precision.

Although you may store your numeric variables as byte, int, long, float, or double, Stata converts all numbers to double before performing any calculations. Consequently, difficulties can arise in comparing numbers that have no finite binary representation.

For example, if the variable x is stored as a float and contains the value 1.1 (a repeating "decimal" in binary), the expression x==1.1 will evaluate to *false* because the literal 1.1 is the double representation of 1.1, which is different from the float representation stored in x. (They differ by 2.384×10^{-8}.) The expression x==float(1.1) will evaluate to *true* because the float() function converts the literal 1.1 to its float representation before it is compared with x. (See [U] **13.10 Precision and problems therein** for more information.)

fmtwidth(*fmtstr*)

Range: strings

Description: returns the output length of the %fmt contained in *fmtstr*.

returns *missing* if *fmtstr* does not contain a valid %fmt. For example, fmtwidth("%9.2f") returns 9 and fmtwidth("%tc") returns 18.

has_eprop(*name*)

Domain: names

Range: 0 or 1

Description: returns 1 if *name* appears as a word in e(properties); otherwise, returns 0.

inlist($z,a,b,\ldots$)

Domain: all reals or all strings

Range: 0 or 1

Description: returns 1 if z is a member of the remaining arguments; otherwise, returns 0. All arguments must be reals or all must be strings. The number of arguments is between 2 and 255 for reals and between 2 and 10 for strings.

inrange(z,a,b)

Domain:	all reals or all strings
Range:	0 or 1
Description:	returns 1 if it is known that $a \leq z \leq b$; otherwise, returns 0.

The following ordered rules apply:

$z \geq$. returns 0.

$a \geq$. and $b =$. returns 1.

$a \geq$. returns 1 if $z \leq b$; otherwise, it returns 0.

$b \geq$. returns 1 if $a \leq z$; otherwise, it returns 0.

Otherwise, 1 is returned if $a \leq z \leq b$.

If the arguments are strings, "." is interpreted as "".

irecode(x,x_1,x_2,x_3,..., x_n)

Domain x:	−8e+307 to 8e+307
Domain x_i:	−8e+307 to 8e+307
Range:	nonnegative integers
Description:	returns *missing* if x is missing or $x_1, \ldots, x_n$ is not weakly increasing.

returns 0 if $x \leq x_1$.

returns 1 if $x_1 < x \leq x_2$.

returns 2 if $x_2 < x \leq x_3$.

. . .

returns n if $x > x_n$.

Also see `autocode()` and `recode()` for other styles of recode functions.

`irecode(3, -10, -5, -3, -3, 0, 15, .) = 5`

matrix(*exp*)

Domain:	any valid expression
Range:	evaluation of *exp*
Description:	restricts name interpretation to scalars and matrices; see `scalar()` function below.

maxbyte()

Range:	one integer number
Description:	returns the largest value that can be stored in storage type `byte`. This function takes no arguments, but the parentheses must be included.

maxdouble()

Range:	one double-precision number
Description:	returns the largest value that can be stored in storage type `double`. This function takes no arguments, but the parentheses must be included.

maxfloat()

Range:	one floating-point number
Description:	returns the largest value that can be stored in storage type `float`. This function takes no arguments, but the parentheses must be included.

maxint()

Range:	one integer number
Description:	returns the largest value that can be stored in storage type `int`. This function takes no arguments, but the parentheses must be included.

maxlong()
> Range: one integer number
> Description: returns the largest value that can be stored in storage type long. This function takes no arguments, but the parentheses must be included.

$\text{mi}(x_1, x_2, \ldots, x_n)$ is a synonym for $\text{missing}(x_1, x_2, \ldots, x_n)$.

minbyte()
> Range: one integer number
> Description: returns the smallest value that can be stored in storage type byte. This function takes no arguments, but the parentheses must be included.

mindouble()
> Range: one double-precision number
> Description: returns the smallest value that can be stored in storage type double. This function takes no arguments, but the parentheses must be included.

minfloat()
> Range: one floating-point number
> Description: returns the smallest value that can be stored in storage type float. This function takes no arguments, but the parentheses must be included.

minint()
> Range: one integer number
> Description: returns the smallest value that can be stored in storage type int. This function takes no arguments, but the parentheses must be included.

minlong()
> Range: one integer number
> Description: returns the smallest value that can be stored in storage type long. This function takes no arguments, but the parentheses must be included.

$\text{missing}(x_1, x_2, \ldots, x_n)$
> Domain x_i: any string or numeric expression
> Range: 0 and 1
> Description: returns 1 if any x_i evaluates to *missing*; otherwise, returns 0.
>
> Stata has two concepts of missing values: a numeric missing value (., .a, .b, ..., .z) and a string missing value (""). missing() returns 1 (meaning *true*) if any expression x_i evaluates to *missing*. If x is numeric, missing(x) is equivalent to $x \geq$ .. If x is string, missing(x) is equivalent to $x==$"".

r(*name*)
> Domain: names
> Range: strings, scalars, matrices, and *missing*
> Description: returns the value of the saved result r(*name*);
> > see [U] **18.8 Accessing results calculated by other programs**
> > r(*name*) = scalar missing if the saved result does not exist
> > r(*name*) = specified matrix if the saved result is a matrix
> > r(*name*) = scalar numeric value if the saved result is a scalar that can be interpreted as a number
> > r(*name*) = a string containing the first 244 characters if the saved result is a string

recode(x,x_1,x_2,...,x_n)
  Domain x: −8e+307 to 8e+307 and *missing*
  Domain x_1:   −8e+307 to 8e+307
  Domain x_2: x_1 to 8e+307
  ...
  Domain x_n: x_{n-1} to 8e+307
  Range: x_1, x_2, ..., x_n and *missing*
  Description: returns *missing* if $x_1,...,x_n$ is not weakly increasing.
           returns x if x is missing.
           returns x_1 if $x \le x_1$; x_2 if $x \le x_2$, ...; otherwise,
               x_n if x is greater than x_1, x_2, ..., x_{n-1}.
               $x_i \ge .$ is interpreted as $x_i = +\infty$.

           Also see autocode() and irecode() for other styles of recode functions.

replay()
  Range: integers 0 and 1, meaning *false* and *true*, respectively
  Description: returns 1 if the first nonblank character of local macro '0' is a comma,
           or if '0' is empty. This is a function for use by programmers writing
           estimation commands; see [P] **ereturn**.

return(*name*)
  Domain: names
  Range: strings, scalars, matrices, and *missing*
  Description: returns the value of the to-be-saved result r(*name*);
           see [U] **18.8 Accessing results calculated by other programs** and
           [U] **18.10 Saving results**
           return(*name*) = scalar missing if the saved result does not exist
           return(*name*) = specified matrix if the saved result is a matrix
           return(*name*) = scalar numeric value if the saved result is a scalar
           return(*name*) = a string containing the first 244 characters
                    if the saved result is a string

s(*name*)
  Domain: names
  Range: strings and *missing*
  Description: returns the value of saved result s(*name*);
           see [U] **18.8 Accessing results calculated by other programs**
           s(*name*) = . if the saved result does not exist
           s(*name*) = a string containing the first 244 characters
                    if the saved result is a string

scalar(*exp*)

    Domain: any valid expression
    Range: evaluation of *exp*
    Description: restricts name interpretation to scalars and matrices.

> Names in expressions can refer to names of variables in the dataset, names of matrices, or names of scalars. Matrices and scalars can have the same names as variables in the dataset. If names conflict, Stata assumes that you are referring to the name of the variable in the dataset.
>
> matrix() and scalar() explicitly state that you are referring to matrices and scalars. matrix() and scalar() are the same function; scalars and matrices may not have the same names and so cannot be confused. Typing scalar(x) makes it clear that you are referring to the scalar or matrix named x and not the variable named x should there happen to be a variable of that name.

Date and time functions

Stata's *date and time functions* are described with examples in [U] **24 Dealing with dates and times** and [D] **dates and times**. What follows is a technical description. We use the following notation:

e_{tc}	%tc	encoded date–time (ms. since 01jan1960 00:00:00.000)
e_{tC}	%tC	encoded date–time (ms. with leap seconds since 01jan1960 00:00:00.000)
e_d	%td	encoded date (days since 01jan1960)
e_w	%tw	encoded weekly date (weeks since 1960w1)
e_m	%tm	encoded monthly date (months since 1960m1)
e_q	%tq	encoded quarterly date (quarters since 1960q1)
e_h	%th	encoded half-yearly date (half years since 1960h1)
e_y	%ty	encoded yearly date (years)
M	month, 1–12	
D	day of month, 1–31	
Y	year, 0100–9999	
h	hour, 0–23	
m	minute, 0–59	
s	second, 0–59 or 60 if leap seconds	
W	week number, 1–52	
Q	quarter number, 1–4	
H	half-year number, 1 or 2	

The date and time functions, where integer arguments are required, allow noninteger values and use the floor() of the value.

A %t date–time is recorded as the milliseconds, days, weeks, etc., depending upon the units from 01jan1960; negative values indicate dates and times before 01jan1960. Allowable dates and times are those between 01jan0100 and 31dec9999, inclusive, but all functions are based on the Gregorian calendar, and values do not correspond to historical dates before Friday, 15oct1582.

Cdhms(e_d,h,m,s)
   Domain e_d: %td dates 01jan0100 to 31dec9999 (integers $-679{,}350$ to $2{,}936{,}549$)
   Domain h: integers 0 to 23
   Domain m: integers 0 to 59
   Domain s: reals 0.000 to 60.999
   Range: dates 01jan0100 00:00:00.000 to 31dec9999 23:59:59.999
                (integers $-58{,}695{,}840{,}000{,}000$ to $>253{,}717{,}919{,}999{,}999$) and *missing*
   Description: returns the e_{tC} date–time (ms. with leap seconds since 01jan1960 00:00:00.000)
                corresponding to e_d, h, m, s.

Chms(h,m,s)
   Domain h: integers 0 to 23
   Domain m: integers 0 to 59
   Domain s: reals 0.000 to 60.999
   Range: dates 01jan0100 00:00:00.000 to 31dec9999 23:59:59.999
                (integers $-58{,}695{,}840{,}000{,}000$ to $>253{,}717{,}919{,}999{,}999$) and *missing*
   Description: returns the e_{tC} date–time (ms. with leap seconds since 01jan1960 00:00:00.000)
                corresponding to h, m, s on 01jan1960.

Clock(s_1,s_2[,Y])
   Domain s_1: strings
   Domain s_2: strings
   Domain Y: integers 1000 to 9998 (but probably 2001 to 2099)
   Range: dates 01jan0100 00:00:00.000 to 31dec9999 23:59:59.999
                (integers $-58{,}695{,}840{,}000{,}000$ to $>253{,}717{,}919{,}999{,}999$) and *missing*
   Description: returns the e_{tC} date–time (ms. with leap seconds since 01jan1960 00:00:00.000)
                corresponding to s_1 based on s_2 and Y.

   Function Clock() works the same as function clock() except that it returns
   a leap second–adjusted %tC value rather than an unadjusted %tc value. Use
   Clock() only if original time values have been adjusted for leap seconds.

clock(s_1,$s_2$$\left[\,,Y\right]$)
  Domain s_1: strings
  Domain s_2: strings
  Domain Y: integers 1000 to 9998 (but probably 2001 to 2099)
  Range: dates 01jan0100 00:00:00.000 to 31dec9999 23:59:59.999
     (integers $-58{,}695{,}840{,}000{,}000$ to $253{,}717{,}919{,}999{,}999$) and *missing*
  Description: returns the e_{tc} date–time (ms. since 01jan1960 00:00:00.000) corresponding to
     s_1 based on s_2 and Y.

     s_1 contains the date and/or time, recorded as a string, in virtually any format.
     Months can be spelled out, abbreviated (to three characters), or indicated as numbers;
     years can include or exclude the century; blanks and punctuation are allowed.

     s_2 is any permutation of M, D, [##]Y, h, m, and s, with their order defining the
     order that month, day, year, hour, minute, and second occur (and whether they
     occur) in s_1. ##, if specified, indicates the default century for 2-digit years in s_1.
     For instance, $s_2 =$ "MD19Y hm" would translate $s_1 =$ "11/15/91 21:14" as
     15nov1991 21:14. The space in "MD19Y hm" was not significant and the string would
     have translated just as well with "MD19Yhm".

     Y provides an alternate way of handling two-digit years. Y specifies the largest
     year that is to be returned when a two-digit year is encountered; see function date()
     below. If neither ## nor Y is specified, clock() returns *missing* when it
     encounters a two-digit year.

Cmdyhms(M,D,Y,h,m,s)
  Domain M: integers 1 to 12
  Domain D: integers 1 to 31
  Domain Y: integers 0100 to 9999 (but probably 1800 to 2100)
  Domain h: integers 0 to 23
  Domain m: integers 0 to 59
  Domain s: reals 0.000 to 60.999
  Range: dates 01jan0100 00:00:00.000 to 31dec9999 23:59:59.999
     (integers $-58{,}695{,}840{,}000{,}000$ to $>253{,}717{,}919{,}999{,}999$) and *missing*
  Description: returns the e_{tC} date–time (ms. with leap seconds since 01jan1960 00:00:00.000)
     corresponding to M, D, Y, h, m, s.

Cofc(e_{tc})
  Domain e_{tc}: dates 01jan0100 00:00:00.000 to 31dec9999 23:59:59.999
     (integers $-58{,}695{,}840{,}000{,}000$ to $253{,}717{,}919{,}999{,}999$)
  Range: dates 01jan0100 00:00:00.000 to 31dec9999 23:59:59.999
     (integers $-58{,}695{,}840{,}000{,}000$ to $>253{,}717{,}919{,}999{,}999$)
  Description: returns the e_{tC} date–time (ms. with leap seconds since 01jan1960 00:00:00.000)
     of e_{tc} (ms. without leap seconds since 01jan1960 00:00:00.000.

cofC(e_{tC})
  Domain e_{tC}: dates 01jan0100 00:00:00.000 to 31dec9999 23:59:59.999
     (integers $-58{,}695{,}840{,}000{,}000$ to $>253{,}717{,}919{,}999{,}999$)
  Range: dates 01jan0100 00:00:00.000 to 31dec9999 23:59:59.999
     (integers $-58{,}695{,}840{,}000{,}000$ to $253{,}717{,}919{,}999{,}999$)
  Description: returns the e_{tc} date–time (ms. without leap seconds since 01jan1960 00:00:00.000)
     of e_{tC} (ms. with leap seconds since 01jan1960 00:00:00.000.

Cofd(e_d)

  Domain e_d: %td dates 01jan0100 to 31dec9999 (integers $-679{,}350$ to $2{,}936{,}549$)
  Range: dates 01jan0100 00:00:00.000 to 31dec9999 23:59:59.999
                (integers $-58{,}695{,}840{,}000{,}000$ to $> 253{,}717{,}919{,}999{,}999$)
  Description: returns the e_{tC} date–time (ms. with leap seconds since 01jan1960 00:00:00.000)
                of date e_d at time 00:00:00.000.

cofd(e_d)

  Domain e_d: %td dates 01jan0100 to 31dec9999 (integers $-679{,}350$ to $2{,}936{,}549$)
  Range: dates 01jan0100 00:00:00.000 to 31dec9999 23:59:59.999
                (integers $-58{,}695{,}840{,}000{,}000$ to $253{,}717{,}919{,}999{,}999$)
  Description: returns the e_{tc} date–time (ms. since 01jan1960 00:00:00.000) of date e_d at time
                00:00:00.000.

date(s_1,s_2[,Y])

  Domain s_1: strings
  Domain s_2: strings
  Domain Y: integers 1000 to 9998 (but probably 2001 to 2099)
  Range: %td dates 01jan0100 to 31dec9999 (integers $-679{,}350$ to $2{,}936{,}549$) and *missing*
  Description: returns the e_d date (days since 01jan1960) corresponding to s_1 based on s_2 and Y.

  > s_1 contains the date, recorded as a string, in virtually any format. Months can
  > be spelled out, abbreviated (to three characters), or indicated as numbers; years can
  > include or exclude the century; blanks and punctuation are allowed.

  > s_2 is any permutation of M, D, and [##]Y, with their order defining the order
  > that month, day, and year occur in s_1. ##, if specified, indicates the default century
  > for two-digit years in s_1. For instance, $s_2 = $ "MD19Y" would translate
  > $s_1 = $ "11/15/91" as 15nov1991.

  > Y provides an alternate way of handling two-digit years. Returned when a
  > two-digit year is encountered is the closest year to Y.

  > > date("1/15/99","MDY",2040) $=$ 15jan1999
  > > date("1/15/99","MDY",2060) $=$ 15jan2099
  > >
  > > date("1/15/51","MDY",2000) $=$ 15jan1951
  > > date("1/15/50","MDY",2000) $=$ 15jan2050
  > > date("1/15/49","MDY",2000) $=$ 15jan2049
  > >
  > > date("1/15/01","MDY",2050) $=$ 15jan2001
  > > date("1/15/00","MDY",2050) $=$ 15jan2000

  > If neither ## nor Y is specified, date() returns *missing* when it encounters
  > a two-digit year.

day(e_d)

  Domain e_d: %td dates 01jan0100 to 31dec9999 (integers $-679{,}350$ to $2{,}936{,}549$)
  Range: integers 1 to 31 and *missing*
  Description: returns the numeric day of the month corresponding to e_d.

dhms(e_d,h,m,s)

    Domain e_d: %td dates 01jan0100 to 31dec9999 (integers $-679,350$ to $2,936,549$)

    Domain h: integers 0 to 23

    Domain m: integers 0 to 59

    Domain s: reals 0.000 to 59.999

    Range: dates 01jan0100 00:00:00.000 to 31dec9999 23:59:59.999

               (integers $-58,695,840,000,000$ to $253,717,919,999,999$) and *missing*

    Description: returns the e_{tc} date–time (ms. since 01jan1960 00:00:00.000) corresponding to e_d, h, m, s.

dofC(e_{tC})

    Domain e_{tC}: dates 01jan0100 00:00:00.000 to 31dec9999 23:59:59.999

               (integers $-58,695,840,000,000$ to $> 253,717,919,999,999$)

    Range: %td dates 01jan0100 to 31dec9999 (integers $-679,350$ to $2,936,549$)

    Description: returns the e_d date (days since 01jan1960) of date–time e_{tC} (ms. with leap seconds since 01jan1960 00:00:00.000).

dofc(e_{tc})

    Domain e_{tc}: dates 01jan0100 00:00:00.000 to 31dec9999 23:59:59.999

               (integers $-58,695,840,000,000$ to $253,717,919,999,999$)

    Range: %td dates 01jan0100 to 31dec9999 (integers $-679,350$ to $2,936,549$)

    Description: returns the e_d date (days since 01jan1960) of date–time e_{tc} (ms. since 01jan1960 00:00:00.000).

dofh(e_h)

    Domain e_h: %th dates 0100h1 to 9999h2 (integers $-3,720$ to $16,079$)

    Range: %td dates 01jan0100 to 01jul9999 (integers $-679,350$ to $2,936,366$)

    Description: returns the e_d date (days since 01jan1960) of the start of half-year e_h.

dofm(e_m)

    Domain e_m: %tm dates 0100m1 to 9999m12 (integers $-22,320$ to $96,479$)

    Range: %td dates 01jan0100 to 01dec9999 (integers $-679,350$ to $2,936,519$)

    Description: returns the e_d date (days since 01jan1960) of the start of month e_m.

dofq(e_q)

    Domain e_q: %tq dates 0100q1 to 9999q4 (integers $-7,440$ to $32,159$)

    Range: %td dates 01jan0100 to 01oct9999 (integers $-679,350$ to $2,936,458$)

    Description: returns the e_d date (days since 01jan1960) of the start of quarter e_q.

dofw(e_w)

    Domain e_w: %tw dates 0100w1 to 9999w52 (integers $-96,720$ to $418,079$)

    Range: %td dates 01jan0100 to 24dec9999 (integers $-679,350$ to $2,936,542$)

    Description: returns the e_d date (days since 01jan1960) of the start of week e_w.

dofy(e_y)

    Domain e_y: %ty dates 0100 to 9999 (integers 0100 to 9999)

    Range: %td dates 01jan0100 to 01jan9999 (integers $-679,350$ to $2,936,185$)

    Description: returns the e_d date (days since 01jan1960) of 01jan in year e_y.

dow(e_d)

    Domain e_d: %td dates 01jan0100 to 31dec9999 (integers $-679,350$ to $2,936,549$)

    Range: integers 0 to 6 and *missing*

    Description: returns the numeric day of the week corresponding to date e_d;

                       0 = Sunday, 1 = Monday, . . . , 6 = Saturday, etc.

doy(e_d)
    Domain e_d: %td dates 01jan0100 to 31dec9999 (integers $-679,350$ to $2,936,549$)
    Range: integers 1 to 366 and *missing*
    Description: returns the numeric day of the year corresponding to date e_d.

halfyear(e_d)
    Domain e_d: %td dates 01jan0100 to 31dec9999 (integers $-679,350$ to $2,936,549$)
    Range: integers 1, 2 and *missing*
    Description: returns the numeric half of the year corresponding to date e_d.

halfyearly(s_1,s_2[,Y])
    Domain s_1:   strings
    Domain s_2:   strings "HY" and "YH"; Y may be prefixed with ##
    Domain Y: integers 1000 to 9998 (but probably 2001 to 2099)
    Range: %th dates 0100h1 to 9999h2 (integers $-3,720$ to $16,079$) and *missing*
    Description: returns the e_h half-yearly date (half years since 1960h1) corresponding to s_1 based
                    on s_2 and Y; Y specifies closest year; see date().

hh(e_{tc})
    Domain e_{tc}: dates 01jan0100 00:00:00.000 to 31dec9999 23:59:59.999
                    (integers $-58,695,840,000,000$ to $253,717,919,999,999$)
    Range: integers 0 through 23, *missing*
    Description: returns the hour corresponding to date–time e_{tc} (ms. since 01jan1960 00:00:00.000).

hhC(e_{tC})
    Domain e_{tC}: dates 01jan0100 00:00:00.000 to 31dec9999 23:59:59.999
                    (integers $-58,695,840,000,000$ to $> 253,717,919,999,999$)
    Range: integers 0 through 23, *missing*
    Description: returns the hour corresponding to date–time e_{tC} (ms. with leap seconds since
                    01jan1960 00:00:00.000).

hms(h,m,s)
    Domain h: integers 0 to 23
    Domain m: integers 0 to 59
    Domain s: reals 0.000 to 59.999
    Range: integers 1 to 31 and *missing*
    Description: returns the e_{tc} date–time (ms. since 01jan1960 00:00:00.000) corresponding to
                    h, m, s on 01jan1960.

hofd(e_d)
    Domain e_d: %td dates 01jan0100 to 31dec9999 (integers $-679,350$ to $2,936,549$)
    Range: %th dates 0100h1 to 9999h2 (integers $-3,720$ to $16,079$)
    Description: returns the e_h half-yearly date (half years since 1960h1) containing date e_d.

hours(ms)
    Domain ms: real; milliseconds
    Range: real and *missing*
    Description: returns $ms/3,600,000$.

mdy(M,D,Y)
    Domain M: integers 1 to 12
    Domain D: integers 1 to 31
    Domain Y: integers 0100 to 9999 (but probably 1800 to 2100)
    Range: %td dates 01jan0100 to 31dec9999 (integers $-679,350$ to $2,936,549$) and *missing*
    Description: returns the e_d date (days since 01jan1960) corresponding to M, D, Y.

mdyhms(M,D,Y,h,m,s)

    Domain M: integers 1 to 12
    Domain D: integers 1 to 31
    Domain Y: integers 0100 to 9999 (but probably 1800 to 2100)
    Domain h: integers 0 to 23
    Domain m: integers 0 to 59
    Domain s: reals 0.000 to 59.999
    Range: dates 01jan0100 00:00:00.000 to 31dec9999 23:59:59.999
               (integers $-58{,}695{,}840{,}000{,}000$ to $253{,}717{,}919{,}999{,}999$) and *missing*
    Description: returns the e_{tc} date–time (ms. since 01jan1960 00:00:00.000) corresponding to M, D, Y, h, m, s.

minutes(ms)

    Domain ms: real; milliseconds
    Range: real and *missing*
    Description: returns $ms/60{,}000$.

mm(e_{tc})

    Domain e_{tc}: dates 01jan0100 00:00:00.000 to 31dec9999 23:59:59.999
               (integers $-58{,}695{,}840{,}000{,}000$ to $253{,}717{,}919{,}999{,}999$)
    Range: integers 0 through 59, *missing*
    Description: returns the minute corresponding to date–time e_{tc} (ms. since 01jan1960 00:00:00.000).

mmC(e_{tC})

    Domain e_{tC}: dates 01jan0100 00:00:00.000 to 31dec9999 23:59:59.999
               (integers $-58{,}695{,}840{,}000{,}000$ to $>253{,}717{,}919{,}999{,}999$)
    Range: integers 0 through 59, *missing*
    Description: returns the minute corresponding to date–time e_{tC} (ms. with leap seconds since 01jan1960 00:00:00.000).

mofd(e_d)

    Domain e_d: %td dates 01jan0100 to 31dec9999 (integers $-679{,}350$ to $2{,}936{,}549$)
    Range: %tm dates 0100m1 to 9999m12 (integers $-22{,}320$ to $96{,}479$)
    Description: returns the e_m monthly date (months since 1960m1) containing date e_d.

month(e_d)

    Domain e_d: %td dates 01jan0100 to 31dec9999 (integers $-679{,}350$ to $2{,}936{,}549$)
    Range: integers 1 to 12 and *missing*
    Description: returns the numeric month corresponding to date e_d.

monthly(s_1,$s_2$$\big[$,$Y$$\big]$)

    Domain s_1:   strings
    Domain s_2:   strings "MY" and "YM"; Y may be prefixed with ##
    Domain Y: integers 1000 to 9998 (but probably 2001 to 2099)
    Range: %tm dates 0100m1 to 9999m12 (integers $-22{,}320$ to $96{,}479$) and *missing*
    Description: returns the e_m monthly date (months since 1960m1) corresponding to s_1 based on s_2 and Y; Y specifies top year; see date() in [D] **dates and times**.

msofhours(h)

    Domain h: real; hours
    Range: real and *missing*; milliseconds
    Description: returns $h \times 3{,}600{,}000$.

msofminutes(m)
   Domain m: real; minutes
   Range: real and *missing*; milliseconds
   Description: returns $m \times 60{,}000$.

msofseconds(s)
   Domain s: real; seconds
   Range: real and *missing*; milliseconds
   Description: returns $s \times 1{,}000$.

qofd(e_d)
   Domain e_d: %td dates 01jan0100 to 31dec9999 (integers $-679{,}350$ to $2{,}936{,}549$)
   Range: %tq dates 0100q1 to 9999q4 (integers $-7{,}440$ to $32{,}159$)
   Description: returns the e_q quarterly date (quarters since 1960q1) containing date e_d.

quarter(e_d)
   Domain e_d: %td dates 01jan0100 to 31dec9999 (integers $-679{,}350$ to $2{,}936{,}549$)
   Range: integers 1 to 4 and *missing*
   Description: returns the numeric quarter of the year corresponding to date e_d.

quarterly(s_1,s_2[,Y])
   Domain s_1:   strings
   Domain s_2:   strings "QY" and "YQ"; Y may be prefixed with ##
   Domain Y: integers 1000 to 9998 (but probably 2001 to 2099)
   Range: %tq dates 0100q1 to 9999q4 (integers $-7{,}440$ to $32{,}159$) and *missing*
   Description: returns the e_q quarterly date (quarters since 1960q1) corresponding to s_1 based on
                   s_2 and Y; Y specifies closest year; see date().

seconds(ms)
   Domain ms: real; milliseconds
   Range: real and *missing*
   Description: returns $ms/1{,}000$.

ss(e_{tc})
   Domain e_{tc}:   dates 01jan0100 00:00:00.000 to 31dec9999 23:59:59.999
                      (integers $-58{,}695{,}840{,}000{,}000$ to $253{,}717{,}919{,}999{,}999$)
   Range: real 0.000 through 59.999, *missing*
   Description: returns the second corresponding to date–time e_{tc} (ms. since 01jan1960 00:00:00.000).

ssC(e_{tC})
   Domain e_{tC}:   dates 01jan0100 00:00:00.000 to 31dec9999 23:59:59.999
                      (integers $-58{,}695{,}840{,}000{,}000$ to $>253{,}717{,}919{,}999{,}999$)
   Range: real 0.000 through 60.999, *missing*
   Description: returns the second corresponding to date–time e_{tC} (ms. with leap seconds since
                      01jan1960 00:00:00.000).

tC(l)
   Domain l: date–time literal strings 01jan0100 00:00:00.000 to 31dec9999 23:59:59.999
   Range: dates 01jan0100 00:00:00.000 to 31dec9999 23:59:59.999
                  (integers $-58{,}695{,}840{,}000{,}000$ to $>253{,}717{,}919{,}999{,}999$)
   Description: convenience function to make typing dates and times in expressions easier;
                  same as tc() except returns leap-second adjusted values; e.g., typing
                  tc(29nov2007 9:15) is equivalent to typing 1511946900000, whereas
                  tC(29nov2007 9:15) is 1511946923000.

tc(l)

　　Domain l: 　　date–time literal strings 01jan0100 00:00:00.000 to 31dec9999 23:59:59.999

　　Range: 　　dates 01jan0100 00:00:00.000 to 31dec9999 23:59:59.999

　　　　　　　(integers −58,695,840,000,000 to 253,717,919,999,999)

　　Description: 　convenience function to make typing dates and times in expressions easier;

　　　　　　　e.g., typing tc(2jan1960 13:42) is equivalent to typing 135720000;

　　　　　　　the date but not the time may be omitted and then 01jan1960 is

　　　　　　　assumed; the seconds portion of the time may be omitted and

　　　　　　　is assumed to be 0.000; tc(11:02) is equivalent to typing 39720000.

td(l)

　　Domain l: 　　date literal strings 01jan0100 to 31dec9999

　　Range: 　　%td dates 01jan0100 to 31dec9999 (integers −679,350 to 2,936,549)

　　Description: 　convenience function to make typing dates in expressions easier;

　　　　　　　e.g., typing td(2jan1960) is equivalent to typing 1.

th(l)

　　Domain l: 　　half-year literal strings 0100h1 to 9999h2

　　Range: 　　%th dates 0100h1 to 9999h2 (integers −3,720 to 16,079)

　　Description: 　convenience function to make typing half-yearly dates in expressions easier;

　　　　　　　e.g., typing th(1960h2) is equivalent to typing 1.

tm(l)

　　Domain l: 　　month literal strings 0100m1 to 9999m12

　　Range: 　　%tm dates 0100m1 to 9999m12 (integers −22,320 to 96,479)

　　Description: 　convenience function to make typing monthly dates in expressions easier;

　　　　　　　e.g., typing tm(1960m2) is equivalent to typing 1.

tq(l)

　　Domain l: 　　quarter literal strings 0100q1 to 9999q4

　　Range: 　　%tq dates 0100q1 to 9999q4 (integers −7,440 to 32,159)

　　Description: 　convenience function to make typing quarterly dates in expressions easier;

　　　　　　　e.g., typing tq(1960q2) is equivalent to typing 1.

tw(l)

　　Domain l: 　　week literal strings 0100w1 to 9999w52

　　Range: 　　%tw dates 0100w1 to 9999w52 (integers −96,720 to 418,079)

　　Description: 　convenience function to make typing weekly dates in expressions easier;

　　　　　　　e.g., typing tw(1960w2) is equivalent to typing 1.

week(e_d)

　　Domain e_d: 　%td dates 01jan0100 to 31dec9999 (integers −679,350 to 2,936,549)

　　Range: 　　integers 1 to 52 and *missing*

　　Description: 　returns the numeric week of the year corresponding to date e_d

　　　　　　　(the first week of a year is the first 7-day period of the year).

weekly(s_1,$s_2$$\left[\,,Y\right]$)

　　Domain s_1: 　strings

　　Domain s_2: 　strings "WY" and "YW"; Y may be prefixed with ##

　　Domain Y: 　integers 1000 to 9998 (but probably 2001 to 2099)

　　Range: 　　%tw dates 0100w1 to 9999w52 (integers −96,720 to 418,079) and *missing*

　　Description: 　returns the e_w weekly date (weeks since 1960w1) corresponding to s_1 based on s_2

　　　　　　　and Y; Y specifies closest year; see date().

wofd(e_d)
    Domain e_d: %td dates 01jan0100 to 31dec9999 (integers −679,350 to 2,936,549)
    Range: %tw dates 0100w1 to 9999w52 (integers −96,720 to 418,079)
    Description: returns the e_w weekly date (weeks since 1960w1) containing date e_d.

year(e_d)
    Domain e_d: %td dates 01jan0100 to 31dec9999 (integers −679,350 to 2,936,549)
    Range: integers 0100 to 9999 (but probably 1800 to 2100)
    Description: returns the numeric year corresponding to date e_d.

yearly(s_1,s_2 [,Y])
    Domain s_1: strings
    Domain s_2: string "Y"; Y may be prefixed with ##
    Domain Y: integers 1000 to 9998 (but probably 2001 to 2099)
    Range: %ty dates 0100 to 9999 (integers 0100 to 9999) and *missing*
    Description: returns the e_y yearly date (year) corresponding to s_1 based on s_2 and Y;
               Y specifies closest year; see date().

yh(Y,H)
    Domain Y: integers 1000 to 9999 (but probably 1800 to 2100)
    Domain H: integers 1, 2
    Range: %th dates 0100h1 to 9999h2 (integers −1,920 to 16,079)
    Description: returns the e_h half-yearly date (half years since 1960h1) corresponding to year Y,
               half-year H.

ym(Y,M)
    Domain Y: integers 1000 to 9999 (but probably 1800 to 2100)
    Domain M: integers 1 to 12
    Range: %tm dates 0100m1 to 9999m12 (integers −11,520 to 96,479)
    Description: returns the e_m monthly date (months since 1960m1) corresponding to year Y,
               month M.

yofd(e_d)
    Domain e_d: %td dates 01jan0100 to 31dec9999 (integers −679,350 to 2,936,549)
    Range: %ty dates 0100 to 9999 (integers 0100 to 9999)
    Description: returns the e_y yearly date (year) containing date e_d.

yq(Y,Q)
    Domain Y: integers 1000 to 9999 (but probably 1800 to 2100)
    Domain Q: integers 1 to 4
    Range: %tq dates 0100q1 to 9999q4 (integers −3,840 to 32,159)
    Description: returns the e_q quarterly date (quarters since 1960q1) corresponding to year Y,
               quarter Q.

yw(Y,W)
    Domain Y: integers 1000 to 9999 (but probably 1800 to 2100)
    Domain W: integers 1 to 52
    Range: %tw dates 0100w1 to 9999w52 (integers −49,920 to 418,079)
    Description: returns the e_w weekly date (weeks since 1960w1) corresponding to year Y,
               week W.

Selecting time spans

tin(d_1,d_2)
 Domain d_1: date or time literals recorded in units of t previously tsset
 Domain d_2: date or time literals recorded in units of t previously tsset
 Range: 0 and 1, 1 $\Rightarrow$ *true*
 Description: *true* if $d_1 \leq t \leq d_2$, where t is the time variable previously tsset.

> You must have previously tsset the data to use tin(); see [TS] **tsset**. When you tsset the data, you specify a time variable t, and the format on t states how it is recorded. You type d_1 and d_2 according to that format.
>
> If t has a %tc format, you could type tin(5jan1992 11:15, 14apr2002 12:25).
>
> If t has a %td format, you could type tin(5jan1992, 14apr2002).
>
> If t has a %tw format, you could type tin(1985w1, 2002w15).
>
> If t has a %tm format, you could type tin(1985m1, 2002m4).
>
> If t has a %tq format, you could type tin(1985q1, 2002q2).
>
> If t has a %th format, you could type tin(1985h1, 2002h1).
>
> If t has a %ty format, you could type tin(1985, 2002).
>
> Otherwise, t is just a set of integers, and you could type tin(12, 38).
>
> The details of the %t format do not matter. If your t is formatted %tdnn/dd/yy so that 5jan1992 displays as 1/5/92, you would still type the date in day–month–year order: tin(5jan1992, 14apr2002).

twithin(d_1,d_2)
 Domain d_1: date or time literals recorded in units of t previously tsset
 Domain d_2: date or time literals recorded in units of t previously tsset
 Range: 0 and 1, 1 $\Rightarrow$ *true*
 Description: *true* if $d_1 < t < d_2$, where t is the time variable previously tsset;
     see the tin() function above; twithin() is similar, except the range is exclusive.

Matrix functions returning a matrix

In addition to the functions listed below, see [P] **matrix svd** for singular value decomposition, [P] **matrix symeigen** for eigenvalues and eigenvectors of symmetric matrices, and [P] **matrix eigenvalues** for eigenvalues of nonsymmetric matrices.

cholesky(M)

    Domain: $n \times n$, positive definite, symmetric matrices

    Range: $n \times n$ lower-triangular matrices

    Description: returns the Cholesky decomposition of the matrix:
           if $R = \text{cholesky}(S)$, then $RR^T = S$.
           R^T indicates the transpose of R.
           Row and column names are obtained from M.

corr(M)

    Domain: $n \times n$ symmetric variance matrices

    Range: $n \times n$ symmetric correlation matrices

    Description: returns the correlation matrix of the variance matrix.
           Row and column names are obtained from M.

diag(v)

    Domain: $1 \times n$ and $n \times 1$ vectors

    Range: $n \times n$ diagonal matrices

    Description: returns the square, diagonal matrix created from the row or column vector.
           Row and column names are obtained from the column names of M if M is
           a row vector or from the row names of M if M is a column vector.

get(*systemname*)

    Domain: existing names of system matrices

    Range: matrices

    Description: returns a copy of Stata internal system matrix *systemname*.

           This function is included for backward compatibility with previous versions
           of Stata.

hadamard(M, N)

    Domain M: $m \times n$ matrices

    Domain N: $m \times n$ matrices

    Range: $m \times n$ matrices

    Description: returns a matrix whose i, j element is $M[i,j] \cdot N[i,j]$ (if M and N
           are not the same size, this function reports a conformability error).

I(n)

    Domain: real scalars 1 to `matsize`

    Range: identity matrices

    Description: returns an $n \times n$ identity matrix if n is an integer; otherwise, this function returns
           the $\text{round}(n) \times \text{round}(n)$ identity matrix.

inv(M)

    Domain: $n \times n$ nonsingular matrices

    Range: $n \times n$ matrices

    Description: returns the inverse of the matrix M. If M is singular, this will result in an error.

           The function `invsym()` should be used in preference to `inv()` because `invsym()`
           is more accurate. The row names of the result are obtained from the column
           names of M, and the column names of the result are obtained from the row names
           of M.

invsym(M)

    Domain: $n \times n$ symmetric matrices

    Range: $n \times n$ symmetric matrices

    Description: returns the inverse of M if M is positive definite. If M is not positive definite, rows will be inverted until the diagonal terms are zero or negative; the rows and columns corresponding to these terms will be set to 0, producing a g2 inverse. The row names of the result are obtained from the column names of M, and the column names of the result are obtained from the row names of M.

J(r,c,z)

    Domain r: integer scalars 1 to `matsize`

    Domain c: integer scalars 1 to `matsize`

    Domain z: scalars –8e+307 to 8e+307

    Range: $r \times c$ matrices

    Description: returns the $r \times c$ matrix containing elements z.

matuniform(r, c)

    Domain r: integer scalars 1 to `matsize`

    Domain c: integer scalars 1 to `matsize`

    Range: $r \times c$ matrices

    Description: returns the $r \times c$ matrices containing uniformly distributed pseudorandom numbers on the interval $[0, 1)$.

nullmat(*matname*)

    Domain: matrix names, existing and nonexisting

    Range: matrices including null if *matname* does not exist

    Description: nullmat() is for use with the row-join (,) and column-join (\\) operators in programming situations. Consider the following code fragment, which is an attempt to create the vector $(1, 2, 3, 4)$:

```
forvalues i = 1/4 {
        mat v = (v, 'i')
}
```

           The above program will not work because, the first time through the loop, v will not yet exist, and thus forming (v, 'i') makes no sense. nullmat() relaxes that restriction:

```
forvalues i = 1/4 {
        mat v = (nullmat(v), 'i')
}
```

           The nullmat() function informs Stata that if v does not exist, the function row-join is to be generalized. Joining nothing with 'i' results in ('i'). Thus the first time through the loop, v = (1) is formed. The second time through, v does exist, so v = $(1, 2)$ is formed, and so on.

           nullmat() can be used only with the , and \\ operators.

`sweep(M,i)`

    Domain M: $n \times n$ matrices

    Domain i: integer scalars 1 to n

    Range: $n \times n$ matrices

    Description: returns matrix M with ith row/column swept. The row and column names of the resultant matrix are obtained from M, except that the nth row and column names are interchanged. If $B = \texttt{sweep}(A, k)$, then

$$B_{kk} = \frac{1}{A_{kk}}$$

$$B_{ik} = -\frac{A_{ik}}{A_{kk}}, \qquad i \neq k$$

$$B_{kj} = \frac{A_{kj}}{A_{kk}}, \qquad j \neq k$$

$$B_{ij} = A_{ij} - \frac{A_{ik}A_{kj}}{A_{kk}}, \qquad i \neq k, j \neq k$$

`vec(M)`

    Domain: matrices

    Range: column vectors ($n \times 1$ matrices)

    Description: returns a column vector formed by listing the elements of M, starting with the first column and proceeding column by column.

`vecdiag(M)`

    Domain: $n \times n$ matrices

    Range: $1 \times n$ vectors

    Description: returns the row vector containing the diagonal of matrix M. `vecdiag()` is the opposite of `diag()`. The row name is set to `r1`; the column names are obtained from the column names of M.

Matrix functions returning a scalar

`colnumb(M,s)`

    Domain M: matrices

    Domain s: strings

    Range: integer scalars 1 to `matsize` and *missing*

    Description: returns the column number of M associated with column name s. returns *missing* if the column cannot be found.

`colsof(M)`

    Domain: matrices

    Range: integer scalars 1 to `matsize`

    Description: returns the number of columns of M.

`det(M)`

    Domain: $n \times n$ (square) matrices

    Range: scalars –8e+307 to 8e+307

    Description: returns the determinant of matrix M.

`diag0cnt(`M`)`
  Domain: $n \times n$ (square) matrices
  Range: integer scalars 0 to n
  Description: returns the number of zeros on the diagonal of M.

`el(`s,i,j`)`
  Domain s: strings containing matrix name
  Domain i: scalars 1 to `matsize`
  Domain j: scalars 1 to `matsize`
  Range: scalars $-8\mathrm{e}{+}307$ to $8\mathrm{e}{+}307$ and *missing*
  Description: returns s`[floor(`i`),floor(`j`)]`, the i,j element of the matrix named s.
               returns *missing* if i or j are out of range or if matrix s does not exist.

`issymetric(`M`)`
  Domain M: matrices
  Range: integers 0 and 1
  Description: returns 1 if the matrix is symmetric; otherwise, returns 0.

`matmissing(`M`)`
  Domain M: matrices
  Range: integers 0 and 1
  Description: returns 1 if any elements of the matrix are missing; otherwise, returns 0.

`mreldif(`X,Y`)`
  Domain X: matrices
  Domain Y: matrices with same number of rows and columns as X
  Range: scalars $-8\mathrm{e}{+}307$ to $8\mathrm{e}{+}307$
  Description: returns the relative difference of X and Y, where the relative difference is
               defined as $\max_{i,j}\big(|x_{ij} - y_{ij}|/(|y_{ij}| + 1)\big)$.

`rownumb(`M,s`)`
  Domain M: matrices
  Domain s: strings
  Range: integer scalars 1 to `matsize` and *missing*
  Description: returns the row number of M associated with row name s.
               returns *missing* if the row cannot be found.

`rowsof(`M`)`
  Domain: matrices
  Range: integer scalars 1 to `matsize`
  Description: returns the number of rows of M.

`trace(`M`)`
  Domain: $n \times n$ (square) matrices
  Range: scalars $-8\mathrm{e}{+}307$ to $8\mathrm{e}{+}307$
  Description: returns the trace of matrix M.

Acknowledgments

We thank George Marsaglia of Florida State University for providing his KISS (keep it simple stupid) random-number generator.

We thank John R. Gleason of Syracuse University for directing our attention to Wichura (1988) for calculating the cumulative normal density accurately, for sharing his experiences about techniques with us, and for providing C code to make the calculations.

Jacques Salomon Hadamard (1865–1963) was born in Versailles, France. He studied at the Ecole Normale Supérieure in Paris and obtained a doctorate in 1892 for a thesis on functions defined by Taylor series. Hadamard taught at Bordeaux for 4 years and in a productive period published an outstanding theorem on prime numbers, proved independently by Charles de la Vallée Poussin, and worked on what are now called Hadamard matrices. In 1897, he returned to Paris, where he held a series of prominent posts. In his later career, his interests extended from pure mathematics toward mathematical physics. Hadamard produced papers and books in many different areas. He campaigned actively against anti-Semitism at the time of the Dreyfus affair. After the fall of France in 1940, he spent some time in the United States and then Great Britain.

References

Abramowitz, M., and I. A. Stegun, ed. 1968. *Handbook of Mathematical Functions with Formulas, Graphs, and Mathematical Tables*, 7th printing. Washington, DC: National Bureau of Standards.

Cox, N. J. 2003. Stata tip 2: Building with floors and ceilings. *Stata Journal* 3: 446–447.

——. 2004. Stata tip 6: Inserting awkward characters in the plot. *Stata Journal* 4: 95–96.

Haynam, G. E., Z. Govindarajulu, and F. C. Leone. 1970. Tables of the cumulative noncentral chi-square distribution. In *Selected Tables in Mathematical Statistics*, vol. 1, ed. H. L. Harter and D. B. Owen, 1–78. Providence, RI: American Mathematical Society.

Johnson, N. L., S. Kotz, and N. Balakrishnan. 1995. *Continuous Univariate Distributions, Vol. 2*. 2nd ed. New York: Wiley.

Kantor, D., and N. J. Cox. 2005. Depending on conditions: A tutorial on the cond() function. *Stata Journal* 5: 413–420.

Marsaglia, G. 1994. Personal communication.

Mazýa, V. and T. Shaposhnikova. 1998. *Jacques Hadamard, A Universal Mathematician*. Providence, RI: American Mathematical Society.

Moore, R. J. 1982. Algorithm AS 187: Derivatives of the gamma integral. *Applied Statistics* 31: 330–335.

Posten, H. O. 1993. An effective algorithm for the noncentral beta distribution function. *American Statistician* 47: 129–131.

Press, W. H., S. A. Teukolsky, W. T. Vetterling, and B. P. Flannery. 1992. *Numerical Recipes in C: The Art of Scientific Computing*. 2nd ed. Cambridge: Cambridge University Press.

Wichura, M. J. 1988. The percentage points of the normal distribution. *Applied Statistics* 37: 454–477.

Also See

[D] **egen** — Extensions to generate

[M-5] **intro** — Mata functions

[U] **13.3 Functions**

[U] **14.8 Matrix functions**

Title

> **generate** — Create or change contents of variable

Syntax

Create new variable

> generate [*type*] *newvar*[*:lblname*] =*exp* [*if*] [*in*]

Replace contents of existing variable

> replace *oldvar* =*exp* [*if*] [*in*] [, nopromote]

Specify initial value of random-number seed

> set seed {#|*code*}

Specify default storage type assigned to new variables

> set type { float|double } [, permanently]

where *type* is one of byte | int | long | float | double | str | str1 | str2 | ... | str244.

See *Description* below for an explanation of str. For the other types, see [U] **12 Data**.

by is allowed with generate and replace; see [D] **by**.

Description

generate creates a new variable. The values of the variable are specified by =*exp*.

If no *type* is specified, the new variable type is determined by the type of result returned by =*exp*. A float variable (or a double, according to set type) is created if the result is numeric, and a string variable is created if the result is a string. In the latter case, a str# variable is created where # is the smallest string that will hold the result.

If a *type* is specified, the result returned by =*exp* must be a string or numeric according to whether *type* is string or numeric. If str is specified, a str# variable is created, where # is the smallest string that will hold the result.

See [D] **egen** for extensions to generate.

replace changes the contents of an existing variable. Because replace alters data, the command cannot be abbreviated.

set seed # specifies the initial value of the random-number seed used by the uniform() function; see [D] **functions**. # should be a positive integer. You may also specify *code* to indicate a previous internal state.

set seed *code* specifies a string that records the state of the random-number generator uniform(). You can obtain the current state of the random-number generator from c(seed).

set type specifies the default storage type assigned to new variables (such as those created by generate) when the storage type is not explicitly specified.

241

Options

nopromote prevents replace from promoting the variable type to accommodate the change. For instance, consider a variable stored as an integer type (byte, int, or long), and assume that you replace some values with nonintegers. By default, replace changes the variable type to a floating point (float or double) and thus correctly stores the changed values. Similarly, replace promotes byte and int variables to longer integers (int and long) if the replacement value is an integer but is too large in absolute value for the current storage type. replace promotes strings to longer strings. nopromote prevents replace from doing this; instead, the replacement values are truncated to fit into the current storage type.

permanently specifies that, in addition to making the change right now, the new limit be remembered and become the default setting when you invoke Stata.

Remarks

Remarks are presented under the following headings:

> *generate and replace*
> *set seed*
> *set type*

generate and replace

generate and replace are used to create new variables and to modify the contents of existing variables, respectively. Although the commands do the same thing, they have different names so that you do not accidentally replace values in your data. Detail descriptions of expressions are given in [U] **13 Functions and expressions**.

Also see [D] **edit**.

▷ Example 1

We have a dataset containing the variable age2, which we have previously defined as age^2 (i.e., age^2). We have changed some of the age data and now want to correct age2 to reflect the new values:

```
. use http://www.stata-press.com/data/r10/genxmpl1
(Wages of women)
. generate age2=age^2
age2 already defined
r(110);
```

When we attempt to re-generate age2, Stata refuses, telling us that age2 is already defined. We could drop age2 and then re-generate it, or we could use the replace command:

```
. replace age2=age^2
(204 real changes made)
```

When we use replace, we are informed of the number of actual changes made to the dataset.

◁

You can explicitly specify the storage type of the new variable being created by putting the *type*, such as byte, int, long, float, double, or str8, in front of the variable name. For example, you could type generate double revenue = qty * price. Not specifying a type is equivalent to specifying `float` if the variable is numeric, or, more correctly, it is equivalent to specifying the default type set by the `set type` command; see below. If the variable is alphanumeric, not specifying a type is equivalent to specifying `str#`, where # is the length of the largest string in the variable.

You may also specify a value label to be associated with the new variable by including "*:lblname*" after the variable name. This is seldom done because you can always associate the value label later by using the `label define` command; see [U] **12.6.3 Value labels**.

▷ Example 2

Among the variables in our dataset is `name`, which contains the first and last name of each person. We wish to create a new variable called `lastname`, which we will then use to sort the data. `name` is a string variable.

```
. use http://www.stata-press.com/data/r10/genxmpl2, clear
. list name
```

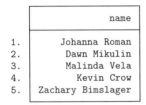

	name
1.	Johanna Roman
2.	Dawn Mikulin
3.	Malinda Vela
4.	Kevin Crow
5.	Zachary Bimslager

```
. generate lastname=word(name,2)
. describe
Contains data from http://www.stata-press.com/data/r10/genxmpl2.dta
  obs:            5
  vars:           2                          18 Jan 2007 12:24
  size:         150 (99.9% of memory free)
```

variable name	storage type	display format	value label	variable label
name	str17	%17s		
lastname	str9	%9s		

```
Sorted by:
     Note:  dataset has changed since last saved
```

Stata is smart. Even though we did not specify the storage type in our `generate` statement, Stata knew to create a `str9 lastname` variable, since the longest last name is `Bimslager`, which has nine characters.

◁

▷ Example 3

We wish to create a new variable `age2` that represents the variable `age` squared. We realize that since `age` is an integer, `age2` will also be an integer and will certainly be less than 32,740. We therefore decide to store `age2` as an `int` to conserve memory:

```
. use http://www.stata-press.com/data/r10/genxmpl3, clear
. generate int age2=age^2
(9 missing values generated)
```

Preceding age2 with int told Stata that the variable was to be stored as an int. After creating the new variable, Stata informed us that nine missing values were generated. generate informs us whenever it produces missing values.

◁

See [U] **13 Functions and expressions** and [U] **25 Dealing with categorical variables** for more information and examples. Also see [D] **recode** for a convenient way to recode categorical variables.

❑ Technical Note

If you specify the if modifier and/or in *range*, the =*exp* is evaluated only for those observations that meet the specified condition and/or are in the specified range. The other observations of the new variable are set to missing:

```
. use http://www.stata-press.com/data/r10/genxmpl3, clear
. generate int age2=age^2 if age>30
(290 missing values generated)
```

❑

▷ Example 4

replace can be used to change just one value, as well as to make sweeping changes to our data. For instance, say that we enter data on the first five odd and even positive integers and then discover that we made a mistake:

```
. use http://www.stata-press.com/data/r10/genxmpl4, clear
. list
```

	odd	even
1.	1	2
2.	3	4
3.	-8	6
4.	7	8
5.	9	10

The third observation is wrong; the value of odd should be 5, not −8. We can use replace to correct the mistake:

```
. replace odd=5 in 3
(1 real change made)
```

We could also have corrected the mistake by typing replace odd=5 if odd==-8.

◁

set seed

set seed {#|*code*} reinitializes the seed for the pseudorandom number generator uniform() and should be specified as a positive integer or as a *code*, indicating a previous internal state obtained from c(seed). (Concerning #s: if a negative number is specified, it will be made positive.) The seed determines the sequence of pseudorandom numbers produced by uniform(). When you start up Stata, it initializes the seed to 123456789, an easy-to-remember nonrandom number that in no way affects the pseudorandomness of the numbers produced by uniform(). Because Stata initializes the seed to a constant, Stata produces the same sequence of random numbers (measured from the start of the session) unless you reinitialize the seed. The seed you select affects only the sequence, not the pseudorandomness. Without loss of pseudorandomness, the seed may be set to small numbers; e.g., set seed 2.

See [D] **functions** for details and formulas.

❏ Technical Note

Suppose that you just invoked Stata and typed the following:

```
. display c(seed)
X075bcd151f123bb5159a55e50022865746ad
```

The c-class macro c(seed) contains the current value of set seed. The code X075bcd151f123bb5159a55e50022865746ad is equivalent to 123456789. Each time you generate a random number, the internal state changes.

```
. generate u = uniform()
. display c(seed)
X167c0bd9c43f462544a474abacbdd93d03c8
```

You can use the previous contents of c(seed) to return uniform() to a previous state.

```
. local curseed = c(seed)
. generate x = uniform()
.  etc.
. set seed `curseed'
```

❏

set type

When you create a new numeric variable and do not specify the storage type for it, say, by typing generate y=x+2, the new variable is made a float if you have not previously issued the set type command. If earlier in your session you typed set type double, the new numeric variable would be made a double.

Methods and Formulas

You can do anything with replace that you can do with generate. The only difference between the commands is that replace requires that the variable already exist, whereas generate requires that the variable be new. In fact, inside Stata, generate and replace have the same code. Since Stata is an interactive system, we force a distinction between replacing existing values and generating new ones so that you do not accidentally replace valuable data while thinking that you are creating a new piece of information.

References

Gleason, J. R. 1997. dm50: Defining variables and recording their definitions. *Stata Technical Bulletin* 40: 9–10. Reprinted in *Stata Technical Bulletin Reprints*, vol. 7, pp. 48–49.

——. 1999. dm50.1: Update to defv. *Stata Technical Bulletin* 51: 2. Reprinted in *Stata Technical Bulletin Reprints*, vol. 9, pp. 14–15.

Newson, R. 2004. dm0008: Stata tip 13: generate and replace use the current sort order. *Stata Journal* 4: 484–485.

Weesie, J. 1997. dm43: Automatic recording of definitions. *Stata Technical Bulletin* 35: 6–7. Reprinted in *Stata Technical Bulletin Reprints*, vol. 6, pp. 18–20.

Also See

Title

> **gsort** — Ascending and descending sort

Syntax

gsort [+|-] *varname* [[+|-] *varname* ...] [, generate(*newvar*) mfirst]

Description

gsort arranges observations to be in ascending or descending order of the specified variables and so differs from sort in that sort produces ascending-order arrangements only; see [D] **sort**.

Each *varname* can be numeric or a string.

The observations are placed in ascending order of *varname* if + or nothing is typed in front of the name and are placed and in descending order if − is typed.

Options

generate(*newvar*) creates *newvar* containing 1, 2, 3, ... for each group denoted by the ordered data. This is useful when using the ordering in a subsequent by operation; see [U] **11.5 by varlist: construct** and examples below.

mfirst specifies that missing values be placed first in descending orderings rather than last.

Remarks

gsort is almost a plug-compatible replacement for sort, except that you cannot specify a general *varlist* with gsort. For instance, sort alpha-gamma means to sort the data in ascending order of alpha, within equal values of alpha; sort on the next variable in the dataset (presumably beta), within equal values of alpha and beta; etc. gsort alpha-gamma would be interpreted as gsort alpha -gamma, meaning to sort the data in ascending order of alpha and, within equal values of alpha, in descending order of gamma.

▷ Example 1

The difference in *varlist* interpretation aside, gsort can be used in place of sort. To list the 10 lowest-priced cars in the data, we might type

```
. use http://www.stata-press.com/data/r10/auto
. gsort price
. list make price in 1/10
```

or, if we prefer,

```
. gsort +price
. list make price in 1/10
```

To list the 10 highest-priced cars in the data, we could type

```
. gsort -price
. list make price in 1/10
```

gsort can also be used with string variables. To list all the makes in reverse alphabetical order, we might type

```
. gsort -make
. list make
```

◁

▷ Example 2

gsort can be used with multiple variables. Given a dataset on hospital patients with multiple observations per patient, typing

```
. use http://www.stata-press.com/data/r10/bp3
. gsort id time
. list id time bp
```

lists each patient's blood pressures in the order the measurements were taken. If we typed

```
. gsort id -time
. list id time bp
```

each patient's blood pressures would be listed in reverse time order.

◁

❑ Technical Note

Say that we wished to attach to each patient's records the lowest and highest blood pressures observed during the hospital stay (let's ignore the easier way to achieve this result, egen's min() and max() functions):

```
. egen lo_bp = min(bp), by(id)
. egen hi_bp = max(bp), by(id)
```

(see [D] **egen**). Here is how we could do it with gsort:

```
. use http://www.stata-press.com/data/r10/bp3, clear
. gsort id bp
. by id: gen lo_bp = bp[1]
. gsort id -bp
. by id: gen hi_bp = bp[1]
. list, sepby(id)
```

This works, even in the presence of missing values of bp, because such missing values are placed last within arrangements, regardless of the direction of the sort.

❑

❑ Technical Note

Assume that we have a dataset containing x for which we wish to obtain the forward and reverse cumulatives. The forward cumulative is defined as $F(X) =$ the fraction of observations such that $x \leq X$. Again let's ignore the easier way to obtain the forward cumulative, which would be to use Stata's cumul command

```
. cumul x, gen(cum)
```

(see [R] **cumul**). Eschewing cumul, we could type

```
. sort x
. by x: gen cum = _N if _n==1
. replace cum = sum(cum)
. replace cum = cum/cum[_N]
```

That is, we first place the data in ascending order of x; we used sort but could have used gsort. Next, for each observed value of x, we generated cum containing the number of observations that take on that value (you can think of this as the discrete density). We summed the density, obtaining the distribution, and finally normalized it to sum to 1.

The reverse cumulative $G(X)$ is defined as the fraction of data such that $x \geq X$. To obtain this, we could try simply reversing the sort:

```
. gsort -x
. by x: gen rcum = _N if _n==1
. replace rcum = sum(rcum)
. replace rcum = rcum/rcum[_N]
```

This would work, except for one detail: Stata will complain that the data are not sorted in the second line. Stata complains because it does not understand descending sorts (gsort is an ado-file). To remedy this problem, gsort's generate() option will create a new grouping variable that is in ascending order (thus satisfying Stata's narrow definition) and that is, in terms of the groups it defines, identical to that of the true sort variables:

```
. gsort -x, gen(revx)
. by revx: gen rcum = _N if _n==1
. replace rcum = sum(rcum)
. replace rcum = rcum/rcum[_N]
```

❏

Methods and Formulas

gsort is implemented as an ado-file.

Also See

[D] **sort** — Sort data

Title

hexdump — Display hexadecimal report on file

Syntax

hexdump *filename* [, *options*]

options	description
analyze	display a report on the dump rather than the dump itself
tabulate	display a full tabulation of the ASCII characters in the analyze report
noextended	do not display printable extended ASCII characters
results	save results containing the frequency with which each character code was observed; programmer's option
from(#)	dump or analyze first byte of the file; default is to start at first byte, from(0)
to(#)	dump or analyze last byte of the file; default is to continue to the end of the file

Description

hexdump displays a hexadecimal dump of a file or, optionally, a report analyzing the dump.

Options

analyze specifies that a report on the dump, rather than the dump itself, be presented.

tabulate specifies in the analyze report that a full tabulation of the ASCII characters also be presented.

noextended specifies that hexdump not display printable extended ASCII characters, characters in the range 161–254, or, equivalently, 0xa1–0xfe. (hexdump does not display characters 128–160 and 255.)

results is for programmers. It specifies that in addition to other saved results, hexdump save r(c0), r(c1), ..., r(c255), containing the frequency with which each character code was observed.

from(#) specifies the first byte of the file to be dumped or analyzed. The default is to start at the first byte of the file, from(0).

to(#) specifies the last byte of the file to be dumped or analyzed. The default is to continue to the end of the file.

Remarks

hexdump is useful when you are having difficulty reading a file with infile, infix, or insheet. Sometimes, the reason for the difficulty is that the file does not contain what you think it contains, or that it does contain the format you have been told, and looking at the file in text mode is either not possible or not revealing enough.

Pretend that we have the file `myfile.raw` containing

```
Datsun 210      4589  35  5  1
VW Scirocco     6850  25  4  1
Merc. Bobcat    3829  22  4  0
Buick Regal     5189  20  3  0
VW Diesel       5397  41  5  1
Pont. Phoenix   4424  19  .  0
Merc. Zephyr    3291  20  3  0
Olds Starfire   4195  24  1  0
BMW 320i        9735  25  4  1
```

We will use `myfile.raw` with `hexdump` to produce output that looks like the following:

```
. hexdump myfile.raw
                                                               character
                           hex representation                representation
          address     0 1  2 3  4 5  6 7  8 9  a b  c d  e f  0123456789abcdef

                0  |  4461 7473 756e 2032 3130 2020 2020 2034  Datsun 210     4
               10  |  3538 3920 2033 3520 2035 2020 310a 5657  589  35  5  1.VW
               20  |  2053 6369 726f 6363 6f20 2020 2036 3835   Scirocco    685
               30  |  3020 2032 3520 2034 2020 310a 4d65 7263  0  25  4  1.Merc

               40  |  2e20 426f 6263 6174 2020 2033 3832 3920  . Bobcat    3829
               50  |  2032 3220 2034 2020 300a 4275 6963 6b20  22  4  0.Buick
               60  |  5265 6761 6c20 2020 2035 3138 3920 2032  Regal     5189 2
               70  |  3020 2033 2020 300a 5657 2044 6965 7365  0  3  0.VW Diese

               80  |  6c20 2020 2020 2035 3339 3720 2034 3120  l       5397 41
               90  |  2035 2020 310a 506f 6e74 2e20 5068 6f65   5  1.Pont. Phoe
               a0  |  6e69 7820 2034 3432 3420 2031 3920 202e  nix  4424  19   .
               b0  |  2020 300a 4d65 7263 2e20 5a65 7068 7972     0.Merc. Zephyr

               c0  |  2020 2033 3239 3120 2032 3020 2033 2020     3291  20  3
               d0  |  300a 4f6c 6473 2053 7461 7266 6972 6520  0.Olds Starfire
               e0  |  2034 3139 3520 2032 3420 2031 2020 300a   4195  24  1  0.
               f0  |  424d 5720 3332 3069 2020 2020 2020 2039  BMW 320i       9
              100  |  3733 3520 2032 3520 2034 2020 310a       735  25  4  1.
```

(*Continued on next page*)

hexdump can also produce output that looks like the following:

```
. hexdump myfile.raw, analyze
  Line-end characters                 Line length (tab=1)
     \r\n        (Windows)       0       minimum                    29
     \r by itself (Mac)          0       maximum                    29
     \n by itself (Unix)         9
  Space/separator characters          Number of lines               9
     [blank]                    99       EOL at EOF?               yes
     [tab]                       0
     [comma] (,)                 0    Length of first 5 lines
  Control characters                     Line 1                     29
     binary 0                    0       Line 2                     29
     CTL excl. \r, \n, \t        0       Line 3                     29
     DEL                         0       Line 4                     29
     Extended (128-159,255)      0       Line 5                     29
  ASCII printable
     A-Z                        20
     a-z                        61    File format               ASCII
     0-9                        77
     Special (!@#$ etc.)         4
     Extended (160-254)          0
                             _____
  Total                        270

  Observed were:
     \n blank . 0 1 2 3 4 5 6 7 8 9 B D M O P R S V W Z a b c d e f g h i k l
     n o p r s t u x y
```

Of the two forms of output, the second is often the more useful because it summarizes the file, and the length of the summary is not a function of the length of the file. Here is the summary for a file that is just over 4 MB long:

```
. hexdump bigfile.raw, analyze
  Line-end characters                 Line length (tab=1)
     \r\n        (Windows)   147,456    minimum                    29
     \r by itself (Mac)          0       maximum                    30
     \n by itself (Unix)         2
  Space/separator characters          Number of lines         147,458
     [blank]              1,622,039      EOL at EOF?               yes
     [tab]                       0
     [comma] (,)                 0    Length of first 5 lines
  Control characters                     Line 1                     30
     binary 0                    0       Line 2                     30
     CTL excl. \r, \n, \t        0       Line 3                     30
     DEL                         0       Line 4                     30
     Extended (128-159,255)      0       Line 5                     30
  ASCII printable
     A-Z                   327,684
     a-z                   999,436    File format               ASCII
     0-9                 1,261,587
     Special (!@#$ etc.)    65,536
     Extended (160-254)          0
                         _____
  Total                4,571,196

  Observed were:
     \n \r blank . 0 1 2 3 4 5 6 7 8 9 B D M O P R S V W Z a b c d e f g h i
     k l n o p r s t u x y
```

Here is the same file but with a subtle problem:

```
. hexdump badfile.raw, analyze
  Line-end characters                     Line length (tab=1)
    \r\n        (Windows)      147,456       minimum                  30
    \r by itself (Mac)               0       maximum                  90
    \n by itself (Unix)              0
  Space/separator characters              Number of lines         147,456
    [blank]                  1,622,016       EOL at EOF?                 yes
    [tab]                            0
    [comma] (,)                      0     Length of first 5 lines
  Control characters                        Line 1                   30
    binary 0                         8       Line 2                   30
    CTL excl. \r, \n, \t             4       Line 3                   30
    DEL                              0       Line 4                   30
    Extended (128-159,255)          24       Line 5                   30
  ASCII printable
    A-Z                        327,683
    a-z                        999,426     File format             BINARY
    0-9                      1,261,568
    Special (!@#$ etc.)         65,539
    Extended (160-254)              16
                            _____
  Total                     4,571,196
  Observed were:
    \0 ^C ^D ^G \n \r ^U blank & . 0 1 2 3 4 5 6 7 8 9 B D E M O P R S U V W
    Z a b c d e f g h i k l n o p r s t u v x y } ~ E^A E^C E^I E^M E^P
    ë é ö 255
```

In the above, the line length varies between 30 and 90 (we were told that each line would be 30 characters long). Also the file contains what hexdump, analyze labeled control characters. Finally, hexdump, analyze declared the file to be BINARY rather than ASCII.

We created the second file by removing two valid lines from bigfile.raw (60 characters) and substituting 60 characters of binary junk. We would defy you to find the problem without using hexdump, analyze. You would succeed, but only after much work. Remember, this file has 147,456 lines, and only two of them are bad. If you print 1,000 lines at random from the file, your chances of listing the bad part is only .013472. To have a 50% chance of finding the bad lines, you would have to list 52,000 lines, which is to say, review about 945 pages of output. On those 945 pages, each line would need to be drawn at random. More likely, you would list lines in groups, and that would greatly reduce your chances of encountering the bad lines.

The situation is not as dire as we make it out to be because, were you to read badfile.raw by using infile, it would complain, and, here it would tell you exactly where it was complaining. Still, at that point you might wonder whether the problem was with how you were using infile or with the data. Moreover, our 60 bytes of binary junk experiment corresponds to transmission error. If the problem were instead that the person who constructed the file constructed two of the lines differently, infile might not complain, but later you would notice some odd values in your data (because obviously you would review the summary statistics, right?). Here hexdump, analyze might be the only way you could prove to yourself and others that the raw data need to be reconstructed.

(Continued on next page)

❑ Technical Note

In the full hexadecimal dump,

```
. hexdump myfile.raw
```

			hex representation						character representation
address	0 1	2 3	4 5	6 7	8 9	a b	c d	e f	0123456789abcdef
0	4461	7473	756e	2032	3130	2020	2020	2034	Datsun 210 4
10	3538	3920	2033	3520	2035	2020	310d	0a56	589 35 5 1..V
20	5720	5363	6972	6f63	636f	2020	2020	3638	W Scirocco 68
30	3530	2020	3235	2020	3420	2031	0d0a	4d65	50 25 4 1..Me

(output omitted)

addresses (listed on the left) are listed in hexadecimal. Above, 10 means decimal 16, 20 means decimal 32, and so on. Sixteen characters are listed across each line.

In some other dump, you might see something like

```
. hexdump myfile2.raw
```

			hex representation						character representation
address	0 1	2 3	4 5	6 7	8 9	a b	c d	e f	0123456789abcdef
0	4461	7473	756e	2032	3130	2020	2020	2034	Datsun 210 4
10	3538	3920	2033	3520	2035	2020	3120	2020	589 35 5 1
20	2020	2020	2020	2020	2020	2020	2020	2020	
*									
160	2020	2020	2020	0a56	5720	5363	6972	6f63	.VW Sciroc
170	636f	2020	2020	3638	3530	2020	3235	2020	co 6850 25

(output omitted)

The * in the address field indicates that the previous line is repeated until we get to hexadecimal address 160 (decimal 352).

❑

Saved Results

hexdump, analyze and hexdump, results save the following in r():

Scalars

r(Windows)	number of \r\n
r(Mac)	number of \r by itself
r(Unix)	number of \n by itself
r(blank)	number of blanks
r(tab)	number of tab characters
r(comma)	number comma (,) characters
r(ctl)	number of binary 0s; A–Z, excluding \r, \n, \t; DELs; and 128–159, 255
r(uc)	number of A–Z
r(lc)	number of a–z
r(digit)	number of 0–9
r(special)	number of printable special characters (!@#, etc.)
r(extended)	number of printable extended characters (160–254)
r(filesize)	number of characters
r(lmin)	minimum line length
r(lmax)	maximum line length
r(lnum)	number of lines
r(eoleof)	1 if EOL at EOF, 0 otherwise
r(l1)	length of 1st line
r(l2)	length of 2nd line
r(l3)	length of 3rd line
r(l4)	length of 4th line
r(l5)	length of 5th line
r(c0)	number of binary 0s (results only)
r(c1)	number of binary 1s (^A) (results only)
r(c2)	number of binary 2s (^B) (results only)
...	...
r(c255)	number of binary 255s (results only)

Macros

r(format)	ASCII, EXTENDED ASCII, or BINARY

Also See

[D] **filefilter** — Convert ASCII text or binary patterns in a file

[D] **type** — Display contents of a file

Title

icd9 — ICD-9-CM diagnostic and procedure codes

Syntax

Verify that variable contains defined codes

{icd9 | icd9p} check *varname* [, any list generate(*newvar*)]

Verify and clean variable

{icd9 | icd9p} clean *varname* [, dots pad]

Generate new variable from existing variable

{icd9 | icd9p} generate *newvar* = *varname* , main

{icd9 | icd9p} generate *newvar* = *varname* , description [long end]

{icd9 | icd9p} generate *newvar* = *varname* , range(*icd9rangelist*)

Display code descriptions

{icd9 | icd9p} lookup *icd9rangelist*

Search for codes from descriptions

{icd9 | icd9p} search ["]*text*["] [["]*text*["] ...] [, or]

Display ICD-9 code source

{icd9 | icd9p} query

where *icd9rangelist* is

icd9code	(the particular code)
*icd9code**	(all codes starting with)
icd9code/*icd9code*	(the code range)

or any combination of the above, such as 001* 018/019 E* 018.02. *icd9codes* must be typed with leading zeros: 1 is an error; type 001 (diagnostic code) or 01 (procedure code).

icd9 is for use with ICD-9 *diagnostic* codes, and icd9p is for use with *procedure* codes. The two commands' syntaxes parallel each other.

Description

The following description from http://www.cdc.gov/nchs/about/otheract/icd9/abticd9.htm explains ICD-9:

> The International Classification of Diseases, Ninth Revision, Clinical Modification (ICD-9-CM) is based on the World Health Organization's Ninth Revision, International Classification of Diseases (ICD-9). ICD-9-CM is the official system of assigning codes to diagnoses and procedures associated with hospital utilization in the United States. The ICD-9 is used to code and classify mortality data from death certificates.

`icd9` and `icd9p` help when working with ICD-9-CM codes.

ICD-9 codes come in two forms: diagnostic codes and procedure codes. In this system, 001 (cholera) and 941.45 (deep 3rd deg burn nose) are examples of diagnostic codes, although some people write (and datasets record) 94145 rather than 941.45. Also, 01 (incise-excis brain/skull) and 55.01 (nephrotomy) are examples of procedure codes, although some people write 5501 rather than 55.01. `icd9` and `icd9p` understand both ways of recording codes.

Important note: what constitutes a valid icd9 code changes over time. For the rest of this entry, a *defined code* is any code that is either currently valid, was valid at some point since version V16 (effective October 1, 1998), or has meaning as a grouping of codes. Some examples would help. The diagnosis code 001, though not valid on its own, is useful because it denotes cholera. It is kept as a defined code whose description ends with an asterisk (*). The diagnosis code 645.01 was deleted between versions V16 and V18. It remains as a defined code, and its description ends with a hash mark (#).

`icd9` and `icd9p` parallel each other; `icd9` is for use with diagnostic codes, and `icd9p` is for use with procedure codes.

`icd9[p]` check verifies that existing variable *varname* contains defined ICD-9 codes. If not, `icd9[p]` check provides a full report on the problems. `icd9[p]` check is useful for tracking down problems when any of the other `icd9[p]` commands tell you that the "variable does not contain ICD-9 codes". `icd9[p]` check verifies that each recorded code actually exists in the defined code list.

`icd9[p]` clean also verifies that existing variable *varname* contains valid ICD-9 codes, and, if it does, `icd9[p]` clean modifies the variable to contain the codes in either of two standard formats. All `icd9[p]` commands work equally well with cleaned or uncleaned codes. There are many ways of writing the same ICD-9 code, and `icd9[p]` clean is designed (1) to ensure consistency and (2) to make subsequent output look better.

`icd9[p]` generate produces new variables based on existing variables containing (cleaned or uncleaned) ICD-9 codes. `icd9[p]` generate, main produces *newvar* containing the main code. `icd9[p]` generate, description produces *newvar* containing a textual description of the ICD-9 code. `icd9[p]` generate, range() produces numeric *newvar* containing 1 if *varname* records an ICD-9 code in the range listed and 0 otherwise.

`icd9[p]` lookup and `icd9[p]` search are utility routines that are useful interactively. `icd9[p]` lookup simply displays descriptions of the codes specified on the command line, so to find out what diagnostic E913.1 means, you can type `icd9 lookup e913.1`. The data that you have in memory are irrelevant—and remain unchanged—when you use `icd9[p]` lookup. `icd9[p]` search is similar to `icd9[p]` lookup, except that it turns the problem around; `icd9[p]` search looks for relevant ICD-9 codes from the description given on the command line. For instance, you could type `icd9 search liver` or `icd9p search liver` to obtain a list of codes containing the word "liver".

`icd9[p]` query displays the identity of the source from which the ICD-9 codes were obtained and the textual description that `icd9[p]` uses.

ICD-9 codes are commonly written in two ways, with and without periods. For instance, with diagnostic codes, you can write 001, 86221, E8008, and V822, or you can write 001., 862.21, E800.8, and V82.2. With procedure codes, you can write 01, 50, 502, and 5021, or 01., 50., 50.2, and 50.21. The icd9[p] command does not care which syntax you use or even whether you are consistent. Case also is irrelevant: v822, v82.2, V822, and V82.2 are all equivalent. Codes may be recorded with or without leading and trailing blanks.

icd9[p] works with V24, V22, V21, V19, V18, and V16 codes.

Options for icd9[p] check

any tells icd9[p] check to verify that the codes fit the format of ICD-9 codes but not to check whether the codes are actually defined. This makes icd9[p] check run faster. For instance, diagnostic code 230.52 (or 23052, if you prefer) looks valid, but there is no such ICD-9 code. Without the any option, 230.52 (or 23052) would be flagged as an error. With any, 230.52 (or 23052) is not an error.

list reports any invalid codes that were found in the data by icd9[p] check. For example, 1, 1.1.1, and perhaps 230.52 if any is not specified, are to be individually listed.

generate(*newvar*) specifies that icd9[p] check creates new variable *newvar* containing, for each observation, 0 if the code is defined and a number from 1 to 10 otherwise. The positive numbers indicate the kind of problem and correspond to the listing produced by icd9[p] check. For instance, 10 means that the code could be valid, but it turns out not to be on the list of defined codes.

Options for icd9[p] clean

dots specifies whether periods are to be included in the final format. Do you want the diagnostic codes recorded, for instance, as 86221 or 862.21? Without the dots option, the 86221 format would be used. With the dots option, the 862.21 format would be used.

pad specifies that the codes are to be padded with spaces, front and back, to make the codes line up vertically in listings. Specifying pad makes the resulting codes look better when used with most other Stata commands.

Options for icd9[p] generate

main, description, and range(*icd9rangelist*) specify what icd9[p] generate is to calculate. *varname* always specifies a variable containing ICD-9 codes.

main specifies that the main code be extracted from the ICD-9 code. For procedure codes, the main code is the first two characters. For diagnostic codes, the main code is usually the first three or four characters (the characters before the dot if the code has dots). In any case, icd9[p] generate does not care whether the code is padded with blanks in front or how strangely it might be written; icd9[p] generate will find the main code and extract it. The resulting variable is itself an ICD-9 code and may be used with the other icd9[p] subcommands. This includes icd9[p] generate, main.

description creates *newvar* containing descriptions of the ICD-9 codes.

long is for use with description. It specifies that the new variable, in addition to containing the text describing the code, contain the code, too. Without long, *newvar* in an observation might contain "bronchus injury-closed". With long, it would contain "862.21 bronchus injury-closed".

end modifies long (specifying end implies long) and places the code at the end of the string: "bronchus injury-closed 862.21".

range(*icd9rangelist*) allows you to create indicator variables equal to 1 when the ICD-9 code is in the inclusive range specified.

Option for icd9[p] search

or specifies that ICD-9 codes be searched for entries that contain any word specified after icd9[p] search. The default is to list only entries that contain all the words specified.

Remarks

Let's begin with the diagnostic codes that icd9 processes. The format of an ICD-9 diagnostic code is

$$\left[\text{blanks}\right]\left\{0\text{--}9,V,v\right\}\left\{0\text{--}9\right\}\left\{0\text{--}9\right\}\left[\,.\,\right]\left[0\text{--}9\left[0\text{--}9\right]\right]\left[\text{blanks}\right]$$

or

$$\left[\text{blanks}\right]\left\{E,e\right\}\left\{0\text{--}9\right\}\left\{0\text{--}9\right\}\left\{0\text{--}9\right\}\left[\,.\,\right]\left[0\text{--}9\right]\left[\text{blanks}\right]$$

icd9 can deal with ICD-9 diagnostic codes written in any of the ways that this format allows. Items in square brackets are optional. The code might start with some number of blanks. Braces, { }, indicate required items. The code then has a digit from 0 to 9, the letter V (uppercase or lowercase, first line), or the letter E (uppercase or lowercase, second line). After that, it has two or more digits, perhaps followed by a period, and then it may have up to two more digits (perhaps followed by more blanks).

All the following codes meet the above definition:

```
001
001.
    001
001.9
        0019
86222
862.22
E800.2
e8002
V82
v82.2
V822
```

Meeting the above definition does not make the code valid. There are 133,100 possible codes meeting the above definition, of which fewer than 20,000 are currently defined.

(Continued on next page)

Examples of currently defined diagnostic codes include

Code	Description
001	cholera*
001.0	cholera d/t vib cholerae
001.1	cholera d/t vib el tor
001.9	cholera nos
. . .	
999	complic medical care nec*
. . .	
V01	communicable dis contact*
V01.0	cholera contact
V01.1	tuberculosis contact
V01.2	poliomyelitis contact
V01.3	smallpox contact
V01.4	rubella contact
V01.5	rabies contact
V01.6	venereal dis contact
V01.7	viral dis contact nec
V01.8	communic dis contact nec
V01.9	communic dis contact nos
. . .	
E800	rr collision nos*
E800.0	rr collision nos-employ
E800.1	rr coll nos-passenger
E800.2	rr coll nos-pedestrian
E800.3	rr coll nos-ped cyclist
E800.8	rr coll nos-person nec
E800.9	rr coll nos-person nos
. . .	

The *main code* refers to the part of the code to the left of the period. 001, 002, . . . , 999; v01, . . . , V82; and E800, . . . , E999 are main codes.

The main code corresponding to a detailed code can be obtained by taking the part of the code to the left of the period, except for codes beginning with 176, 764, 765, v29, and v69. Those main codes are not defined, yet there are more detailed codes under them:

Code	Description
176	CODE DOES NOT EXIST, but 8 codes starting with 176 do exist:
176.0	skin - kaposi's sarcoma
176.1	sft tisue - kpsi's srcma
. . .	
764	CODE DOES NOT EXIST, but 44 codes starting with 764 do exist:
764.0	lt-for-dates w/o fet mal*
764.00	light-for-dates wtnos
. . .	
765	CODE DOES NOT EXIST, but 22 codes starting with 765 do exist:
765.0	extreme immaturity*
765.00	extreme immatur wtnos
. . .	
V29	CODE DOES NOT EXIST, but 6 codes starting with V29 do exist:
V29.0	nb obsrv suspct infect
V29.1	nb obsrv suspct neurlgcl
. . .	
V69	CODE DOES NOT EXIST, but 6 codes starting with V69 do exist:
V69.0	lack of physical exercise
V69.1	inapprt diet eat habits
. . .	

Our solution is to define four new codes:

Code	Description
176	kaposi's sarcoma (Stata)*
764	light-for-dates (Stata)*
765	immat & preterm (Stata)*
V29	nb suspct cnd (Stata)*
V69	lifestyle (Stata)*

Things are less confusing with respect to the procedure codes processed by icd9p. The format of ICD-9 procedure codes is

$$\left[\,\text{blanks}\,\right]\left\{0\text{--}9\right\}\left\{0\text{--}9\right\}\left[\,.\,\right]\left[\,0\text{--}9\left[\,0\text{--}9\,\right]\,\right]\left[\,\text{blanks}\,\right]$$

Thus there are 10,000 possible procedure codes, of which fewer than 5,000 are currently valid. The first two digits represent the main code, of which 100 are feasible and 98 are currently used (00 and 17 are not used).

Descriptions

The description given for each of the codes is as found in the original source. The procedure codes contain the addition of five new codes created by Stata. An asterisk on the end of a description indicates that the corresponding ICD-9 diagnostic code has subcategories. A hash mark (#) at the end of a description denotes a code that is not valid in the most current version but that was valid at some time between version V16 and the present version.

icd9[p] query reports the original source of the information on the codes:

```
. icd9 query
_dta:
  1.  ICD9 Diagnostic Code Mapping Data for use with Stata, History
  2.  ──── V16 ─────────────────────────────────────────────
  3.  Dataset obtained 24aug1999 from http://www.hcfa.gov/stats/pufiles.htm,
      file http://www.hcfa.gov/stats/icd9v16.exe
  4.  Codes 176, 764, 765, V29, and V69 defined by StataCorp: 176 [kaposi's
      sarcoma (Stata)*], 765 [immat & preterm (Stata)*], 764 [light-for-dates
      (Stata)*], V29 [nb suspct cnd (Stata)*], V69 [lifestyle (Stata)*]
  5.  ──── V18 ─────────────────────────────────────────────
(output omitted)
 12.  ──── V19 ─────────────────────────────────────────────
 13.  Dataset obtained 3jan2002 from http://www.hcfa.gov/stats/pufiles.htm,
      file http://www.hcfa.gov/stats/icd9v19.zip, file 9v19diag.txt
 14.  27feb2002: V19 put into Stata distribution
(output omitted)
. icd9p query
_dta:
  1.  ICD9 Procedure Code Mapping Data for use with Stata, History
  2.  ──── V16 ─────────────────────────────────────────────
  3.  Dataset obtained 24aug1999 from http://www.hcfa.gov/stats/pufiles.htm,
      file http://www.hcfa.gov/stats/icd9v16.exe
  4.  ──── V18 ─────────────────────────────────────────────
  5.  Dataset obtained 10may2001 from http://www.hcfa.gov/stats/pufiles.htm,
      file http://www.hcfa.gov/stats/icd9v18.zip, file V18SURG.TXT
  6.  11jun2001: V18 data put into Stata distribution
  7.  BETWEEN V16 and V18: 9 codes added: 3971 3979 4107 4108 4109 4697 6096
      6097 9975
(output omitted)
```

▷ Example 1

We have a dataset containing up to three diagnostic codes and up to two procedures on a sample of 1,000 patients:

```
. use http://www.stata-press.com/data/r10/patients
. list in 1/10
```

	patid	diag1	diag2	diag3	proc1	proc2
1.	1	65450			9383	
2.	2	23v.6	37456		8383	17
3.	3	V10.02				
4.	4	102.6			629	
5.	5	861.01				
6.	6	38601	2969		9337	
7.	7	705			7309	8385
8.	8	v53.32			7878	951
9.	9	20200	7548	E8247	0479	
10.	10	464.11	20197		4641	

Do not try to make sense of these data because, in constructing this example, the diagnostic and procedure codes were randomly selected.

First, variable `diag1` is recorded sloppily—sometimes the dot notation is used and sometimes not, and sometimes there are leading blanks. That does not matter. We decide to begin by using `icd9 clean` to clean up this variable:

```
. icd9 clean diag1
diag1 contains invalid ICD-9 codes
r(459);
```

`icd9 clean` refused because there are invalid codes among the 1,000 observations. We can use `icd9 check` to find and flag the problem observations (or observation, as here):

```
. icd9 check diag1, gen(prob)
diag1 contains invalid codes:
      1.  Invalid placement of period                     0
      2.  Too many periods                                0
      3.  Code too short                                  0
      4.  Code too long                                   0
      5.  Invalid 1st char (not 0-9, E, or V)             0
      6.  Invalid 2nd char (not 0-9)                      0
      7.  Invalid 3rd char (not 0-9)                      1
      8.  Invalid 4th char (not 0-9)                      0
      9.  Invalid 5th char (not 0-9)                      0
     10.  Code not defined                                0
                                                      _____
          Total                                          1
```

```
. list patid diag1 prob if prob
```

	patid	diag1	prob
2.	2	23v.6	7

Let's assume that we go back to the patient records and determine that this should have been coded 230.6:

```
. replace diag1 = "230.6" if patid==2
(1 real change made)
. drop prob
```

We now try again to clean up the formatting of the variable:

```
. icd9 clean diag1
(643 changes made)
. list in 1/10
```

	patid	diag1	diag2	diag3	proc1	proc2
1.	1	65450			9383	
2.	2	2306	37456		8383	17
3.	3	V1002				
4.	4	1026			629	
5.	5	86101				
6.	6	38601	2969		9337	
7.	7	705			7309	8385
8.	8	V5332			7878	951
9.	9	20200	7548	E8247	0479	
10.	10	46411	20197		4641	

Perhaps we prefer the dot notation. `icd9 clean` can be used again on `diag1`, and now we will clean up `diag2` and `diag3`:

```
. icd9 clean diag1, dots
(936 changes made)

. icd9 clean diag2, dots
(551 changes made)

. icd9 clean diag3, dots
(100 changes made)

. list in 1/10
```

	patid	diag1	diag2	diag3	proc1	proc2
1.	1	654.50			9383	
2.	2	230.6	374.56		8383	17
3.	3	V10.02				
4.	4	102.6			629	
5.	5	861.01				
6.	6	386.01	296.9		9337	
7.	7	705			7309	8385
8.	8	V53.32			7878	951
9.	9	202.00	754.8	E824.7	0479	
10.	10	464.11	201.97		4641	

We now turn to cleaning the procedure codes. We use `icd9p` (emphasis on the *p*) to clean these codes:

```
. icd9p clean proc1, dots
(816 changes made)
. icd9p clean proc2, dots
(140 changes made)
. list in 1/10
```

	patid	diag1	diag2	diag3	proc1	proc2
1.	1	654.50			93.83	
2.	2	230.6	374.56		83.83	17
3.	3	V10.02				
4.	4	102.6			62.9	
5.	5	861.01				
6.	6	386.01	296.9		93.37	
7.	7	705			73.09	83.85
8.	8	V53.32			78.78	95.1
9.	9	202.00	754.8	E824.7	04.79	
10.	10	464.11	201.97		46.41	

Both `icd9 clean` and `icd9p clean` verify only that the variable being cleaned follows the construction rules for the code; it does not check that the code is itself valid. `icd9[p] check` does that:

```
. icd9p check proc1
(proc1 contains valid ICD-9 procedure codes; 168 missing values)
. icd9p check proc2

proc2 contains invalid codes:
     1.   Invalid placement of period            0
     2.   Too many periods                       0
     3.   Code too short                         0
     4.   Code too long                          0
     5.   Invalid 1st char (not 0-9)             0
     6.   Invalid 2nd char (not 0-9)             0
     7.   Invalid 3rd char (not 0-9)             0
     8.   Invalid 4th char (not 0-9)             0
    10.   Code not defined                       1
                                            _____
          Total                                  1
```

proc2 has an invalid code. We could find it by using `icd9p check, generate()`, just as we did above with `icd9 check, generate()`.

`icd9[p]` can create new variables containing textual descriptions of our diagnostic and procedure codes:

```
. icd9 generate td1 = diag1, description
. sort patid
```

```
. list patid diag1 td1 in 1/10
```

	patid	diag1	td1
1.	1	654.50	cerv incompet preg-unsp
2.	2	230.6	ca in situ anus nos
3.	3	V10.02	hx-oral/pharynx malg nec
4.	4	102.6	yaws of bone & joint
5.	5	861.01	heart contusion-closed
6.	6	386.01	meniere dis cochlvestib
7.	7	705	disorders of sweat gland*
8.	8	V53.32	ftng autmtc dfibrillator
9.	9	202.00	ndlr lym unsp xtrndl org
10.	10	464.11	ac tracheitis w obstruct

icd9[p] generate, description does not preserve the sort order of the data (and neither does icd9[p] check, unless you specify the any option).

Procedure code proc2 had an invalid code. Even so, icd9p generate, description is willing to create a textual description variable:

```
. icd9p gen tp2 = proc2, description
(1 nonmissing value invalid and so could not be labeled)
. sort patid
. list patid proc2 tp2 in 1/10
```

	patid	proc2	tp2
1.	1		
2.	2	17	
3.	3		
4.	4		
5.	5		
6.	6		
7.	7	83.85	musc/tend lng change nec
8.	8	95.1	form & structur eye exam*
9.	9		
10.	10		

tp2 contains nothing when proc2 is 17 because 17 is not a valid procedure code.

icd9[p] generate can also create variables containing main codes:

(Continued on next page)

```
. icd9 generate main1 = diag1, main
. list patid diag1 main1 in 1/10
```

	patid	diag1	main1
1.	1	654.50	654
2.	2	230.6	230
3.	3	V10.02	V10
4.	4	102.6	102
5.	5	861.01	861
6.	6	386.01	386
7.	7	705	705
8.	8	V53.32	V53
9.	9	202.00	202
10.	10	464.11	464

`icd9p generate, main` can similarly generate main procedure codes.

Sometimes we might merely be examining an observation:

```
. list diag* if patid==563
```

	diag1	diag2	diag3
563.	526.4		

If we wondered what 526.4 was, we could type

```
. icd9 lookup 526.4
1 match found:
     526.4     inflammation of jaw
```

`icd9[p] lookup` can list ranges of codes:

```
. icd9 lookup 526/526.99
15 matches found:
     526       jaw diseases*
     526.0     devel odontogenic cysts
     526.1     fissural cysts of jaw
     526.2     cysts of jaws nec
     526.3     cent giant cell granulom
     526.4     inflammation of jaw
     526.5     alveolitis of jaw
     526.61    perfor root canal space
     526.62    endodontic overfill
     526.63    endodontic underfill
     526.69    periradicular path nec
     526.8     other jaw diseases*
     526.81    exostosis of jaw
     526.89    jaw disease nec
     526.9     jaw disease nos
```

The same result could be found by typing

```
. icd9 lookup 526*
```

`icd9[p] search` can find a code from the description:

```
. icd9 search jaw disease
4 matches found:
    526       jaw diseases*
    526.8     other jaw diseases*
    526.89    jaw disease nec
    526.9     jaw disease nos
```

◁

Saved Results

`icd9 check` and `icd9p check` save the following in `r()`:

Scalars
    `r(e#)`   number of errors of type #
    `r(esum)` total number of errors

`icd9 clean` and `icd9p clean` save the following in `r()`:

Scalars
    `r(N)` number of changes

Methods and Formulas

`icd9` and `icd9p` are implemented as ado-files.

Reference

Gould, W. 2000. dm76: ICD-9 diagnostic and procedure codes. *Stata Technical Bulletin* 54: 8–16. Reprinted in *Stata Technical Bulletin Reprints*, vol. 9, pp. 77–87.

Title

> **impute** — Fill in missing values

Syntax

impute *depvar indepvars* [*if*] [*in*] [*weight*] , <u>g</u>enerate(*newvar$_1$*) [*options*]

options	description
Main	
*<u>g</u>enerate(*newvar$_1$*)	generate *newvar$_1$* to contain the imputations
<u>nomiss</u>ings(*varlist*)	include *varlist* without missing values in the best-subset regressions
all	estimate using all observations (regardless of *if* and *in*)
<u>reg</u>sample(*exp*)	estimate using observations specified in *exp*
<u>copyrest</u>	copy out-of-sample values of dependent variable to generated variable
<u>var</u>p(*newvar$_2$*)	create new variable to contain the variance of the prediction

* generate(*newvar$_1$*) is required.

indepvars and *varlist* may contain time-series operators; see [U] **11.4.3 Time-series varlists**.

aweights and fweights are allowed; see [U] **11.1.6 weight**.

Description

impute fills in missing values; *depvar* is the variable whose missing values are to be imputed. *indepvars* is the list of variables on which the imputations are to be based, and *newvar$_1$* is the new variable that contains the imputations.

impute organizes the cases by patterns of missing data so that the missing-value regressions can be conducted efficiently; this necessitates a limit of 31 variables in *indepvars*.

if *exp* and in *range* restrict the sample in which missings are imputed and, unless regsample() or all is specified, also the sample used in the regressions.

Options

> Main

generate(*newvar$_1$*) specifies the name of the new variable to be created. generate() is required.

nomissings(*varlist*) specifies the variables to include in the best-subset regressions. This option requires that the specified variables be free of missing values within the sample of observations used in the regressions.

all specifies that all observations be used in the regression sample. Thus all is equivalent to regsample(_n<=_N) or regsample(1).

regsample(*exp*) specifies the sample used to fit regressions. Don't confuse the if and in clauses with the regsample() option. If regsample() is not specified, the regression sample defaults to all observations if if and in are not specified or to all selected observations otherwise.

copyrest specifies that out-of-sample values of *depvar* be copied to the generate()-d variable. The default is to set out-of-sample values to missing (.).

varp(*newvar₂*) specifies the name of a new variable to contain the variance (*not* the standard error) of the prediction.

Remarks

In observations in which *depvar* is not equal to missing, *newvar₁* is set equal to *depvar*, and *newvar₂* (if specified) is set to zero. Where *depvar* is missing, *newvar₁* is imputed using the prediction from the best available subset of present data. *newvar₂* (if specified) is set to the variance of the prediction. This variance is in the sense of `predict`'s `stdp` option, although it is squared; see [R] **predict**. It is an estimate of how far the prediction of the mean would differ from the actual data point were it known.

This is not the only method for coping with missing data, but it is often much better than deleting cases with any missing data, which is the default. For a discussion of different methods of imputation, see Little and Rubin (2002).

▷ Example 1

`impute` may be used in conjunction with, for instance, regression (or any estimation technique) to avoid the loss of an unacceptable number of cases because of missing data. However, the later estimates may be biased because any variable imputed by `impute` is only an estimate of the unknown, true value. For linear regression, a reasonable bound (in fractional terms) for the bias is given by the ratio of the mean of *newvar₂* to the variance of *newvar₁*. Usually, the bias is toward zero, meaning that the effect of the variable will be underestimated.

Say that we have been hired by the restaurant industry to study expenditures on eating and drinking. We have data on 898 U.S. cities:

```
. use http://www.stata-press.com/data/r10/emd
(1980 City Data)

. describe

Contains data from http://www.stata-press.com/data/r10/emd.dta
  obs:           898                          1980 City Data
  vars:            9                          10 Jan 2007 20:25
  size:        34,124 (96.6% of memory free)
```

variable name	storage type	display format	value label	variable label
fips	long	%10.0g		state/place code
ln_eat	float	%9.0g		ln(Dining sales per capita)
income_pc	int	%8.0g		Per capita money income
ln_rsales_pc	float	%9.0g		ln(retail sales per capita)
jantemp	float	%9.0g		Median January temperature (Fahrenheit)
precipitation	float	%9.0g		Annual precipitation (in.)
ln_income	float	%9.0g		ln(median per capita income)
median_age	float	%9.0g		Median age
hhsize	float	%9.0g		Persons per household

```
Sorted by:
```

We begin by running the regression:

```
. regress ln_eat ln_rsales jantemp precip ln_inc median_age hhsize
```

Source	SS	df	MS		Number of obs =	664
					F(6, 657) =	212.55
Model	87.7285014	6	14.6214169		Prob > F =	0.0000
Residual	45.1948678	657	.068789753		R-squared =	0.6600
					Adj R-squared =	0.6569
Total	132.923369	663	.200487736		Root MSE =	.26228

ln_eat	Coef.	Std. Err.	t	P>|t|	[95% Conf.	Interval]
ln_rsales_pc	.6611241	.026623	24.83	0.000	.6088476	.7134006
jantemp	.0019624	.0007601	2.58	0.010	.0004698	.003455
precipitat~n	-.0014311	.0008433	-1.70	0.090	-.0030869	.0002247
ln_income	.1158486	.056352	2.06	0.040	.0051969	.2265003
median_age	-.0010863	.0002823	-3.85	0.000	-.0016407	-.0005319
hhsize	-.0050407	.0004243	-11.88	0.000	-.0058739	-.0042076
_cons	-1.377592	.4777641	-2.88	0.004	-2.31572	-.439463

Despite having data on 898 cities, our regression was performed on only 664 cities—74% of the original 898. Some 234 observations were unused because of missing data. Here when we type summarize, we discover that each of the independent variables has missing values, so the problem is not that one variable is missing in 26% of the observations but that each variable is missing in some observations. In fact, summarize revealed that each variable is missing in roughly 5% of the observations. We lost 26% of our data because, in aggregate, 26% of the observations have one or more missing variables.

Thus we impute each independent variable on the basis of the other independent variables:

```
. impute ln_rsales jantemp precip ln_inc median_age hhsize, gen(i_ln_rsales)
4.90% (44) observations imputed
. impute jantemp ln_rsales precip ln_inc median_age hhsize, gen(i_jantemp)
5.90% (53) observations imputed
. impute precip ln_rsales jantemp ln_inc median_age hhsize, gen(i_precip)
4.57% (41) observations imputed
. impute ln_inc ln_rsales jantemp precip median_age hhsize, gen(i_ln_inc)
4.34% (39) observations imputed
. impute median_age ln_rsales jantemp precip ln_inc hhsize, gen(i_median_age)
4.45% (40) observations imputed
. impute hhsize ln_rsales jantemp precip ln_inc median_age, gen(i_hhsize)
5.23% (47) observations imputed
```

That done, we can now refit the model on the imputed variables:

```
. regress ln_eat i_ln_rsales i_jantemp i_precip i_ln_inc i_median_age i_hhsize
```

Source	SS	df	MS		Number of obs =	898
					F(6, 891) =	253.41
Model	108.85923	6	18.1432051		Prob > F =	0.0000
Residual	63.7929145	891	.071596986		R-squared =	0.6305
					Adj R-squared =	0.6280
Total	172.652145	897	.192477308		Root MSE =	.26758

ln_eat	Coef.	Std. Err.	t	P>|t|	[95% Conf. Interval]	
i_ln_rsales	.660906	.0245827	26.89	0.000	.6126593	.7091528
i_jantemp	.0021019	.0006932	3.03	0.002	.0007414	.0034625
i_precip	-.0013268	.0007646	-1.74	0.083	-.0028275	.0001739
i_ln_inc	.095863	.0510231	1.88	0.061	-.0042764	.1960024
i_median_age	-.0011234	.0002584	-4.35	0.000	-.0016304	-.0006163
i_hhsize	-.0052508	.0003953	-13.28	0.000	-.0060267	-.004475
_cons	-1.143142	.4304284	-2.66	0.008	-1.987914	-.2983702

The regression model is now fitted on all 898 observations.

◁

❑ Technical Note

impute runs regressions by best-subset regression, looking at the pattern of missing values in the predictors. impute may have to run a regression for each combination of the predictor variables, depending upon the pattern of missingness.

To count all best-subset combinations, impute looks at the 0s and 1s in the binary representation of a long integer. In Stata, a long integer contains 32 bits—one of which is used for the sign. Thus the remaining bits are used to identify whether to include a predictor variable in a given regression. Increasing this limit beyond 31 will not have the desired result (even though the modified impute will not exit with an error).

To illustrate how impute determines which variables to include in a regression, suppose that there are three predictors and that the pattern of missing values among them requires a regression for each combination. In this worst-case scenario, there are $2^3 = 8$ regressions to run. You can determine which predictor to include in a regression by looking at the binary representation of the regression index (starting from 0):

integer (base 10)	integer (binary)
0	000
1	001
2	010
3	011
4	100
5	101
6	110
7	111

If the names of the predictor variables are x1, x2, and x3, you can interpret the binary number like this:

x3	x2	x1
<digit>	<digit>	<digit>

Thus 001 means include x1, 011 means include x1 and x2, ...

Given this implementation, there has to be a limit on the number of predictors allowed by the `impute` command before the generated `long integer` variable is automatically `recast` as a `float` or `double`, thus breaking the implementation.

If you increase the limit, all variables beyond the first 31 (possibly fewer) will not be used in any of the regressions.

One way to get around this limit would be to add `impute`'s `nomissings`(*varlist*) option. These variables will be assumed to have no missing values so that they could be present in all regressions.

❏

Methods and Formulas

`impute` is implemented as an ado-file.

Consider the command `impute` y x_1 x_2 ... x_k, `gen`($\widehat{y}$) `varp`($\widehat{v}$).

When y is not missing, $\widehat{y} = y$ and $\widehat{v} = 0$.

Let y_j be an observation for which y is missing. A regressor list is formed from x_1, x_2, ..., x_k containing all x's for which x_{ij} is not missing. If the resulting list is empty, $\widehat{y}_j$ and $\widehat{v}_j$ are set to missing. Otherwise, a regression of y on the list is fitted (see [R] **regress**), and $\widehat{y}_j$ is defined as the predicted value of y_j (see [R] **predict**). $\widehat{v}_j$ is defined as the square of the standard error of the prediction, as calculated by `predict, stdp`; see [R] **predict**.

References

Goldstein, R. 1996a. sed10: Patterns of missing data. *Stata Technical Bulletin* 32: 12–13. Reprinted in *Stata Technical Bulletin Reprints*, vol. 6, p. 115.

———. 1996b. sed10.1: Patterns of missing data, update. *Stata Technical Bulletin* 33: 2. Reprinted in *Stata Technical Bulletin Reprints*, vol. 6, pp. 115–116.

Little, R. J. A., and D. B. Rubin. 2002. *Statistical Analysis with Missing Data*. 2nd ed. New York: Wiley.

Mander, A., and D. Clayton. 1999. sg116: Hotdeck imputation. *Stata Technical Bulletin* 51: 32–34. Reprinted in *Stata Technical Bulletin Reprints*, vol. 9, pp. 196–199.

———. 2000. sg116.1: Update to hotdeck imputation. *Stata Technical Bulletin* 54: 26. Reprinted in *Stata Technical Bulletin Reprints*, vol. 9, p. 199.

Royston, P. 2004. st0067: Multiple imputation of missing values. *Stata Journal* 4: 227–241.

———. 2005a. st0067_1: Multiple imputation of missing values: update. *Stata Journal* 5: 188–201.

———. 2005b. st0067_2: Multiple imputation of missing values: Update of ice. *Stata Journal* 5: 527–536.

Also See

[R] **predict** — Obtain predictions, residuals, etc., after estimation

[D] **ipolate** — Linearly interpolate (extrapolate) values

[R] **regress** — Linear regression

Title

> **infile** — Overview of reading data into Stata

Description

This entry provides a quick reference for determining which method to use for reading non-Stata data into memory. See [U] **21 Inputting data** for more details.

Remarks

Remarks are presented under the following headings:

> *Summary of the different methods*
> > *insheet*
> > *infile (free format)—infile without a dictionary*
> > *infix (fixed format)*
> > *infile (fixed format)—infile with a dictionary*
> > *fdause*
> > *haver (Windows only)*
> > *odbc*
> > *xmluse*
> *Examples*

Summary of the different methods

insheet

o `insheet` reads text (ASCII) files created by a spreadsheet or a database program.

o The data must be tab separated or comma separated, but not both simultaneously, and cannot be space separated.

o An observation must be on only one line.

o The first line in the file can optionally contain the names of the variables.

infile (free format)—infile without a dictionary

o The data can be space separated, tab separated, or comma separated.

o Strings with embedded spaces or commas must be enclosed in quotes (even if tab- or comma separated).

o An observation can be on more than one line, or there can even be multiple observations per line.

infix (fixed format)

o The data must be in fixed-column format.

o An observation can be on more than one line.

o `infix` has simpler syntax than `infile` (fixed format).

infile (fixed format)—infile with a dictionary

○ The data may be in fixed-column format.

○ An observation can be on more than one line.

○ `infile` (fixed format) has the most capabilities for reading data.

fdause

○ `fdause` reads SAS XPORT Transport format files—the file format required by the U.S. Food and Drug Administration (FDA).

○ `fdause` will also read value label information from a `formats.xpf` XPORT file, if available.

haver (Windows only)

○ `haver` reads Haver Analytics (http://www.haver.com/) database files.

○ `haver` is available only for Windows and requires a corresponding DLL (`DLXAPI32.DLL`) available from Haver Analytics.

odbc

○ ODBC, an acronym for Open DataBase Connectivity, is a standard for exchanging data between programs. Stata supports the ODBC standard for importing data via the `odbc` command and can read from any ODBC data source on your computer. See [D] **odbc**.

xmluse

○ `xmluse` reads extensible markup language (XML) files—highly adaptable text-format files derived from ground station markup language (GSML).

○ `xmluse` can read either an Excel-format XML or a Stata-format XML file into Stata.

Examples

▷ Example 1: Tab-separated data

```
                                                         top of examp1.raw
    1        0        1       John Smith      m
    0        0        1       Paul Lin        m
    0        1        0       Jan Doe f
    0        0        .       Julie McDonald  f
                                                         end of examp1.raw
```

contains tab-separated data. The `type` command with the `showtabs` option shows the tabs:

```
. type examp1.raw, showtabs

1<T>0<T>1<T>John Smith<T>m
0<T>0<T>1<T>Paul Lin<T>m
0<T>1<T>0<T>Jan Doe<T>f
0<T>0<T>.<T>Julie McDonald<T>f
```

It could be read in by

```
. insheet a b c name gender using examp1
```

◁

▷ Example 2: Comma-separated data

──────────────────────────────────── top of examp2.raw ───────────

```
a,b,c,name,gender
1,0,1,John Smith,m
0,0,1,Paul Lin,m
0,1,0,Jan Doe,f
0,0,,Julie McDonald,f
```
──────────────────────────────────── end of examp2.raw ───────────

could be read in by

```
. insheet using examp2
```

◁

▷ Example 3: Tab-separated data with double-quoted strings

──────────────────────────────────── top of examp3.raw ───────────

```
1       0       1       "John Smith"      m
0       0       1       "Paul Lin"        m
0       1       0       "Jan Doe"         f
0       0       .       "Julie McDonald"        f
```
──────────────────────────────────── end of examp3.raw ───────────

contains tab-separated data with strings in double quotes.

```
. type examp3.raw, showtabs
1<T>0<T>1<T>"John Smith"<T>m
0<T>0<T>1<T>"Paul Lin"<T>m
0<T>1<T>0<T>"Jan Doe"<T>f
0<T>0<T>.<T>"Julie McDonald"<T>f
```

It could be read in by

```
. infile byte (a b c) str15 name str1 gender using examp3
```

or

```
. insheet a b c name gender using examp3
```

or

```
. infile using dict3
```

where the dictionary `dict3.dct` contains

──────────────────────────────────── top of dict3.dct ───────────

```
infile dictionary using examp3 {
        byte    a
        byte    b
        byte    c
        str15   name
        str1    gender
}
```
──────────────────────────────────── end of dict3.dct ───────────

◁

▷ Example 4: Space-separated data with double-quoted strings

```
──────────────────────────────────────── top of examp4.raw ────────────
  1 0 1 "John Smith" m
  0 0 1 "Paul Lin" m
  0 1 0 "Jan Doe" f
  0 0 . "Julie McDonald" f
──────────────────────────────────────── end of examp4.raw ────────────
```

could be read in by

```
. infile byte (a b c) str15 name str1 gender using examp4
```

or

```
. infile using dict4
```

where the dictionary dict4.dct contains

```
──────────────────────────────────────── top of dict4.dct ────────────
infile dictionary using examp4 {
        byte    a
        byte    b
        byte    c
        str15   name
        str1    gender
}
──────────────────────────────────────── end of dict4.dct ────────────
```

◁

▷ Example 5: Fixed-column format

```
──────────────────────────────────────── top of examp5.raw ────────────
  101mJohn Smith
  001mPaul Lin
  010fJan Doe
  00 fJulie McDonald
──────────────────────────────────────── end of examp5.raw ────────────
```

could be read in by

```
. infix a 1 b 2 c 3 str gender 4 str name 5-19 using examp5
```

or

```
. infix using dict5a
```

where dict5a.dct contains

```
──────────────────────────────────────── top of dict5a.dct ────────────
infix dictionary using examp5 {
                a       1
                b       2
                c       3
        str     gender  4
        str     name    5-19
}
──────────────────────────────────────── end of dict5a.dct ────────────
```

or

```
. infile using dict5b
```

where `dict5b.dct` contains

```
                                              ──────── top of dict5b.dct ────────
     infile dictionary using examp5 {
             byte      a       %1f
             byte      b       %1f
             byte      c       %1f
             str1      gender  %1s
             str15     name    %15s
     }
                                              ──────── end of dict5b.dct ────────
```
◁

▷ Example 6: Fixed-column format with headings

```
                                              ──────── top of examp6.raw ────────
     line 1 : a heading
     There are a total of 4 lines of heading.
     The next line contains a useful heading:
     ----+----1----+----2----+----3----+----4----+-
     1       0       1       m       John Smith
     0       0       1       m       Paul Lin
     0       1       0       f       Jan Doe
     0       0               f       Julie McDonald
                                              ──────── end of examp6.raw ────────
```

could be read in by

        . infile using dict6a

where `dict6a.dct` contains

```
                                              ──────── top of dict6a.dct ────────
     infile dictionary using examp6 {
     _firstline(5)
                   byte      a
                   byte      b
     _column(17)   byte      c       %1f
                   str1      gender
     _column(33)   str15     name    %15s
     }
                                              ──────── end of dict6a.dct ────────
```

or could be read in by

        . infix 5 first a 1 b 9 c 17 str gender 25 str name 33-46 using examp6

or could be read in by

        . infix using dict6b

where `dict6b.dct` contains

```
                                              ──────── top of dict6b.dct ────────
     infix dictionary using examp6 {
     5 first
                   a         1
                   b         9
                   c         17
             str   gender    25
             str   name      33-46
     }
                                              ──────── end of dict6b.dct ────────
```
◁

▷ Example 7: Fixed-column format with observations spanning multiple lines

```
——————————————————————————— top of examp7.raw ———————
a b c gender name
1 0 1
m
John Smith
0 0 1
m
Paul Lin
0 1 0
f
Jan Doe
0 0
f
Julie McDonald
——————————————————————————— end of examp7.raw ———————
```

could be read in by

    . infile using dict7a

where `dict7a.dct` contains

```
——————————————————————————————— top of dict7a.dct ———————
infile dictionary using examp7 {
_firstline(2)
            byte    a
            byte    b
            byte    c
_line(2)
            str1    gender
_line(3)
            str15   name    %15s
}
——————————————————————————————— end of dict7a.dct ———————
```

or, if we wanted to include variable labels,

    . infile using dict7b

where `dict7b.dct` contains

```
——————————————————————————————— top of dict7b.dct ———————
infile dictionary using examp7 {
_firstline(2)
            byte    a            "Question 1"
            byte    b            "Question 2"
            byte    c            "Question 3"
_line(2)
            str1    gender       "Gender of subject"
_line(3)
            str15   name    %15s
}
——————————————————————————————— end of dict7b.dct ———————
```

`infix` could also read these data,

    . infix 2 first 3 lines a 1 b 3 c 5 str gender 2:1 str name 3:1-15 using examp7

or the data could be read in by

    . infix using dict7c

where `dict7c.dct` contains

```
──────────────────────────────────────────── top of dict7c.dct ────────
    infix dictionary using examp7 {
    2 first
                a       1
                b       3
                c       5
        str     gender  2:1
        str     name    3:1-15
    }
──────────────────────────────────────────── end of dict7c.dct ────────
```

or the data could be read in by

    . infix using dict7d

where `dict7d.dct` contains

```
──────────────────────────────────────────── top of dict7d.dct ────────
    infix dictionary using examp7 {
    2 first
                a       1
                b       3
                c       5
    /
        str     gender  1
    /
        str     name    1-15
    }
──────────────────────────────────────────── end of dict7d.dct ────────
```
◁

Also See

[D] **edit** — Edit and list data with Data Editor

[D] **infile (fixed format)** — Read ASCII (text) data in fixed format with a dictionary

[D] **infile (free format)** — Read unformatted ASCII (text) data

[D] **infix (fixed format)** — Read ASCII (text) data in fixed format

[D] **input** — Enter data from keyboard

[D] **insheet** — Read ASCII (text) data created by a spreadsheet

[D] **fdasave** — Save and use datasets in FDA (SAS XPORT) format

[D] **odbc** — Load, write, or view data from ODBC sources

[D] **xmlsave** — Save and use datasets in XML format

[TS] **haver** — Load data from Haver Analytics database

[U] **21 Inputting data**

Title

---
infile (fixed format) — Read ASCII (text) data in fixed format with a dictionary
---

Syntax

infile using *dfilename* [*if*] [*in*] [, *options*]

options	description
Main	
using(*filename*)	ASCII dataset filename
clear	replace data in memory
Options	
automatic	create value labels from nonnumeric data

The syntax for a dictionary (a file created with an editor or word processor outside Stata) is

———————————————————————————— top of dictionary file ————

```
[infile] dictionary [using filename] {
    * comments may be included freely
    _lrecl(#)
    _firstlineoffile(#)
    _lines(#)
    _line(#)
    _newline[(#)]
    _column(#)
    _skip[(#)]
    [type] varname [:lblname] [% infmt] ["variable label"]
}
(your data might appear here)
```

———————————————————————————— end of dictionary file ————

where % *infmt* is { %[#[.#]]{f|g|e} | %[#]s | %[#]S }

Description

infile using reads from a disk a dataset that is not in Stata format. infile using does this by first reading *dfilename*—a "dictionary" that describes the format of the data file—and then reads the file containing the data. The dictionary is a file you create in an editor or word processor outside Stata. If *dfilename* is specified without an extension, .dct is assumed.

If using *filename* is not specified, the data are assumed to begin on the line following the closing brace. If using *filename* is specified, the data are assumed to be located in *filename*. If *filename* is specified without an extension, .raw is assumed.

Note for Stata for Windows and Stata for Macintosh users: if *dfilename* or *filename* contains embedded spaces, remember to enclose it in double quotes.

The data may be in the same file as the dictionary or in another file.

Another variation on `infile` omits the intermediate dictionary; see [D] **infile (free format)**. This variation is easier to use but will not read fixed-format files. On the other hand, although `infile using` will read free-format files, `infile` without a dictionary is even better at it.

An alternative to `infile using` for reading fixed-format files is `infix`; see [D] **infix (fixed format)**. `infix` provides fewer features than `infile using` but is easier to use.

Stata has other commands for reading data. If you are not certain that `infile using` will do what you are looking for, see [D] **infile** and [U] **21 Inputting data**.

Options

        ⌐ Main ⌐

using(*filename*) specifies the name of a file containing the data. If `using()` is not specified, the data are assumed to follow the dictionary in *dfilename*, or, if the dictionary specifies the name of some other file, that file is assumed to contain the data. If `using`(*filename*) is specified, *filename* is used to obtain the data, even if the dictionary says otherwise. If *filename* is specified without an extension, `.raw` is assumed.

    Note for Stata for Windows and Stata for Macintosh users: if *filename* contains embedded spaces, remember to enclose it in double quotes.

clear specifies that it is okay for the new data to replace what is currently in memory. To ensure that you do not lose something important, `infile using` will refuse to read new data if other data are already in memory. `clear` allows `infile using` to replace the data in memory. You can also drop the data yourself by typing `drop _all` before reading new data.

        ⌐ Options ⌐

automatic causes Stata to create value labels from the nonnumeric data it reads. It also automatically widens the display format to fit the longest label.

Dictionary directives

* marks comment lines. Wherever you wish to place a comment, begin the line with a *. Comments can appear many times in the same dictionary.

_lrecl(#) is used only for reading datasets that do not have end-of-line delimiters (carriage return, line feed, or some combination of these). Such files are often produced by mainframe computers and have been poorly translated from EBCDIC into ASCII. `_lrecl()` specifies the logical record length. `_lrecl()` requests that `infile` act as if a line ends every # characters.

    _lrecl() appears only once, and typically not at all, in a dictionary.

_firstlineoffile(#) (abbreviation _first()) is also rarely specified. It states the line of the file where the data begin. You do not need to specify `_first()` when the data follow the dictionary; Stata can figure that out for itself. However, you might specify `_first()` when reading data from another file in which the first line does not contain data because of headers or other markers.

    _first() appears only once, and typically not at all, in a dictionary.

_lines(#) states the number of lines per observation in the file. Simple datasets typically have `_lines(1)`. Large datasets often have many lines (sometimes called records) per observation. `_lines()` is optional, even when there is more than one line per observation because `infile` can sometimes figure it out for itself. Still, if `_lines(1)` is not right for your data, it is best to specify the correct number through `_lines(#)`.

_lines() appears only once in a dictionary.

_line(#) tells infile to jump to line # of the observation. _lines() is not the same as _line(). Consider a file with _lines(4), meaning four lines per observation. _line(2) says to jump to the second line of the observation. _line(4) says to jump to the fourth line of the observation. You may jump forward or backward. infile does not care, and there is no inefficiency in going forward to _line(3), reading a few variables, jumping back to _line(1), reading another variable, and jumping forward again to _line(3).

You need not ensure that, at the end of your dictionary, you are on the last line of the observation. infile knows how to get to the next observation because it knows where you are and it knows _lines(), the total number of lines per observation.

_line() may appear many times in a dictionary.

_newline[(#)] is an alternative to _line(). _newline(1), which may be abbreviated _newline, goes forward one line. _newline(2) goes forward two lines. We do not recommend using _newline() because _line() is better. If you are currently on line 2 of an observation and want to get to line 6, you could type _newline(4), but your meaning is clearer if you type _line(6).

_newline() may appear many times in a dictionary.

_column(#) jumps to column # on the current line. You may jump forward or backward within a line. _column() may appear many times in a dictionary.

_skip(#) jumps forward # columns on the current line. _skip() is just an alternative to _column(). _skip() may appear many times in a dictionary.

[type] varname [:lblname] [% infmt] ["variable label"] instructs infile to read a variable. The simplest form of this instruction is the variable name itself: varname.

At all times, infile is on some column of some line of an observation. infile starts on column 1 of line 1, so pretend that is where we are. Given the simplest directive 'varname', infile goes through the following logic:

If the current column is blank, it skips forward until there is a nonblank column (or until the end of the line). If it just skipped all the way to the end of the line, it stores a missing value in varname. If it skipped to a nonblank column, it begins collecting what is there until it comes to a blank column or the end of the line. These are the data for varname. Then it sets the current column to wherever it is.

The logic is a bit more complicated. For instance, when skipping forward to find the data, infile might encounter a quote. If so, it then collects the characters for the data by skipping forward until it finds the matching quote. If you specified a % infmt, infile skips the skipping-forward step and simply collects the specified number of characters. If you specified a %S infmt, infile does not skip leading or trailing blanks. Nevertheless, the general logic is (optionally) skip, collect, and reset.

Remarks

Remarks are presented under the following headings:

Introduction
Reading free-format files
Reading fixed-format files
Numeric formats
String formats
Specifying column and line numbers
Examples of reading fixed-format files
Reading fixed-block files

Introduction

`infile using` follows a two-step process to read your data. You type something like `infile using descript`, and

1. `infile using` reads the file `descript.dct`, which tells `infile` about the format of the data; and

2. `infile using` then reads the data according to the instructions recorded in `descript.dct`.

`descript.dct` (the file could be named anything) is called a dictionary, and `descript.dct` is just a text file that you create with an editor or word processor outside Stata.

As for the data, they can be in the same file as the dictionary or in a different file. It does not matter.

Reading free-format files

Another variation of `infile` for reading free-format files is described in [D] **infile (free format)**. We will refer to this variation as `infile` without a dictionary. The distinction between the two variations is in the treatment of line breaks. `infile` without a dictionary does not consider them significant. `infile` with a dictionary does.

A line, also known as a record, physical record, or physical line (as opposed to observations, logical records, or logical lines), is a string of characters followed by the line terminator. If you were to type the file, a line is what would appear on your screen if your screen were infinitely wide. Your screen would have to be infinitely wide so that there would be no possibility that one line could take more than one line of your screen, thus fooling you into thinking there are multiple lines when there is only one.

A logical line, on the other hand, is a sequence of one or more physical lines that represents one observation of your data. `infile` with a dictionary does not spontaneously go to new physical lines; it goes to a new line only between observations and when you tell it to. `infile` without a dictionary, on the other hand, goes to a new line whenever it needs to, which can be right in the middle of an observation. Thus consider the following little bit of data, which is for three variables:

```
5 4
1 9 3
2
```

How do you interpret these data?

Here is one interpretation: There are 3 observations. The first is 5, 4, and missing. The second is 1, 9, and 3. The third is 2, missing, and missing. That is the interpretation that `infile` with a dictionary makes.

Here is another interpretation: There are 2 observations. The first is 5, 4, and 1. The second is 9, 3, and 2. That is the interpretation that `infile` without a dictionary makes.

Which is right? You would have to ask the person who entered these data. The question is, are the line breaks significant? Do they mean anything? If the line breaks are significant, you use `infile` with a dictionary. If the line breaks are not significant, you use `infile` without a dictionary.

The other distinction between the two `infiles` is that `infile` with a dictionary does not process comma-separated-value format. If your data are comma separated, see [D] **infile (free format)** or [D] **insheet**.

▷ Example 1

Outside Stata, we have typed into the file `highway.dct` information on the accident rate per million vehicle miles along a stretch of highway, the speed limit on that highway, and the number of access points (on-ramps and off-ramps) per mile. Our file contains

```
───────────────────────────────────────── top of highway.dct, example 1 ─────────
infile dictionary {
        acc_rate  spdlimit acc_pts
}
4.58 55 4.6
2.86  60 4.4
1.61 . 2.2
3.02 60 4.7
───────────────────────────────────────── end of highway.dct, example 1 ─────────
```

This file can be read by typing `infile using highway`. Stata displays the dictionary and reads the data:

```
. infile using highway
infile dictionary {
        acc_rate  spdlimit acc_pts
}
(4 observations read)
. list
```

	acc_rate	spdlimit	acc_pts
1.	4.58	55	4.6
2.	2.86	60	4.4
3.	1.61	.	2.2
4.	3.02	60	4.7

◁

▷ Example 2

We can include variable labels in a dictionary so that after we `infile` the data, the data will be fully labeled. We could change `highway.dct` to read

```
───────────────────────────────────────── top of highway.dct, example 2 ─────────
infile dictionary {
* This is a comment and will be ignored by Stata
* You might type the source of the data here.
        acc_rate  "Acc. Rate/Million Miles"
        spdlimit   "Speed Limit (mph)"
        acc_pts   "Access Pts/Mile"
}
4.58 55 4.6
2.86  60 4.4
1.61 . 2.2
3.02 60 4.7
───────────────────────────────────────── end of highway.dct, example 2 ─────────
```

Now when we type `infile using highway`, Stata not only reads the data but also labels the variables.

◁

▷ Example 3

We can indicate the variable types in the dictionary. For instance, if we wanted to store `acc_rate` as a `double` and `spdlimit` as a `byte`, we could change `highway.dct` to read

———————————————————————————————— top of highway.dct, example 3 ————————

```
infile dictionary {
* This is a comment and will be ignored by Stata
* You might type the source of the data here.
 double acc_rate  "Acc. Rate/Million Miles"
 byte   spdlimit  "Speed Limit (mph)"
        acc_pts   "Access Pts/Mile"
}
4.58 55 4.6
2.86  60 4.4
1.61 . 2.2
3.02 60 4.7
```

———————————————————————————————— end of highway.dct, example 3 ————————

Because we do not indicate the variable type for `acc_pts`, it is given the default variable type `float` (or the type specified by the `set type` command).

◁

▷ Example 4

By specifying the types, we can read string variables as well as numeric variables. For instance,

—— top of emp.dct ————————

```
infile dictionary {
* data on employees
  str20 name      "Name"
        age       "Age"
    int sex       "Sex coded 0 male 1 female"
}
"Lisa Gilmore" 25 1
Branton 32 1
'Bill Ross' 27 0
```

—— end of emp.dct ————————

The strings can be delimited by single or double quotes, and quotes may be omitted altogether if the string contains no blanks or other special characters.

◁

❏ Technical Note

You may attach value labels to variables in the dictionary by using the colon notation:

—— top of emp2.dct ————————

```
infile dictionary {
* data on name, sex, and age
  str16 name      "Name"
        sex:sexlbl "Sex"
    int age       "Age"
}
"Arthur Doyle" Male 22
"Mary Hope" Female 37
"Guy Fawkes" Male 48
"Sherry Crooks" Female 25
```

—— end of emp2.dct ————————

If you want the value labels to be created automatically, you must specify the `automatic` option on the `infile` command. These data could be read by typing `infile using person2, automatic`, assuming the dictionary and data are stored in the file `person2.dct`.

❏

▷ Example 5

The data need not be in the same file as the dictionary. We might leave the highway data in `highway.raw` and write a dictionary called `highway.dct` describing the data:

```
──────────────── top of highway.dct, example 4 ────────────
infile dictionary using highway {
* This dictionary reads the file highway.raw.  If the
* file were called highway.txt, the first line would
* read "dictionary using highway.txt"
        acc_rate   "Acc. Rate/Million Miles"
        spdlimit   "Speed Limit (mph)"
        acc_pts    "Access Pts/Mile"
}
──────────────── end of highway.dct, example 4 ────────────
```

◁

▷ Example 6

The `firstlineoffile()` directive allows us to ignore lines at the top of the file. Consider the following raw dataset:

```
──────────────── top of mydata.raw ────────────
The following data was entered by Marsha Martinez.  It was checked by
Helen Troy.
id income educ sex age
1024 25000 HS Male 28
1025 27000 C Female 24
──────────────── end of mydata.raw ────────────
```

Our dictionary might read

```
──────────────── top of mydata.dct ────────────
infile dictionary using mydata {
        _first(4)
        int id "Identification Number"
        income "Annual income"
        str2 educ "Highest educ level"
        str6 sex
        byte age
}
──────────────── end of mydata.dct ────────────
```

◁

▷ Example 7

The `_line()` and `_lines()` directives tell Stata how to read our data when there are multiple records per observation. We have the following in `mydata2.raw`:

```
——————————————————————————————————————————————————— top of mydata2.raw ———————————
id income educ sex age
1024 25000 HS
Male
28
1025 27000 C
Female
24
1035 26000 HS
Male
32
1036 25000 C
Female
25
——————————————————————————————————————————————————— end of mydata2.raw ———————————
```

We can read this with a dictionary `mydata2.dct`, which we will just let Stata list as it simultaneously reads the data:

```
. infile using mydata2, clear

infile dictionary using mydata2 {
    _first(2)                               * Begin reading on line 2
    _lines(3)                               * Each observation takes 3 lines.
    int id "Identification Number"          * Since _line is not specified, Stata
    income "Annual income"                  * assumes that it is 1.
    str2 educ "Highest educ level"
    _line(2)                                * Go to line 2 of the observation.
    str6 sex                                * (values for sex are located on line 2)
    _line(3)                                * Go to line 3 of the observation.
    int age                                 * (values for age are located on line 3)
}
(4 observations read)

. list
```

	id	income	educ	sex	age
1.	1024	25000	HS	Male	28
2.	1025	27000	C	Female	24
3.	1035	26000	HS	Male	32
4.	1036	25000	C	Female	25

Here is the really good part: we read these variables in order, but that was not necessary. We could just as well have used the dictionary:

```
—————————————————————————————————————————————————— top of mydata2p.dct ———————————
infile dictionary using mydata2 {
        _first(2)
        _lines(3)
        _line(1)    int    id      "Identification number"
                    income "Annual income"
                    str2   educ    "Highest educ level"
        _line(3)    int    age
        _line(2)    str6   sex
}
—————————————————————————————————————————————————— end of mydata2p.dct ———————————
```

We would have obtained the same results just as quickly, the only difference being that our variables in the final dataset would be in the order specified: id, income, educ, age, and sex.

◁

❑ Technical Note

You can use _newline to specify where breaks occur, if you prefer:

─────────────────────────────────── top of highway.dct, example 5 ───────────

```
infile dictionary {
         acc_rate  "Acc. Rate/Million Miles"
         spdlimit  "Speed Limit (mph)"
_newline acc_pts   "Access Pts/Mile"
}
4.58 55
4.6
2.86  60
 4.4
1.61 .
2.2
3.02 60
 4.7
```

─────────────────────────────────── end of highway.dct, example 5 ───────────

The line reading '1.61 .' could have been read 1.61 (without the period), and the results would have been unchanged. Since dictionaries do not go to new lines automatically, a missing value is assumed for all values not found in the record.

❑

Reading fixed-format files

Values in formatted data are sometimes packed one against the other with no intervening blanks. For instance, the highway data might appear as

─────────────────────────────────── top of highway.raw, example 6 ───────────

```
4.58554.6
2.86604.4
1.61  2.2
3.02604.7
```

─────────────────────────────────── end of highway.raw, example 6 ───────────

The first four columns of each record represent the accident rate; the next two columns, the speed limit; and the last three columns, the number of access points per mile.

To read these data, you must specify the % *infmt* in the dictionary. Numeric % *infmt*s are denoted by a leading percent sign (%) followed optionally by a string of the form w or $w.d$, where w and d stand for two integers. The first integer, w, specifies the width of the format. The second integer, d, specifies the number of digits that are to follow the decimal point. d must be less than or equal to w. Finally, a character denoting the format type (f, g, or e) is appended. For example, %9.2f specifies an f format that is nine characters wide and has two digits following the decimal point.

Numeric formats

The f format indicates that `infile` is to attempt to read the data as a number. When you do not specify the % *infmt* in the dictionary, `infile` assumes the %f format. The missing width w means that `infile` is to attempt to read the data in free format.

As it starts reading each observation, `infile` reads a record into its buffer and sets a column pointer to 1, indicating that it is currently on the first column. When `infile` processes a %f format, it moves the column pointer forward through white space. It then collects the characters up to the next occurrence of white space and attempts to interpret those characters as a number. The column pointer is left at the first occurrence of white space following those characters. If the next variable is also free format, the logic repeats.

When you explicitly specify the field width w, as in %wf, `infile` does not skip leading white space. Instead, it collects the next w characters starting at the column pointer and attempts to interpret the result as a number. The column pointer is left at the old value of the column pointer plus w, that is, on the first character following the specified field.

▷ Example 8

If the data above were stored in `highway.raw`, we could create the following dictionary to read the data:

```
――――――――――――――――――――――――――――― top of highway.dct, example 6 ―――――――
    infile dictionary using highway {
            acc_rate    %4f  "Acc. Rate/Million Miles"
            spdlimit    %2f  "Speed Limit (mph)"
            acc_pts     %3f  "Access Pts/Mile
    }
――――――――――――――――――――――――――――― end of highway.dct, example 6 ―――――――
```

When we explicitly indicate the field width, `infile` does not skip intervening characters. The first four columns are used for the variable `acc_rate`, the next two for `spdlimit`, and the last three for `acc_pts`.

◁

❏ Technical Note

The d specification in the %$w.d$f indicates the number of *implied* decimal places in the data. For instance, the string 212 read in a %3.2f format represents the number 2.12. Do *not* specify d unless your data have elements of this form. The w alone is sufficient to tell `infile` how to read data in which the decimal point is explicitly indicated.

When you specify d, Stata takes it only as a suggestion. If the decimal point is explicitly indicated in the data, that decimal point always overrides the d specification. Decimal points are also not implied if the data contain an E, e, D, or d, indicating scientific notation.

Fields are right justified before implying decimal points. Thus '2 ', ' 2 ', and ' 2' are all read as 0.2 by the %3.1f format.

❏

❑ Technical Note

The g and e formats are the same as the f format. You can specify any of these letters interchangeably. The letters g and e are included as a convenience to those familiar with Fortran, in which the e format indicates scientific notation. For example, the number 250 could be indicated as 2.5E+02 or 2.5D+02. Fortran programmers would refer to this as an E7.5 format, and in Stata, this format would be indicated as %7.5e. In Stata, however, you need specify only the field width w, so you could read this number by using %7f, %7g, or %7e.

The g format is really a Fortran output format that indicates a freer format than f. In Stata, the two formats are identical.

Throughout this section, you may freely substitute the g or e formats for the f format.

❑

❑ Technical Note

Be careful to distinguish between % *fmts* and % *infmts*. % *fmts* are also known as *display* formats—they describe how a variable is to look when it is displayed; see [U] **12.5 Formats: controlling how data are displayed**. % *infmts* are also known as *input* formats—they describe how a variable looks when you input it. For instance, there is an output date format %td, but there is no corresponding input format. (See [U] **24 Dealing with dates and times** for recommendations on how to read dates.) For the other formats, we have attempted to make the input and output definitions as similar as possible. Thus we include g, e, and f % *infmts*, even though they all mean the same thing, since g, e, and f are also % *fmts*.

❑

String formats

The s and S formats are used for reading strings. The syntax is %ws or %wS, where the w is optional. If you do not specify the field width, your strings must either be enclosed in quotes (single or double) or not contain any characters other than letters, numbers, and "_".

This may surprise you, but the s format can be used for reading numeric variables, and the f format can be used for reading string variables! When you specify the field width w in the %wf format, all embedded blanks in the field are removed before the result is interpreted. They are not removed by the %ws format.

For instance, the %3f format would read "- 2", "-2 ", or " -2" as the number −2. The %3s format would not be able to read "- 2" as a number, since the sign is separated from the digit, but it could read " -2" or "-2 ". The %wf format removes blanks; datasets written by some Fortran programs separate the sign from the number.

There are, however, some side effects of this practice. The string "2 2" will be read as 22 by a %3f format. Most Fortran compilers would read this number as 202. The %3s format would issue a warning and store a *missing* value.

Now consider reading the string "a b" into a string variable. Using a %3s format, Stata will store it as it appears: a b. Using a %3f format, however, it will be stored as ab—the middle blank will be removed.

%wS is a special case of %ws. A string read with %ws will have leading and trailing blanks removed, but a string read with %wS will not have them removed.

Examples using the %s format are provided below, right after we discuss specifying column and line numbers.

Specifying column and line numbers

_column() jumps to the specified column. For instance, the documentation of some dataset indicates that the variable age is recorded as a two-digit number in column 47. You could read this by coding

```
_column(47) age %2f
```

After typing this, you are now at column 49, so if immediately following age there were a one-digit number recording sex as 0 or 1, you could code

```
_column(47) age %2f
            sex %1f
```

or, if you wanted to be explicit about it, you could instead code

```
_column(47) age %2f
_column(49) sex %1f
```

It makes no difference. If at column 50 there were a one-digit code for race, and you wanted to read it but skip reading the sex code, you could code

```
_column(47) age %2f
_column(50) race %1f
```

You could equivalently skip forward using _skip():

```
_column(47) age %2f
_skip(1)    race %1f
```

One advantage of column() over _skip is that it lets you jump forward or backward in a record. If you wanted to read race and then age, you could code

```
_column(50) race %1f
_column(47) age %2f
```

If the data you are reading have multiple lines per observation (sometimes said as multiple records per observation), you can tell infile how many lines per record there are by using _lines():

```
_lines(4)
```

_lines() appears only once in a dictionary. Good style says that it should be placed near the top of the dictionary, but Stata does not care.

When you want to go to a particular line, include the _line() directive. In our example, let's assume that race, sex, and age are recorded on the second line of each observation:

```
_lines(4)
_line(2)
    _column(47) age %2f
    _column(50) race %1f
```

Let's assume that id is recorded on line 1.

```
_lines(4)
_line(1)
    _column(1)  id  %4f
_line(2)
    _column(47) age %2f
    _column(50) race %1f
```

_line() works like _column() in that you can jump forward or backward, so these data could just as well be read by

```
_lines(4)
_line(2)
    _column(47) age %2f
    _column(50) race %1f
_line(1)
    _column(1)  id  %4f
```

Remember that this dataset has four lines per observation, and yet we have never referred to line(3) or line(4). That is okay. Also, at the end of our dictionary, we are on line 1, not line 4. That is okay, too. infile will still get to the next observation correctly.

❏ Technical Note

Another way to move between records is _newline(). _newline() is to _line() as _skip() is to _column(), which is to say, _newline() can go only forward. There is one difference: _skip() has its uses, whereas _newline() is useful only for backward capability with older versions of Stata.

_skip() has its uses because sometimes we think in columns and sometimes we think in widths. Some data documentation might include the sentence, "At column 54 are recorded the answers to the 25 questions, with one column allotted to each." If we want to read the answers to questions 1 and 5, it would indeed be natural to code

```
_column(54) q1 %1f
_skip(3)
            q5 %1f
```

Nobody has ever read data documentation with the statement, "Demographics are recorded on record 2, and, 2 records after that, are the income values." The documentation would instead say, "Record 2 contains the demographic information and record 4, income." The _newline() way of thinking is based on what is convenient for the computer, which does, after all, have to eject a certain number of records. That, however, is no reason for making you think that way.

Before that thought occurred to us, Stata users specified _newline() to go forward a number of records. They still can, so their old dictionaries will work. When you use _newline() and do not specify _lines(), you must eject the right number of records so that, at the end of the dictionary, you are on the last record. In this mode, when Stata reexecutes the dictionary to process the next observation, it goes forward one record.

❏

Examples of reading fixed-format files

▷ Example 9

In this example, each observation occupies two lines. The first 2 observations in the dataset are

```
John Dunbar            10001   101 North 42nd Street
1010111111
Sam K. Newey, Jr.      10002   15663 Roustabout Boulevard
0101000000
```

The first observation tells us that the name of the respondent is John Dunbar; that his ID is 10001; that his address is 101 North 42nd Street; and that his answers to questions 1–10 are yes, no, yes, no, yes, yes, yes, yes, yes, and yes.

The second observation tells us that the name of the respondent is Sam K. Newey, Jr.; that his ID is 10002; that his address is 15663 Roustabout Boulevard; and that his answers to questions 1–10 were no, yes, no, yes, no, no, no, no, no, and no.

To see the layout within the file, we can temporarily add two rulers to show the appropriate columns:

```
----+----1----+----2----+----3----+----4----+----5----+----6----+----7----+----8
John Dunbar                     10001   101 North 42nd Street
1010111111
Sam K. Newey, Jr.               10002   15663 Roustabout Boulevard
0101000000
----+----1----+----2----+----3----+----4----+----5----+----6----+----7----+----8
```

Each observation in the data appears in two physical lines within our text file. We had to check in our editor to be sure that there really were new-line characters (e.g., "hard returns") after the address. This is important because some programs will wrap output for you so that one line may appear as many lines. The two seemingly identical files will differ in that one has a hard return and the other has a soft return added only for display purposes.

In our data, the name occupies columns 1–32; a person identifier occupies columns 33–37; and the address occupies columns 40–80. Our worksheet revealed that the widest address ended in column 80.

The text file containing these data is called `fname.txt`. Our dictionary file looks like this:

```
───────────────────────────────────────────────── top of fname.dct ───────────
infile dictionary using fname.txt {
*
* Example reading in data where observations extend across more
* than one line.  The next line tells infile there are 2 lines/obs:
*
_lines(2)
*
            str50    name    %32s        "Name of respondent"
_column(33) long     id      %5f         "Person id"
_skip(2)    str50    addr    %41s        "Address"
_line(2)
_column(1)  byte     q1      %1f         "Question 1"
            byte     q2      %1f         "Question 2"
            byte     q3      %1f         "Question 3"
            byte     q4      %1f         "Question 4"
            byte     q5      %1f         "Question 5"
            byte     q6      %1f         "Question 6"
            byte     q7      %1f         "Question 7"
            byte     q8      %1f         "Question 8"
            byte     q9      %1f         "Question 9"
            byte     q10     %1f         "Question 10"
}
─────────────────────────────────────────────────── end of fname.dct ──────────
```

Up to five pieces of information may be supplied in the dictionary for each variable: the location of the data, the storage type of the variable, the name of the variable, the input format, and the variable label.

Thus the `str50` line says that the first variable is to be given a storage type of `str50`, called `name`, and is to have the variable label "Name of respondent". The `%32s` is the input format, which tells Stata how to read the data. The s tells Stata not to remove any embedded blanks; the 32 tells Stata to go across 32 columns when reading the data.

The next line says that the second variable is to be assigned a storage type of long, named id, and be labeled "Person id". Stata should start reading the information for this variable in column 33. The f tells Stata to remove any embedded blanks, and the 5 says to read across five columns.

The third variable is to be given a storage type of str50, called addr, and be labeled "Address". The _skip(2) directs Stata to skip two columns before beginning to read the data for this variable, and the %41s instructs Stata to read across 41 columns and not to remove embedded blanks.

line(2) instructs Stata to go to line 2 of the observation.

The remainder of the data is 0/1 coded, indicating the answers to the questions. It would be convenient if we could use a shorthand to specify this portion of the dictionary, but we must supply explicit directives.

◁

❑ Technical Note

In the preceding example, there were two pieces of information about location: where the data begin for each variable (the _column(), _skip(), _line()) and how many columns the data span (the %32s, %5f, %41s, %1f). In our dictionary, some of this information was redundant. After reading name, Stata had finished with 32 columns of information. Unless instructed otherwise, Stata would proceed to the next column—column 33—to begin reading information about id. The _column(33) was unnecessary.

The _skip(2) was necessary, however. Stata had read 37 columns of information and was ready to look at column 38. Although the address information does not begin until column 40, columns 38 and 39 contain blanks. Since these are leading blanks instead of embedded blanks, Stata would just ignore them without any trouble. The problem is with the %41s. If Stata begins reading the address information from column 38 and reads 41 columns, Stata would stop reading in column 78 ($78 - 41 + 1 = 38$), but the widest address ends in column 80. We could have omitted the _skip(2) if we had specified an input format of %43s.

The _line(2) was necessary, although we could have read the second line by coding _newline instead.

The _column(1) could have been omitted. After the _line(), Stata begins in column 1.

See the next example for a dataset in which both pieces of location information are required.

❑

▷ Example 10

The following file contains six variables in a variety of formats. In the dictionary we read the variables fifth and sixth out of order by forcing the column pointer.

```
─────────────────────────────────────────────────────── top of example.dct ───────────
    infile dictionary {
                            first    %3f
                   double   second   %2.1f
                            third    %6f
        _skip(2)     str4   fourth   %4s
        _column(21)         sixth %4.1f
        _column(18)         fifth %2f
    }
    1.2125.7e+252abcd 1 .232
    1.3135.7   52efgh2    5
    1.41457    52abcd 3 100.
    1.5155.7D+252efgh04 1.7
    16 16 .57  52abcd 5 1.71
─────────────────────────────────────────────────────── end of example.dct ───────────
```

Assuming that the above is stored in a file called `example.dct`, we can infile and list it by typing

```
. infile using example
infile dictionary {
                        first    %3f
               double   second   %2.1f
                        third    %6f
    _skip(2)     str4   fourth   %4s
    _column(21)         sixth %4.1f
    _column(18)         fifth %2f
}
(5 observations read)

. list
```

	first	second	third	fourth	sixth	fifth
1.	1.2	1.2	570	abcd	.232	1
2.	1.3	1.3	5.7	efgh	.5	2
3.	1.4	1.4	57	abcd	100	3
4.	1.5	1.5	570	efgh	1.7	4
5.	16	1.6	.57	abcd	1.71	5

◁

Reading fixed-block files

❏ Technical Note

The `_lrecl(#)` directive is used for reading datasets that do not have end-of-line delimiters (carriage return, line feed, or some combination of these). Such datasets are typical of IBM mainframes, where they are known as fixed block, or FB. The word LRECL is IBM mainframe jargon for logical record length.

In a fixed-block dataset, each # characters are to be interpreted as a record. For instance, consider the data

```
1 21
2 42
3 63
```

In fixed-block format, these data might be recorded as

```
─────────────────────────────────────────── top of mydata.ibm ───────────
     1 212 423 63
─────────────────────────────────────────── end of mydata.ibm ───────────
```

and you would be told, on the side, that the LRECL is 4. If you then pass along that information to `infile`, it can read the data:

```
─────────────────────────────────────────── top of mydata.dct ───────────
     infile dictionary using mydata.ibm {
            _lrecl(4)
            int     id
            int     age
     }
─────────────────────────────────────────── end of mydata.dct ───────────
```

When you do not specify the `_lrecl(#)` directive, `infile` assumes that each line ends with the standard ASCII delimiter (which can be a line feed, a carriage return, a line feed followed by carriage return, or a carriage return followed by line feed). When you specify `_lrecl(#)`, `infile` reads the data in blocks of # characters and then acts as if that is a line.

A common mistake in processing fixed-block datasets is to use an incorrect LRECL value, such as 160 when it is really 80. To understand what can happen, pretend that you thought the LRECL in your data was 6 rather than 4. Taking the characters in groups of 6, the data appear as

```
     1 212
     423 63
```

Stata cannot verify that you have specified the correct LRECL, so, if the data appear incorrect, verify that you have the correct number.

The maximum LRECL `infile` allows is 524,275.

❏

References

Gleason, J. R. 1998. dm54: Capturing comments from data dictionaries. *Stata Technical Bulletin* 42: 3–4. Reprinted in *Stata Technical Bulletin Reprints*, vol. 7, pp. 55–57.

Gould, W. W. 1992. dm10: Infiling data: Automatic dictionary creation. *Stata Technical Bulletin* 9: 4–8. Reprinted in *Stata Technical Bulletin Reprints*, vol. 2, pp. 28–34.

Nash, J. D. 1994. dm19: Merging raw data and dictionary files. *Stata Technical Bulletin* 20: 3–5. Reprinted in *Stata Technical Bulletin Reprints*, vol. 4, pp. 22–25.

Also See

Title

> **infile (free format)** — Read unformatted ASCII (text) data

Syntax

<u>inf</u>ile *varlist* [_skip[(#)]] [*varlist* [_skip[(#)]] ...]]] <u>using</u> *filename* [*if*] [*in*]

[, *options*]

options	description
Main	
clear	replace data in memory
Options	
<u>auto</u>matic	create value labels from nonnumeric data
<u>byv</u>ariable(#)	organize external file by variables; # is number of observations

Description

infile reads into memory from a disk a dataset that is not in Stata format. If *filename* is specified without an extension, .raw is assumed.

Note for Stata for Windows and Stata for Macintosh users: if your *filename* contains embedded spaces, remember to enclose it in double quotes.

Here we discuss using infile to read free-format data, meaning datasets in which Stata does not need to know the formatting information. Another variation on infile allows reading fixed-format data; see [D] **infile (fixed format)**. Yet another alternative is insheet, which is easier to use if your data are tab- or comma separated and contain 1 observation per line. Stata has other commands for reading data, too. If you are not certain that infile will do what you are looking for, see [D] **infile** and [U] **21 Inputting data**.

After the data are read into Stata, they can be saved in a Stata-format dataset; see [D] **save**.

Options

> ⌐ **Main** ⌐

clear specifies that it is okay for the new data to replace the data that are currently in memory. To ensure that you do not lose something important, infile will refuse to read new data if data are already in memory. clear allows infile to replace the data in memory. You can also drop the data yourself by typing drop _all before reading new data.

> ⌐ **Options** ⌐

automatic causes Stata to create value labels from the nonnumeric data it reads. It also automatically widens the display format to fit the longest label.

byvariable(#) specifies that the external data file is organized by variables rather than by observations. All the observations on the first variable appear, followed by all the observations on the second variable, and so on. Time-series datasets sometimes come in this format.

Remarks

This section describes infile features for reading data in free or comma-separated-value format. Remarks are presented under the following headings:

> *Reading free-format data*
> *Reading comma-separated data*
> *Specifying variable types*
> *Reading string variables*
> *Skipping variables*
> *Skipping observations*
> *Reading time-series data*

Reading free-format data

In free format, data are separated by one or more white-space characters—blanks, tabs, or new lines (carriage return, line feed, or carriage-return/line feed combinations). Thus one observation may span any number of lines.

Numeric missing values are indicated by single periods (".").

▷ Example 1

In the file highway.raw, we have information on the accident rate per million vehicle miles along a stretch of highway, the speed limit on that highway, and the number of access points (on-ramps and off-ramps) per mile. Our file contains

────────────────────────────────── top of highway.raw, example 1 ──────────
```
4.58 55 4.6
2.86  60 4.4
1.61  . 2.2
3.02 60
4.7
```
────────────────────────────────── end of highway.raw, example 1 ──────────

We can read these data by typing

```
. infile acc_rate spdlimit acc_pts using highway
(4 observations read)

. list
```

	acc_rate	spdlimit	acc_pts
1.	4.58	55	4.6
2.	2.86	60	4.4
3.	1.61	.	2.2
4.	3.02	60	4.7

The spacing of the numbers in the original file is irrelevant.

◁

❑ Technical Note

Missing values need not be indicated by one period. The third observation on the speed limit is missing in example 1. The raw data file indicates this by recording one period. Let's assume, instead, that the missing value was indicated by the word `unknown`. Thus the raw data file appears as

—— top of highway.raw, example 2 ————————
```
4.58 55 4.6
2.86  60 4.4
1.61 unknown 2.2
3.02 60
4.7
```
—— end of highway.raw, example 2 ————————

Here is the result of infiling these data:
```
. infile acc_rate spdlimit acc_pts using highway
'unknown' cannot be read as a number for spdlimit[3]
(4 observations read)
```

`infile` warned us that it could not read the word `unknown`, stored a *missing*, and then continued to read the rest of the dataset. Thus aside from the warning message, results are unchanged.

Because not all packages indicate missing data in the same way, this feature can be useful when reading data. Whenever `infile` sees something that it does not understand, it warns you, records a *missing*, and continues. If, on the other hand, the missing values were recorded not as `unknown` but as, say, 99, Stata would have had no difficulty reading the number, but it would also have stored 99 rather than missing. To convert such coded missing values to true missing values, see [D] **mvencode**.

❑

Reading comma-separated data

In comma-separated-value format, data are separated by commas. You may mix comma-separated-value and free format. Missing values are indicated either by single periods or by multiple commas that serve as placeholders, or both. As with free format, 1 observation may span any number of input lines.

▷ Example 2

We can modify the format of `highway.raw` used in example 1 without affecting `infile`'s ability to read it. The dataset can be read with the same command, and the results would be the same if the file instead contained

—— top of highway.raw, example 3 ————————
```
4.58,55 4.6
2.86, 60,4.4
1.61,,2.2
3.02,60
4.7
```
—— end of highway.raw, example 3 ————————

◁

(Continued on next page)

Specifying variable types

The variable names you type after the word `infile` are new variables. The syntax for a new variable is

$$\left[\,type\,\right]\ new\_varname\left[\,:label\_name\,\right]$$

A full discussion of this syntax can be found in [U] **11.4 varlists**. As a quick review, new variables are, by default, of type `float`. This default can be overridden by preceding the variable name with a storage type (`byte`, `int`, `long`, `float`, `double`, or `str#`) or by using the `set type` command. A list of variables placed in parentheses will be given the same type. For example,

`double`(*first_var second_var ... last_var*)

causes *first_var second_var ... last_var* to all be of type `double`.

There is also a shorthand syntax for variable names with numeric suffixes. The varlist `var1-var4` is equivalent to specifying `var1 var2 var3 var4`.

▷ Example 3

In the highway example, we could `infile` the data `acc_rate`, `spdlimit`, and `acc_pts` and force the variable `spdlimit` to be of type `int` by typing

```
. infile acc_rate int spdlimit acc_pts using highway, clear
(4 observations read)
```

We could force all variables to be of type `double` by typing

```
. infile double(acc_rate spdlimit acc_pts) using highway, clear
(4 observations read)
```

We could call the three variables `v1`, `v2`, and `v3` and make them all of type `double` by typing

```
. infile double(v1-v3) using highway, clear
(4 observations read)
```

◁

Reading string variables

By explicitly specifying the types, you can read string variables, as well as numeric variables.

▷ Example 4

Typing `infile str20 name age sex using myfile` would read

── top of myfile.raw ────────────
```
"Sherri Holliday" 25 1
Branton 32 1
"Bill Ross" 27,0
```
── top of myfile.raw ────────────

or even

──────────────────────────────────── top of myfile.raw, variation 2 ────────
```
'Sherri Holliday' 25,1 "Branton" 32
1,'Bill Ross', 27,0
```
──────────────────────────────────── end of myfile.raw, variation 2 ────────

The spacing is irrelevant, and either single or double quotes may be used to delimit strings. The quotes do not count when calculating the length of strings. Quotes may be omitted altogether if the string contains no blanks or other special characters (anything other than letters, numbers, or underscores).

Typing

```
. infile str20 name age sex using myfile, clear
(3 observations read)
```

makes `name` a `str20` and `age` and `sex` `floats`. We might have typed

```
. infile str20 name age int sex using myfile, clear
(3 observations read)
```

to make `sex` an `int` or

```
. infile str20 name int(age sex) using myfile, clear
(3 observations read)
```

to make both `age` and `sex` `ints`.   ◁

❏ Technical Note

`infile` can also handle nonnumeric data by using *value labels*. We will briefly review value labels, but you should see [U] **12.6.3 Value labels** for a complete description.

A value label is a mapping from the set of integers to words. For instance, if we had a variable called `sex` in our data that represented the sex of the individual, we might code 0 for male and 1 for female. We could then just remember that every time we see a value of 0 for `sex`, that observation refers to a male, whereas 1 refers to a female.

Even better, we could inform Stata that 0 represents males and 1 represents females by typing

```
. label define sexfmt 0 "Male" 1 "Female"
```

Then we must tell Stata that this coding scheme is to be associated with the variable `sex`. This is typically done by typing

```
. label values sex sexfmt
```

Thereafter, Stata will print `Male` rather than 0 and `Female` rather than 1 for this variable.

Stata has the ability to turn a value label around. Not only can it go from numeric codes to words such as "Male" and "Female", it can also go from the words to the numeric code. We tell `infile` the value label that goes with each variable by placing a colon (`:`) after the variable name and typing the name of the value label. Before we do that, we use the `label define` command to inform Stata of the coding.

Let's assume that we wish to `infile` a dataset containing the words `Male` and `Female` and that we wish to store numeric codes rather than the strings themselves. This will result in considerable data compression, especially if we store the numeric code as a `byte`. We have a dataset named `persons.raw` that contains name, sex, and age:

———————————————————————————————————— top of persons.raw ————————

```
"Arthur Doyle" Male 22
"Mary Hope" Female 37
"Guy Fawkes" Male 48
"Carrie House" Female 25
```

———————————————————————————————————— end of persons.raw ————————

Here is how we read and encode it at the same time:

```
. label define sexfmt 0 "Male" 1 "Female"
. infile str16 name sex:sexfmt age using persons
(4 observations read)
. list
```

	name	sex	age
1.	Arthur Doyle	Male	22
2.	Mary Hope	Female	37
3.	Guy Fawkes	Male	48
4.	Carrie House	Female	25

The `str16` in the `infile` command applies only to the `name` variable; `sex` is a numeric variable, which we can prove by typing

```
. list, nolabel
```

	name	sex	age
1.	Arthur Doyle	0	22
2.	Mary Hope	1	37
3.	Guy Fawkes	0	48
4.	Carrie House	1	25

❑

❑ Technical Note

When `infile` is directed to use a value label and it finds an entry in the file that does not match any of the codings recorded in the label, it prints a warning message and stores *missing* for the observation. By specifying the `automatic` option, you can instead have `infile` automatically add new entries to the value label.

Say that we have a dataset containing three variables. The first, region of the country, is a character string; the remaining two variables, which we will just call `var1` and `var2`, contain numbers. We have stored the data in a file called `geog.raw`:

```
─────────────────────────────────────── top of geog.raw ───────────
    "NE"       31.23        87.78
    'NCntrl'   29.52        98.92
    South      29.62       114.69
    West       28.28       218.92
    NE         17.50        44.33
    NCntrl     22.51        55.21
─────────────────────────────────────── end of geog.raw ───────────
```

The easiest way to read this dataset is to type

```
. infile str6 region var1 var2 using geog
```

making `region` a string variable. We do not want to do this, however, because we are practicing for reading a dataset like this containing 20,000 observations. If `region` were numerically encoded and stored as a `byte`, there would be a 5-byte saving per observation, reducing the size of the data by 100,000 bytes. We also do not want to bother with first creating the value label. Using the `automatic` option, `infile` creates the value label automatically as it encounters new regions.

```
. infile byte region:regfmt var1 var2 using geog, automatic clear
(6 observations read)
. list, sep(0)
```

	region	var1	var2
1.	NE	31.23	87.78
2.	NCntrl	29.52	98.92
3.	South	29.62	114.69
4.	West	28.28	218.92
5.	NE	17.5	44.33
6.	NCntrl	22.51	55.21

infile automatically created and defined a new value label called `regfmt`. We can use the `label list` command to view its contents:

```
. label list regfmt
regfmt:
              1 NE
              2 NCntrl
              3 South
              4 West
```

The value label need not be undefined before we use `infile` with the `automatic` option. If the value label `regfmt` had been previously defined as

```
. label define regfmt 2 "West"
```

the result of `label list` after the `infile` would have been

```
regfmt:
              2 West
              3 NE
              4 NCntrl
              5 South
```

The `automatic` option is convenient, but there is one reason for using it. Suppose that we had a dataset containing, among other things, information about an individual's sex. We know that the sex variable is supposed to be coded `male` and `female`. If we read the data by using the `automatic` option and if one of the records contains `fmlae`, `infile` will blindly create a third sex rather than print a warning.

❏

Skipping variables

Specifying `_skip` instead of a variable name directs `infile` to ignore the variable in that location. This feature makes it possible to extract manageable subsets from large disk datasets. A number of contiguous variables can be skipped by specifying `_skip(#)`, where # is the number of variables to ignore.

▷ Example 5

In the highway example that started this section, the data file contained three variables: `acc_rate`, `spdlimit`, and `acc_pts`. We can read the first two variables by typing

```
. infile acc_rate spdlimit _skip using highway
(4 observations read)
```

We can read the first and last variables by typing

```
. infile acc_rate _skip acc_pts using highway, clear
(4 observations read)
```

We can read the first variable by typing

```
. infile acc_rate _skip(2) using highway, clear
(4 observations read)
```

_skip may be specified more than once. If we had a dataset containing four variables—say, a, b, c, and d—and we wanted to read just a and c, we could type infile a _skip c _skip using *filename*.

◁

Skipping observations

Subsets of observations can be extracted by specifying if *exp*, which also makes it possible to extract manageable subsets from large disk datasets. Do not, however, use the *_variable* _N in *exp*. Use the in *range* modifier to refer to observation numbers within the disk dataset.

▷ Example 6

Again referring to the highway example, if we type

```
. infile acc_rate spdlimit acc_pts if acc_rate>3 using highway, clear
(2 observations read)
```

only observations for which acc_rate is greater than 3 will be infiled. We can type

```
. infile acc_rate spdlimit acc_pts in 2/4 using highway, clear
(eof not at end of obs)
(3 observations read)
```

to read only the second, third, and fourth observations.

◁

Reading time-series data

If you are dealing with time-series data, you may receive datasets organized by variables rather than by observations. All the observations on the first variable appear, followed by all the observations on the second variable, and so on. The byvariable(#) option specifies that the external data file is organized in this way. You specify the number of observations in the parentheses, since infile needs to know that number to read the data properly. You can also mark the end of one variable's data and the beginning of another's data by placing a semicolon (";") in the raw data file. You may then specify a number larger than the number of observations in the dataset and leave it to infile to determine the actual number of observations. This method can also be used to read unbalanced data.

▷ Example 7

We have time-series data on 4 years recorded in the file time.raw. The dataset contains information on year, amount, and cost, and is organized by variable:

```
─────────────────────────────────────────── top of time.raw ───────────
1980 1981 1982 1983
14 17 25 30
120 135 150
180
─────────────────────────────────────────── end of time.raw ───────────
```

We can read these data by typing

```
. infile year amount cost using time, byvariable(4) clear
(4 observations read)

. list
```

	year	amount	cost
1.	1980	14	120
2.	1981	17	135
3.	1982	25	150
4.	1983	30	180

If the data instead contained semicolons marking the end of each series and had no information for amount in 1983, the raw data might appear as

```
1980 1981 1982 1983 ;
14 17 25 ;
120 135 150
180 ;
```

We could read these data by typing

```
. infile year amount cost using time, byvariable(100) clear
(4 observations read)

. list
```

	year	amount	cost
1.	1980	14	120
2.	1981	17	135
3.	1982	25	150
4.	1983	.	180

◁

Also See

[D] **outfile** — Write ASCII-format dataset

[D] **outsheet** — Write spreadsheet-style dataset

[D] **save** — Save datasets

[D] **infile (fixed format)** — Read ASCII (text) data in fixed format with a dictionary

[D] **input** — Enter data from keyboard

[D] **insheet** — Read ASCII (text) data created by a spreadsheet

[U] **21 Inputting data**

[D] **infile** — Overview of reading data into Stata

Title

> **infix (fixed format)** — Read ASCII (text) data in fixed format

Syntax

infix using *dfilename* $\left[if \right]$ $\left[in \right]$ $\left[, \text{ using}(filename_2) \text{ clear} \right]$

infix *specification* using *filename* $\left[if \right]$ $\left[in \right]$ $\left[, \text{ clear} \right]$

where *dfilename*, if it exists, contains

———————————————————————————— top of dictionary file ————————

```
    infix dictionary [using filename] {
            * comments preceded by asterisk may appear freely
            specifications
    }
    (your data might appear here)
```

———————————————————————————— end of dictionary file ————————

and where *specification* is

```
# firstlineoffile
# lines
#:
/
[ byte | int | float | long | double | str ] varlist [#:]#[-#]
```

Description

infix reads into memory from a disk dataset that is *not* in Stata format. infix requires that the data be in fixed-column format.

If *dfilename* is specified without an extension, .dct is assumed. If *filename* is specified without an extension, .raw is assumed. If *dfilename* contains embedded spaces, remember to enclose it in double quotes.

In the first syntax, if using *filename₂* is not specified on the command line and using *filename* is not specified in the dictionary, the data are assumed to begin on the line following the closing brace.

infile and insheet are alternatives to infix. infile can also read data in fixed format—see [D] **infile (fixed format)**—and it can read data in free format—see [D] **infile (free format)**. Most people think that infix is easier to use for reading fixed-format data, but infile has more features. If your data are not fixed format, you can use insheet; see [D] **insheet**. If you are not certain that infix will do what you are looking for, see [D] **infile** and [U] **21 Inputting data**.

In its first syntax, infix reads the data in a two-step process. You first create a disk file describing how the data are recorded. You tell infix to read that file—called a dictionary—and from there, infix reads the data. The data can be in the same file as the dictionary or in a different file.

In its second syntax, you tell infix how to read the data right on the command line with no intermediate file.

306

Options

Main

using(*filename₂*) specifies the name of a file containing the data. If using() is not specified, the data are assumed to follow the dictionary in *dfilename*, or, if the dictionary specifies the name of some other file, that file is assumed to contain the data. If using(*filename₂*) is specified, *filename₂* is used to obtain the data, even if the dictionary says otherwise. If *filename₂* is specified without an extension, .raw is assumed. If *filename₂* contains embedded spaces, remember to enclose it in double quotes.

clear specifies that it is okay for the new data to replace what is currently in memory. To ensure that you do not lose something important, infix will refuse to read new data if data are already in memory. clear allows infix to replace the data in memory. You can also drop the data yourself by typing drop _all before reading new data.

Specifications

firstlineoffile (abbreviation first) is rarely specified. It states the line of the file at which the data begin. You need not specify first when the data follow the dictionary; infix can figure that out for itself. You can specify first when only the data appear in a file and the first few lines of that file contain headers or other markers.

first appears only once in the specifications.

lines states the number of lines per observation in the file. Simple datasets typically have "1 lines". Large datasets often have many lines (sometimes called records) per observation. lines is optional, even when there is more than one line per observation, because infix can sometimes figure it out for itself. Still, if 1 lines is not right for your data, it is best to specify the appropriate number of lines.

lines appears only once in the specifications.

#: tells infix to jump to line # of the observation. Consider a file with 4 lines, meaning four lines per observation. 2: says to jump to the second line of the observation. 4: says to jump to the fourth line of the observation. You may jump forward or backward: infix does not care, and there is no inefficiency in going forward to 3:, reading a few variables, jumping back to 1:, reading another variable, and jumping back again to 3:.

You need not ensure that, at the end of your specification, you are on the last line of the observation. infix knows how to get to the next observation because it knows where you are and it knows lines, the total number of lines per observation.

#: may appear many times in the specifications.

/ is an alternative to #:. / goes forward one line. // goes forward two lines. We do not recommend using / because #: is better. If you are currently on line 2 of an observation and want to get to line 6, you could type ////, but your meaning is clearer if you type 6:.

/ may appear many times in the specifications.

[byte | int | float | long | double | str] *varlist* [#:]#[-#] instructs infix to read a variable and, sometimes, more than one.

The simplest form of this is *varname* #, such as sex 20. That says that variable *varname* be read from column # of the current line; that variable sex be read from column 20; and that here, sex is a one-digit number.

varname #-#, such as `age 21-23`, says that *varname* be read from the column range specified; that `age` be read from columns 21 through 23; and that here, age is a three-digit number.

You can prefix the variable with a storage type. `str name 25-44` means to read the string variable `name` from columns 25 through 44. If you do not specify `str`, the variable is assumed to be numeric. You can specify the numeric subtype if you wish.

You can specify more than one variable, with or without a type. `byte q1-q5 51-55` means read variables q1, q2, q3, q4, and q5 from columns 51 through 55 and store the five variables as `bytes`.

Finally, you can specify the line on which the variable(s) appear. `age 2:21-23` says that age is to be obtained from the second line, columns 21 through 23. Another way to do this is to put together the `#:` directive with the input-variable directive: `2: age 21-23`. There is a difference, but not with respect to reading the variable `age`. Let's consider two alternatives:

```
1:  str name 25-44    age 2:21-23   q1-q5 51-55
1:  str name 25-44  2: age 21-23    q1-q5 51-55
```

The difference is that the first directive says that variables q1 through q5 are on line 1, whereas the second says that they are on line 2.

When the colon is put in front, it indicates the line on which variables are to be found when we do not explicitly say otherwise. When the colon is put inside, it applies only to the variable under consideration.

Remarks

Remarks are presented under the following headings:

> *Two ways to use infix*
> *Reading string variables*
> *Reading data with multiple lines per observation*
> *Reading subsets of observations*

Two ways to use infix

There are two ways to use `infix`. One is to type the specifications that describe how to read the fixed-format data on the command line:

```
. infix acc_rate 1-4  spdlimit 6-7  acc_pts 9-11  using highway.raw
```

The other is to type the specifications into a file

```
─────────────────────────────────────── top of highway.dct, example 1 ───────────
infix dictionary using highway.raw {
        acc_rate 1-4
        spdlimit 6-7
        acc_pts  9-11
}
─────────────────────────────────────── end of highway.dct, example 1 ───────────
```

and then, Stata, type

```
. infix using highway.dct
```

The method you use makes no difference to Stata. The first method is more convenient if there are only a few variables, and the second method is less prone to error if you are reading a big, complicated file.

The second method allows two variations, the one we just showed—where the data are in another file—and one where the data are in the same file as the dictionary:

```
──────────────────────────────────────── top of highway.dct, example 2 ────────────
    infix dictionary {
            acc_rate 1-4
            spdlimit 6-7
            acc_pts  9-11
    }
    4.58 55 .46
    2.86 60 4.4
    1.61    2.2
    3.02 60 4.7
──────────────────────────────────────── end of highway.dct, example 2 ────────────
```

Note that, in the first example, the top line of the file read infix dictionary using highway.raw, whereas in the second, the line reads simply infix dictionary. When you do not say where the data are, Stata assumes that the data follow the dictionary.

▷ Example 1

So, let's complete the example we started. We have a dataset on the accident rate per million vehicle miles along a stretch of highway, the speed limit on that highway, and the number of access points per mile. We have created the dictionary file, highway.dct, which contains the dictionary and the data:

```
──────────────────────────────────────── top of highway.dct, example 2 ────────────
    infix dictionary {
            acc_rate 1-4
            spdlimit 6-7
            acc_pts  9-11
    }
    4.58 55 .46
    2.86 60 4.4
    1.61    2.2
    3.02 60 4.7
──────────────────────────────────────── end of highway.dct, example 2 ────────────
```

We created this file outside Stata by using an editor or word processor. In Stata, we now read the data. infix lists the dictionary so that we will know the directives it follows:

```
. infix using highway
infix dictionary {
        acc_rate 1-4
        spdlimit 6-7
        acc_pts  9-11
}
(4 observations read)
. list
```

	acc_rate	spdlimit	acc_pts
1.	4.58	55	.46
2.	2.86	60	4.4
3.	1.61	.	2.2
4.	3.02	60	4.7

We simply typed infix using highway rather than infix using highway.dct. When we do not specify the file extension, infix assumes that we mean .dct.

◁

Reading string variables

When you do not say otherwise in your specification—either in the command line or in the dictionary—infix assumes that variables are numeric. You specify that a variable is a string by placing str in front of its name:

```
. infix id 1-6  str name 7-36  age 38-39  str sex 41  using employee.raw
```

or

```
——————————————————————————————— top of employee.dct ————————
    infix dictionary using employee.raw {
            id          1-6
            str name    7-36
            age         38-39
            str sex     40
    }
——————————————————————————————— end of employee.dct ————————
```

Reading data with multiple lines per observation

When a dataset has multiple lines per observation—sometimes called multiple records per observation—you specify the number of lines per observation by using lines, and you specify the line on which the elements appear by using #:.

```
. infix  2 lines  1: id 1-6  str name 7-36  2: age 1-2  str sex 4  using emp2.raw
```

or

```
——————————————————————————————— top of emp2.dct ————————
    infix dictionary using emp2.raw {
        2 lines
        1:
            id          1-6
            str name    7-36
        2:
            age         1-2
            str sex     4
    }
——————————————————————————————— end of emp2.dct ————————
```

There are many different ways to do the same thing.

▷ Example 2

Consider the following raw data:

```
                                                        top of mydata.raw ————
  id income educ / sex age / rcode, answers to questions 1-5
  1024 25000 HS
       Male   28
        1 1 9 5 0 3
  1025 27000 C
       Female 24
        0 2 2 1 1 3
  1035 26000 HS
       Male   32
        1 1 0 3 2 1
  1036 25000 C
       Female 25
        1 3 1 2 3 2
                                                         end of mydata.raw ————
```

This dataset has three lines per observation, and the first line is just a comment. One possible method for reading these data is

```
                                                        top of mydata1.dct ————
  infix dictionary using mydata {
      2 first
      3 lines
      1:    id        1-4
            income    6-10
            str educ 12-13
      2:    str sex   6-11
            int age  13-14
      3:    rcode     6
            q1-q5     7-16
  }
                                                        end of mydata1.dct ————
```

although we prefer

```
                                                        top of mydata2.dct ————
  infix dictionary using mydata {
      2 first
      3 lines
            id        1: 1-4
            income    1: 6-10
            str educ 1:12-13
            str sex  2: 6-11
            age      2:13-14
            rcode    3: 6
            q1-q5    3: 7-16
  }
                                                        end of mydata2.dct ————
```

Either method will read these data, so we will use the first and then explain why we prefer the second.

```
. infix using mydata1
infix dictionary using mydata {
    2 first
    3 lines
    1:      id        1-4
            income    6-10
            str educ 12-13
    2:      str sex   6-11
            int age  13-14
    3:      rcode     6
            q1-q5     7-16
}
(4 observations read)
. list in 1/2
```

	id	income	educ	sex	age	rcode	q1	q2	q3	q4	q5
1.	1024	25000	HS	Male	28	1	1	9	5	0	3
2.	1025	27000	C	Female	24	0	2	2	1	1	3

What is better about the second is that the location of each variable is completely documented on each line—the line number and column. Since `infix` does not care about the order in which we read the variables, we could take the dictionary and jumble the lines, and it would still work. For instance,

———————————————————————————————— top of mydata3.dct ————————

```
infix dictionary using mydata {
    2 first
    3 lines
            str sex   2: 6-11
            rcode     3: 6
            str educ 1:12-13
            age       2:13-14
            id        1: 1-4
            q1-q5     3: 7-16
            income    1: 6-10
}
```

———————————————————————————————— end of mydata3.dct ————————

will also read these data even though, for each observation, we start on line 2, go forward to line 3, jump back to line 1, and end up on line 1. It is not inefficient to do this because `infix` does not really jump to record 2, then record 3, then record 1 again, etc. `infix` takes what we say and organizes it efficiently. The order in which we say it makes no difference, except that the order of the variables in the resulting Stata dataset will be the order we specify.

Here the reordering is senseless, but in real datasets, reordering variables is often desirable. Moreover, we often construct dictionaries, realize that we omitted a variable, and then go back and modify them. By making each line complete, we can add new variables anywhere in the dictionary and not worry that, because of our addition, something that occurs later will no longer read correctly.

◁

Reading subsets of observations

If you wanted to read only the information about males from some raw data file, you might type

```
. infix id 1-6  str name 7-36  age 38-39  str sex 41  using employee.raw
> if sex=="M"
```

If your specification was instead recorded in a dictionary, you could type

```
. infix using employee.dct if sex=="M"
```

In another dataset, if you wanted to read just the first 100 observations, you could type

```
. infix 2 lines  1: id 1-6  str name 7-36  2: age 1-2  str sex 4  using emp2.raw
> in 1/100
```

or, if the specification was instead recorded in a dictionary and you wanted observations 101–573, you could type

```
. infix using emp2.dct in 101/573
```

Also See

[D] **outfile** — Write ASCII-format dataset

[D] **outsheet** — Write spreadsheet-style dataset

[D] **save** — Save datasets

[D] **infile (fixed format)** — Read ASCII (text) data in fixed format with a dictionary

[D] **insheet** — Read ASCII (text) data created by a spreadsheet

[U] **21 Inputting data**

[D] **infile** — Overview of reading data into Stata

Title

---
input — Enter data from keyboard

Syntax

input [*varlist*] [, <u>au</u>tomatic <u>la</u>bel]

Description

input allows you to type data directly into the dataset in memory. See also [D] **edit** for a windowed alternative to input.

Options

<u>au</u>tomatic causes Stata to create value labels from the nonnumeric data it encounters. It also automatically widens the display format to fit the longest label. Specifying automatic implies label, even if you do not explicitly type the label option.

label allows you to type the labels (strings) instead of the numeric values for variables associated with value labels. New value labels are not automatically created unless automatic is specified.

Remarks

If no data are in memory, you must specify a *varlist* when you type input. Stata will then prompt you to enter the new observations until you type end.

▷ Example 1

We have data on the accident rate per million vehicle miles along a stretch of highway, along with the speed limit on that highway. We wish to type these data directly into Stata:

```
. input
nothing to input
r(104);
```

Typing input by itself does not provide enough information about our intentions. Stata needs to know the names of the variables we wish to create.

```
. input acc_rate spdlimit

      acc_rate    spdlimit
  1. 4.58 55
  2. 2.86 60
  3. 1.61 .
  4. end

. _
```

We typed input acc_rate spdlimit, and Stata responded by repeating the variable names and prompting us for the first observation. We entered the values for the first two observations, pressing *Return* after each value was entered. For the third observation, we entered the accident rate (1.61), but we entered a period (.) for missing because we did not know the corresponding speed limit for the highway. After entering data for the fourth observation, we typed end to let Stata know that there were no more observations.

We can now list the data to verify that we have entered the data correctly:

```
. list
```

	acc_rate	spdlimit
1.	4.58	55
2.	2.86	60
3.	1.61	.

◁

If you have data in memory and type input without a *varlist*, you will be prompted to enter more information on *all* the variables. This continues until you type end.

▷ Example 2: Adding observations

We now have another observation that we wish to add to the dataset. Typing input by itself tells Stata that we wish to add new observations:

```
. input

        acc_rate   spdlimit
4. 3.02 60
5. end

. _
```

Stata reminded us of the names of our variables and prompted us for the fourth observation. We entered the numbers 3.02 and 60 and pressed *Return*. Stata then prompted us for the fifth observation. We could add as many new observations as we wish. Since we needed to add only 1 observation, we typed end. Our dataset now has 4 observations.

◁

You may add new variables to the data in memory by typing input followed by the names of the new variables. Stata will begin by prompting you for the first observation, then the second, and so on, until you type end or enter the last observation.

▷ Example 3: Adding variables

In addition to the accident rate and speed limit, we now obtain data on the number of access points (on-ramps and off-ramps) per mile along each stretch of highway. We wish to enter the new data.

```
. input acc_pts

        acc_pts
1. 4.6
2. 4.4
3. 2.2
4. 4.7

. _
```

When we typed `input acc_pts`, Stata responded by prompting us for the first observation. There are 4.6 access points per mile for the first highway, so we entered 4.6. Stata then prompted us for the second observation, and so on. We entered each of the numbers. When we entered the final observation, Stata automatically stopped prompting us—we did not have to type `end`. Stata knows that there are 4 observations in memory, and since we are adding a new variable, it stops automatically.

We can, however, type `end` anytime we wish, and Stata fills the remaining observations on the new variables with *missing*. To illustrate this, we enter one more variable to our data and then `list` the result:

```
. input junk

        junk
1. 1
2. 2
3. end

. list
```

	acc_rate	spdlimit	acc_pts	junk
1.	4.58	55	4.6	1
2.	2.86	60	4.4	2
3.	1.61	.	2.2	.
4.	3.02	60	4.7	.

◁

You can input string variables by using `input`, but you must remember to indicate explicitly that the variables are strings by specifying the type of the variable before the variable's name.

▷ Example 4: Inputting string variables

String variables are indicated by the types `str#`, where `#` represents the storage length, or maximum length, of the variable. For instance, a `str4` variable has a maximum length of 4, meaning that it can contain the strings a, ab, abc, and abcd, but not abcde. Strings shorter than the maximum length can be stored in the variable, but strings longer than the maximum length cannot. You can create variables up to `str244`.

Although a `str80` variable can store strings shorter than 80 characters, you should not make all your string variables `str80` because Stata allocates space for strings on the basis of their *maximum* length. Thus doing so would waste the computer's memory.

Let's assume that we have no data in memory and wish to enter the following data:

```
. input str16 name age str6 sex

                name         age          sex
1. "Arthur Doyle" 22 male
2. "Mary Hope" 37 "female"
3. Guy Fawkes 48 male
'Fawkes' cannot be read as a number
3. "Guy Fawkes" 48 male
4. "Kriste Yeager" 25 female
5. end

. _
```

We first typed `input str16 name age str6 sex`, meaning that `name` is to be a `str16` variable and `sex` a `str6` variable. Since we did not specify anything about `age`, Stata made it a numeric variable.

Stata then prompted us to enter our data. On the first line, the name is Arthur Doyle, which we typed in double quotes. The double quotes are not really part of the string; they merely delimit the beginning and end of the string. We followed that with Mr. Doyle's age, 22, and his sex, male. We did not bother to type double quotes around the word `male` because it contained no blanks or special characters. For the second observation, we typed the double quotes around `female`; it changed nothing.

In the third observation, we omitted the double quotes around the name, and Stata informed us that Fawkes could not be read as a number and reprompted us for the observation. When we omitted the double quotes, Stata interpreted `Guy` as the name, `Fawkes` as the age, and 48 as the sex. This would have been okay with Stata, except for one problem: `Fawkes` looks nothing like a number, so Stata complained and gave us another chance. This time, we remembered to put the double quotes around the name.

Stata was satisfied, and we continued. We entered the fourth observation and typed `end`. Here is our dataset:

```
. list
```

	name	age	sex
1.	Arthur Doyle	22	male
2.	Mary Hope	37	female
3.	Guy Fawkes	48	male
4.	Kriste Yeager	25	female

◁

▷ Example 5: Specifying numeric storage types

Just as we indicated the string variables by placing a storage type in front of the variable name, we can indicate the storage type of our numeric variables as well. Stata has five numeric storage types: `byte`, `int`, `long`, `float`, and `double`. When you do not specify the storage type, Stata assumes that the variable is a `float`. See the definitions of numbers in [U] **12 Data**.

There are two reasons for explicitly specifying the storage type: to induce more precision or to conserve memory. The default type `float` has plenty of precision for most circumstances because Stata performs all calculations in double precision, no matter how the data are stored. If you were storing nine-digit Social Security numbers, however, you would want to use a different storage type, or the last digit would be rounded. `long` would be the best choice; `double` would work equally well, but it would waste memory.

Sometimes you do not need to store a variable as `float`. If the variable contains only integers between $-32,767$ and $32,740$, it can be stored as an `int` and would take only half the space. If a variable contains only integers between -127 and 100, it can be stored as a `byte`, which would take only half again as much space. For instance, in example 4 we entered data for `age` without explicitly specifying the storage type; hence, it was stored as a `float`. It would have been better to store it as a `byte`. To do that, we would have typed

```
. input str16 name byte age str6 sex
              name          age       sex
  1. "Arthur Doyle" 22 male
  2. "Mary Hope" 37 "female"
  3. "Guy Fawkes" 48 male
  4. "Kriste Yeager" 25 female
  5. end

  .  _
```

Stata understands several shorthands. For instance, typing

. `input int(a b) c`

allows you to input three variables—a, b, and c—and makes both a and b `ints` and c a `float`. Remember, typing

. `input int a b c`

would make a an `int` but both b and c `floats`. Typing

. `input a long b double(c d) e`

would make a a `float`, b a `long`, c and d `doubles`, and e a `float`.

Stata has a shorthand for variable names with numeric suffixes. Typing `v1-v4` is equivalent to typing `v1 v2 v3 v4`. Thus typing

. `input int(v1-v4)`

inputs four variables and stores them as `ints`.

◁

❏ Technical Note

The rest of this section deals with using `input` with value labels. If you are not familiar with value labels, see [U] **12.6.3 Value labels**.

Value labels map numbers into words and vice versa. There are two aspects to the process. First, we must define the association between numbers and words. We might tell Stata that 0 corresponds to `male` and 1 corresponds to `female` by typing `label define sexlbl 0 "male" 1 "female"`. The correspondences are named, and here we have named the 0↔male 1↔female correspondence `sexlbl`.

Next, we must associate this value label with a variable. If we had already entered the data and the variable were called `sex`, we would do this by typing `label values sex sexlbl`. We would have entered the data by typing 0s and 1s, but at least now when we `list` the data, we would see the words rather than the underlying numbers.

We can do better than that. After defining the value label, we can associate the value label with the variable at the time we `input` the data and tell Stata to use the value label to interpret what we type:

```
. label define sexlbl 0 "male" 1 "female"
. input str16 name byte(age sex:sexlbl), label
              name       age        sex
  1. "Arthur Doyle" 22 male
  2. "Mary Hope" 37 "female"
  3. "Guy Fawkes" 48 male
  4. "Kriste Yeager" 25 female
  5. end

. _
```

After defining the value label, we typed our `input` command. We added the `label` option at the end of the command, and we typed `sex:sexlbl` for the name of the sex variable. The `byte(...)` around `age` and `sex:sexlbl` was not really necessary; it merely forced both `age` and `sex` to be stored as `bytes`.

Let's first decipher sex:sexlbl. sex is the name of the variable we want to input. The :sexlbl part tells Stata that the new variable is to be associated with the value label named sexlbl. The label option tells Stata to look up any strings we type for labeled variables in their corresponding value label and substitute the number when it stores the data. Thus when we entered the first observation of our data, we typed male for Mr. Doyle's sex, even though the corresponding variable is numeric. Rather than complaining that ""male" could not be read as a number", Stata accepted what we typed, looked up the number corresponding to male, and stored that number in the data.

That Stata has actually stored a number rather than the words male or female is almost irrelevant. Whenever we list the data or make a table, Stata will use the words male and female just as if those words were actually stored in the dataset rather than their numeric codings:

```
. list
```

	name	age	sex
1.	Arthur Doyle	22	male
2.	Mary Hope	37	female
3.	Guy Fawkes	48	male
4.	Kriste Yeager	25	female

```
. tabulate sex
```

sex	Freq.	Percent	Cum.
male	2	50.00	50.00
female	2	50.00	100.00
Total	4	100.00	

It is only almost irrelevant because we can use the underlying numbers in statistical analyses. For instance, if we were to ask Stata to calculate the mean of sex by typing summarize sex, Stata would report 0.5. We would interpret that to mean that one-half of our sample is female.

Value labels are permanently associated with variables, so once we associate a value label with a variable, we never have to do so again. If we wanted to add another observation to these data, we could type

```
. input, label
                 name        age        sex
  5. "Mark Esman" 26 male
  6. end

. _
```

❑

❑ Technical Note

The automatic option automates the definition of the value label. In the previous example, we informed Stata that male corresponds to 0 and female corresponds to 1 by typing label define sexlbl 0 "male" 1 "female". It was not necessary to explicitly specify the mapping. Specifying the automatic option tells Stata to interpret what we type as follows:

First, see if the value is a number. If so, store that number and be done with it. If it is not a number, check the value label associated with the variable in an attempt to interpret it. If an interpretation exists, store the corresponding numeric code. If one does not exist, add a new numeric code corresponding to what was typed. Store that new number and update the value label so that the new correspondence is never forgotten.

We can use these features to reenter our age and sex data. Before reentering the data, we drop
_all and label drop _all to prove that we have nothing up our sleeve:

```
. drop _all

. label drop _all

. input str16 name byte(age sex:sexlbl), automatic
                    name        age       sex
  1. "Arthur Doyle" 22 male
  2. "Mary Hope" 37 "female"
  3. "Guy Fawkes" 48 male
  4. "Kriste Yeager" 25 female
  5. end

.
```

We previously defined the value label sexlbl so that male corresponded to 0 and female corresponded
to 1. The label that Stata automatically created is slightly different, but is just as good:

```
. label list sexlbl
sexlbl:
               1 male
               2 female
```

❏

Reference

Kohler, U. 2005. Stata tip 16: Using input to generate variables. *Stata Journal* 5: 134.

Also See

[D] **save** — Save datasets

[D] **edit** — Edit and list data with Data Editor

[D] **infile** — Overview of reading data into Stata

[U] **21 Inputting data**

Title

> **insheet** — Read ASCII (text) data created by a spreadsheet

Syntax

insheet [*varlist*] using *filename* [, *options*]

options	description
[no]double	override default storage type
tab	tab-delimited data
comma	comma-delimited data
delimiter("*char*")	use *char* as delimiter
clear	replace data in memory
case	preserve variable name's case
† [no]names	variable names are included on the first line of the file

† [no]names is not shown in the dialog box.

Description

insheet reads into memory from a disk a dataset that is not in Stata format. insheet is intended for reading files created by a spreadsheet or database program. Regardless of the creator of the file, insheet reads text (ASCII) files in which there is 1 observation per line and the values are separated by tabs or commas. Also the first line of the file can contain the variable names. If you type

    . insheet using *filename*

insheet reads your data; that's all there is to it.

If *filename* is specified without an extension, .raw is assumed. If your *filename* contains embedded spaces, remember to enclose it in double quotes.

Stata has other commands for reading data. If you are not sure that insheet will do what you are looking for, see [D] **infile** and [U] **21 Inputting data**. If you want to save your data in spreadsheet-style format, see [D] **outsheet**.

Options

[no]double affects the way Stata handles the storage of floating-point variables. If the default storage type (see [D] **generate**) is set to float, specifying the double option forces Stata to store floating-point variables as doubles rather than floats. If the default storage type has been set to double, you must specify nodouble to have floating-point variables stored as floats rather than doubles; see [U] **12.2.2 Numeric storage types**.

tab tells Stata that the values are tab separated. Specifying this option will speed insheet's processing, assuming that you are right. insheet can determine for itself whether the separation character is a tab or a comma.

comma tells Stata that the values are comma separated. Specifying this option will speed insheet's processing, assuming that you are right. insheet can determine for itself whether the separation character is a comma or a tab.

delimiter("*char*") allows you to specify other separation characters. For instance, if values in the file are separated by a semicolon, specify delimiter(";").

clear specifies that it is okay for the new data to replace the data that are currently in memory. To ensure that you do not lose something important, insheet will refuse to read new data if data are already in memory. clear allows insheet to replace the data in memory. You can also drop the data yourself by typing drop _all before reading new data.

case preserves the variable name's case. By default, all variable names are imported as lowercase.

The following option is available with insheet but is not shown in the dialog box:

[no]names informs Stata whether variable names are included on the first line of the file. Specifying this option will speed insheet's processing, assuming that you are right. insheet can determine for itself whether the file includes variable names.

Remarks

insheet is easy. You type

    . insheet using *filename*

and insheet reads your data. That is, it reads your data if

1. it can find the file and

2. the file meets insheet's expectations as to its format.

Assuring 1 is easy enough; just realize that if you type insheet using myfile, Stata interprets this as an instruction to read myfile.raw. If your file is called myfile.txt, type insheet using myfile.txt.

As for the file's format, most spreadsheets and some database programs write data in the form insheet expects. It is easy enough to look—as we will show you—and it is even easier simply to try and see what happens. If typing

    . insheet using *filename*

does not produce the desired result, try one of Stata's other infile commands; see [D] **infile**.

▷ Example 1

We have a raw data file on automobiles called auto.raw. This file was saved by a spreadsheet and can be read by typing

    . insheet using auto
    (5 vars, 10 obs)

    . _

That done, we can now look at what we just loaded:

```
. describe
Contains data
  obs:             10
  vars:             5
  size:            310 (99.8% of memory free)

                storage  display    value
variable name    type    format     label     variable label

make             str13   %13s
price            int     %8.0g
mpg              byte    %8.0g
rep78            byte    %8.0g
foreign          str10   %10s

Sorted by:
     Note:  dataset has changed since last saved

. list

            make    price   mpg   rep78    foreign

  1.   AMC Concord     4099    22       3   Domestic
  2.     AMC Pacer     4749    17       3   Domestic
  3.    AMC Spirit     3799    22       .   Domestic
  4.  Buick Century    4816    20       3   Domestic
  5.  Buick Electra    7827    15       4   Domestic

  6.  Buick LeSabre    5788    18       3   Domestic
  7.    Buick Opel     4453    26       .   Domestic
  8.    Buick Regal    5189    20       3   Domestic
  9.  Buick Riviera   10372    16       3   Domestic
 10.  Buick Skylark    4082    19       3   Domestic
```

These data contain a combination of string and numeric variables. insheet figured all that out by itself.

◁

❑ Technical Note

Now let's back up and look at the auto.raw file. Stata's type command will display files on the screen:

```
. type auto.raw
make      price   mpg    rep78    foreign
AMC Concord    4099   22     3        Domestic
AMC Pacer      4749   17     3        Domestic
AMC Spirit     3799   22     .        Domestic
Buick Century  4816   20     3        Domestic
Buick Electra  7827   15     4        Domestic
Buick LeSabre  5788   18     3        Domestic
Buick Opel     4453   26     .        Domestic
Buick Regal    5189   20     3        Domestic
Buick Riviera  10372  16     3        Domestic
Buick Skylark  4082   19     3        Domestic
```

These data have tab characters between values. Tab characters are invisible and are indistinguishable from blanks. type's showtabs option makes the tabs visible:

```
. type auto.raw, showtabs
make<T>price<T>mpg<T>rep78<T>foreign
AMC Concord<T>4099<T>22<T>3<T>Domestic
AMC Pacer<T>4749<T>17<T>3<T>Domestic
AMC Spirit<T>3799<T>22<T>.<T>Domestic
Buick Century<T>4816<T>20<T>3<T>Domestic
Buick Electra<T>7827<T>15<T>4<T>Domestic
Buick LeSabre<T>5788<T>18<T>3<T>Domestic
Buick Opel<T>4453<T>26<T>.<T>Domestic
Buick Regal<T>5189<T>20<T>3<T>Domestic
Buick Riviera<T>10372<T>16<T>3<T>Domestic
Buick Skylark<T>4082<T>19<T>3<T>Domestic
```

This is an example of the kind of data insheet is willing to read. The first line contains the variable names, although that is not necessary. What is necessary is that the data values have tab characters between them.

insheet would be just as happy if the data values were separated by commas. Here is another variation on auto.raw that insheet can read:

```
. type auto2.raw
make,price,mpg,rep78,foreign
AMC Concord,4099,22,3,Domestic
AMC Pacer,4749,17,3,Domestic
AMC Spirit,3799,22,,Domestic
Buick Century,4816,20,3,Domestic
Buick Electra,7827,15,4,Domestic
Buick LeSabre,5788,18,3,Domestic
Buick Opel,4453,26,,Domestic
Buick Regal,5189,20,3,Domestic
Buick Riviera,10372,16,3,Domestic
Buick Skylark,4082,19,3,Domestic
```

It is easier for us human beings to see the commas rather than the tabs, but computers do not care one way or the other.

❏

▷ Example 2

The file does not have to contain variable names. Here is another variation on auto.raw without the first line, this time with commas rather than tabs separating the values:

```
. type auto3.raw
AMC Concord,4099,22,3,Domestic
AMC Pacer,4749,17,3,Domestic
 (output omitted )
Buick Skylark,4082,19,3,Domestic
```

Here is what happens when we read it:

```
. insheet using auto3
you must start with an empty dataset
r(18);

. _
```

Oops! We still have the data from the last example in memory. We need to clear the old data before reading the new data.

```
. insheet using auto3, clear
(5 vars, 10 obs)

. describe

Contains data
  obs:             10
  vars:             5
  size:           310 (99.8% of memory free)
```

variable name	storage type	display format	value label	variable label
v1	str13	%13s		
v2	int	%8.0g		
v3	byte	%8.0g		
v4	byte	%8.0g		
v5	str10	%10s		

```
Sorted by:
     Note:  dataset has changed since last saved

. list
```

	v1	v2	v3	v4	v5
1.	AMC Concord	4099	22	3	Domestic
2.	AMC Pacer	4749	17	3	Domestic
	(output omitted)				
10.	Buick Skylark	4082	19	3	Domestic

The only difference in this dataset is that rather than the variables being nicely named make, price, mpg, rep78, and foreign, they are named v1, v2, ..., v5. We could now give our variables nicer names:

```
. rename v1 make

. rename v2 price

. _
```

We can also specify the variable names when reading the data:

```
. insheet make price mpg rep78 foreign using auto3, clear
(5 vars, 10 obs)

. list
```

	make	price	mpg	rep78	foreign
1.	AMC Concord	4099	22	3	Domestic
2.	AMC Pacer	4749	17	3	Domestic
	(output omitted)				
10.	Buick Skylark	4082	19	3	Domestic

If we use this approach, we must not specify too few variables

```
. insheet make price mpg rep78 using auto3, clear
too few variables specified
error in line 11 of file
r(102);
```

or too many:

```
. insheet make price mpg rep78 foreign weight using auto3, clear
too many variables specified
error in line 11 of file
r(103);
```

We recommend typing

```
. insheet using filename
```

It is not difficult to rename your variables afterward, should that be necessary.

◁

▷ Example 3

The data may not always be appropriate for reading by insheet. Here is yet another version of the automobile data:

```
. type auto4.raw, showtabs
"AMC Concord"     4099  22  3  Domestic
"AMC Pacer"       4749  17  3  Domestic
"AMC Spirit"      3799  22  .  Domestic
"Buick Century"   4816  20  3  Domestic
"Buick Electra"   7827  15  4  Domestic
"Buick LeSabre"   5788  18  3  Domestic
"Buick Opel"      4453  26  .  Domestic
"Buick Regal"     5189  20  3  Domestic
"Buick Riviera"  10372  16  3  Domestic
"Buick Skylark"   4082  19  3  Domestic
```

We specified type's showtabs option, and no tabs are shown. These data are not tab delimited or comma delimited and are not the kind of data that insheet is designed to read. Let's try insheet anyway:

```
. insheet using auto4, clear
(1 var, 10 obs)
. describe
Contains data
  obs:            10
  vars:            1
  size:          430 (99.8% of memory free)
```

variable name	storage type	display format	value label	variable label
v1	str39	%39s		

```
Sorted by:
     Note:  dataset has changed since last saved
. list
```

	v1
1.	AMC Concord 4099 22 3 Domestic
2.	AMC Pacer 4749 17 3 Domestic
	(output omitted)
10.	Buick Skylark 4082 19 3 Domestic

When `insheet` tries to read data that have no tabs or commas, it is fooled into thinking that the data contain just one variable. If we had these data, we would have to read the data with one of Stata's other commands, such as `infile` (free format).

◁

Also See

[D] **outfile** — Write ASCII-format dataset

[D] **outsheet** — Write spreadsheet-style dataset

[D] **rename** — Rename variable

[D] **save** — Save datasets

[D] **infile (free format)** — Read unformatted ASCII (text) data

[U] **21 Inputting data**

[D] **infile** — Overview of reading data into Stata

Title

inspect — Display simple summary of data's attributes

Syntax

<u>insp</u>ect [*varlist*] [*if*] [*in*]

by is allowed; see [D] **by**.

Description

The inspect command quickly summarizes a numeric variable that differs from that provided by summarize or tabulate. It reports the number of negative, zero, and positive values; the number of integers and nonintegers; the number of unique values; and the number of *missing*; and it produces a small histogram. Its purpose is not analytical but is to allow you to quickly gain familiarity with unknown data.

Remarks

Typing inspect by itself produces an inspection for all the variables in the dataset. If you specify the *varlist*, an inspection of just those variables is presented.

▷ Example 1

inspect is not a replacement or substitute for summarize and tabulate. It is instead a data management or information tool that lets us quickly gain insight into the values stored in a variable.

For instance, we receive data that purport to be on automobiles, and among the variables in the dataset is one called mpg. Its variable label is Mileage (mpg), which is surely suggestive. We inspect the variable

```
. use http://www.stata-press.com/data/r10/auto
(1978 Automobile Data)

. inspect mpg
mpg:  Mileage (mpg)                          Number of Observations

                                         Total    Integers   Nonintegers
          #                   Negative      -          -           -
          #                   Zero          -          -           -
          #                   Positive     74         74           -
      #   #
      #   #   #               Total        74         74           -
      #   #   #   #     .     Missing       -

    12              41                      74
      (21 unique values)
```

and we discover that the variable is never *missing*; all 74 observations in the dataset have some value for mpg. Moreover, the values are all positive and are all integers, as well. Among those 74 observations are 21 unique (different) values. The variable ranges from 12 to 41, and we are provided with a small histogram that suggests that the variable appears to be what it claims.

◁

▷ Example 2

Bob, a coworker, presents us with some census data. Among the variables in the dataset is one called `region`, which is labeled Census Region and is evidently a numeric variable. We `inspect` this variable.

```
. use http://www.stata-press.com/data/r10/bobsdata
(1980 Census data by state)
. inspect region
region:  Census region                             Number of Observations

                                              Total   Integers   Nonintegers
          #                      Negative        -        -           -
          #   #                  Zero            -        -           -
      #   #   #                  Positive       50       50           -
  #   #   #   #                                -----    -----       -----
  #   #   #   #                  Total          50       50           -
  #   #   #   #    .             Missing         -
 +----------------
 1               5                              -----
                                                 50
     (5 unique values)
         region is labeled but 1 value is NOT documented in the label.
```

In this dataset something may be wrong. `region` takes on five unique values. The variable has a value label, however, and one of the observed values is not documented in the label. Perhaps there is a typographical error.

◁

▷ Example 3

There was indeed an error. Bob fixes it and returns the data to us. Here is what `inspect` produces now:

```
. use http://www.stata-press.com/data/r10/census
(1980 Census data by state)
. inspect region
region:  Census region                             Number of Observations

                                              Total   Integers   Nonintegers
          #                      Negative        -        -           -
          #                      Zero            -        -           -
      #   #   #                  Positive       50       50           -
  #   #   #   #                                -----    -----       -----
  #   #   #   #                  Total          50       50           -
  #   #   #   #                  Missing         -
 +----------------
 1               4                              -----
                                                 50
     (4 unique values)
         region is labeled and all values are documented in the label.
```

◁

▷ Example 4

We receive data on the climate in 956 U.S. cities. The variable `tempjan` records the Average January `temperature` in degrees Fahrenheit. The results of `inspect` are

```
. use http://www.stata-press.com/data/r10/citytemp
(City Temperature Data)
. inspect tempjan
```

tempjan: Average January temperature   Number of Observations

						Total	Integers	Nonintegers
	#				Negative	–	–	–
	#				Zero	–	–	–
	#				Positive	954	78	876
	#	#	#					
	#	#	#		Total	954	78	876
.	#	#	#	.	Missing	2		
2.2			72.6			956		

(More than 99 unique values)

In two of the 956 observations, `tempjan` is *missing*. Of the 954 cities that have a recorded `tempjan`, all are positive, and 78 of them are integer values. `tempjan` varies between 2.2 and 72.6. There are more than 99 unique values of `tempjan` in the dataset. (Stata stops counting unique values after 99.)

◁

Saved Results

`inspect` saves the following in `r()`:

Scalars

`r(N)`	number of observations
`r(N_neg)`	number of negative observations
`r(N_0)`	number of observations equal to 0
`r(N_pos)`	number of positive observations
`r(N_negint)`	number of negative integer observations
`r(N_posint)`	number of positive integer observations
`r(N_unique)`	number of unique values or . if more than 99
`r(N_undoc)`	number of undocumented values or . if not labeled

Also See

[D] **codebook** — Describe data contents

[D] **compare** — Compare two variables

[D] **describe** — Describe data in memory or in file

[D] **isid** — Check for unique identifiers

[R] **lv** — Letter-value displays

[R] **summarize** — Summary statistics

[R] **table** — Tables of summary statistics

[R] **tabulate oneway** — One-way tables of frequencies

[R] **tabulate, summarize()** — One- and two-way tables of summary statistics

[R] **tabulate twoway** — Two-way tables of frequencies

Title

> **ipolate** — Linearly interpolate (extrapolate) values

Syntax

> ipolate *yvar xvar* [*if*] [*in*] , g̲enerate(*newvar*) [e̲polate]

by is allowed; see [D] **by**.

Description

> ipolate creates in *newvar* a linear interpolation of *yvar* on *xvar* for missing values of *yvar*.

> Because interpolation requires that *yvar* be a function of *xvar*, *yvar* is also interpolated for tied values of *xvar*. When *yvar* is not missing and *xvar* is neither missing nor repeated, the value of *newvar* is just *yvar*.

Options

> generate(*newvar*) is required and specifies the name of the new variable to be created.

> epolate specifies that values be both interpolated and extrapolated. Interpolation only is the default.

Remarks

▷ Example 1

> We have data points on y and x, although sometimes the observations on y are missing. We believe that y is a function of x, justifying filling in the missing values by linear interpolation:

```
. use http://www.stata-press.com/data/r10/ipolxmpl1
. list, sep(0)
```

	x	y
1.	0	.
2.	1	3
3.	1.5	.
4.	2	6
5.	3	.
6.	3.5	.
7.	4	18

```
. ipolate y x, gen(y1)
(1 missing value generated)
. ipolate y x, gen(y2) epolate
```

331

. list, sep(0)

	x	y	y1	y2
1.	0	.	.	0
2.	1	3	3	3
3.	1.5	.	4.5	4.5
4.	2	6	6	6
5.	3	.	12	12
6.	3.5	.	15	15
7.	4	18	18	18

◁

▷ Example 2

We have a dataset of circulations for 10 magazines from 1980 through 2003. The identity of the magazines is recorded in `magazine`, circulation is recorded in `circ`, and the year is recorded in `year`. In a few of the years, the circulation is not known, so we want to fill it in by linear interpolation.

```
. use http://www.stata-press.com/data/r10/ipolxmpl2, clear
. by magazine: ipolate circ year, gen(icirc)
```

When the by prefix is specified, interpolation is performed separately for each group.

◁

Methods and Formulas

`ipolate` is implemented as an ado-file.

The value y at x is found by finding the closest points (x_0, y_0) and (x_1, y_1), such that $x_0 < x$ and $x_1 > x$ where y_0 and y_1 are observed, and calculating

$$y = \frac{y_1 - y_0}{x_1 - x_0}(x - x_0) + y_0$$

If `epolate` is specified and if (x_0, y_0) and (x_1, y_1) cannot be found on both sides of x, the two closest points on the same side of x are found, and the same formula is applied.

If there are multiple observations with the same value for x_0, y_0 is taken as the average of the corresponding y values for those observations. (x_1, y_1) is handled in the same way.

Also See

[R] **lowess** — Lowess smoothing

[D] **impute** — Fill in missing values

Title

isid — Check for unique identifiers

Syntax

isid *varlist* [using *filename*] [, <u>s</u>ort <u>m</u>issok]

Description

isid checks whether the specified variables uniquely identify the observations.

Options

sort specifies that the dataset be sorted by *varlist*.

missok indicates that missing values are permitted in *varlist*.

Remarks

▷ Example 1

Suppose that we want to check whether the mileage ratings (mpg) uniquely identify the observations in our auto dataset.

```
. use http://www.stata-press.com/data/r10/auto
(1978 Automobile Data)
. isid mpg
variable mpg does not uniquely identify the observations
r(459);
```

isid returns an error and reports that there are multiple observations with the same mileage rating. We can locate those observations manually:

```
. sort mpg
. by mpg: generate nobs = _N
. list make mpg if nobs >1, sepby(mpg)
```

	make	mpg
1.	Linc. Mark V	12
2.	Linc. Continental	12
	(output omitted)	
68.	Mazda GLC	30
69.	Dodge Colt	30
72.	Datsun 210	35
73.	Subaru	35

◁

▷ Example 2

isid is useful for checking a time-series panel dataset. For this type of dataset, we usually need two variables to identify the observations: one that labels the individual IDs and another that labels the periods. Before we set the data using `tsset`, we want to make sure that there are no duplicates with the same panel ID and time. Suppose that we have a dataset that records the yearly gross investment of 10 companies for 20 years. The panel and time variables are `company` and `year`.

```
. use http://www.stata-press.com/data/r10/grunfeld, clear
. isid company year
```

isid reports no error, so the two variables `company` and `year` uniquely identify the observations. Therefore, we should be able to `tsset` the data successfully:

```
. tsset company year
       panel variable:  company (strongly balanced)
        time variable:  year, 1935 to 1954
               delta:  1 year
```

◁

❑ Technical Note

The `sort` option is a convenient shortcut, especially when combined with `using`. The following command

```
. isid patient_id date using newdata, sort
```

is equivalent to

```
. preserve
. use newdata, clear
. sort patient_id date
. isid patient_id date
. save, replace
. restore
```

❑

Methods and Formulas

isid is implemented as an ado-file.

Also See

[D] **describe** — Describe data in memory or in file

[D] **duplicates** — Report, tag, or drop duplicate observations

[D] **lookfor** — Search for string in variable names and labels

[D] **codebook** — Describe data contents

[D] **inspect** — Display simple summary of data's attributes

Title

joinby — Form all pairwise combinations within groups

Syntax

joinby [*varlist*] using *filename* [, *options*]

options	description
Options	
When observations match:	
update	replace missing data in memory with values from *filename*
replace	replace all data in memory with values from *filename*
When observations do not match:	
unmatched(none)	ignore all; the default
unmatched(both)	include from both datasets
unmatched(master)	include from data in memory
unmatched(using)	include from data in *filename*
_merge(*varname*)	*varname* marks source of resulting observation; default is _merge
nolabel	do not copy value label definitions from *filename*

Description

joinby joins, within groups formed by *varlist*, observations of the dataset in memory with *filename*, a Stata-format dataset. By *join* we mean to form all pairwise combinations. *filename* is required to be sorted by *varlist*. If *filename* is specified without an extension, .dta is assumed.

If *varlist* is not specified, joinby takes as *varlist* the set of variables common to the dataset in memory and in *filename*.

Observations unique to one or the other dataset are ignored unless unmatched() specifies differently. Whether you load one dataset and join the other or vice versa makes no difference in the number of resulting observations.

If there are common variables between the two datasets, however, the combined dataset will contain the values from the master data for those observations. This behavior can be modified with the update and replace options.

Options

_____ Options _____

update varies the action that joinby takes when an observation is matched. By default, values from the master data are retained when the same variables are found in both datasets. If update is specified, however, the values from the using dataset are retained where the master dataset contains missing.

335

replace, allowed with update only, specifies that nonmissing values in the master dataset be replaced with corresponding values from the using dataset. A nonmissing value, however, will never be replaced with a missing value.

unmatched(none | both | master | using) specifies whether observations unique to one of the datasets are to be kept, with the variables from the other dataset set to missing. Valid values are

none	ignore all unmatched observations (default)
both	include unmatched observations from the master and using data
master	include unmatched observations from the master data
using	include unmatched observations from the using data

_merge(*varname*) specifies the name of the variable that will mark the source of the resulting observation. The default name is _merge(_merge). To preserve compatibility with earlier versions of joinby, _merge is generated only if unmatched is specified.

nolabel prevents Stata from copying the value-label definitions from the dataset on disk into the dataset in memory. Even if you do not specify this option, label definitions from the disk dataset do not replace label definitions already in memory.

Remarks

The following, admittedly artificial, example illustrates joinby.

▷ Example 1

We have two datasets: child.dta and parent.dta. Both contain a family_id variable, which identifies the people who belong to the same family.

```
. use http://www.stata-press.com/data/r10/child
(Data on Children)
. describe
Contains data from http://www.stata-press.com/data/r10/child.dta
  obs:           5                          Data on Children
  vars:          4                          11 Dec 2006 21:08
  size:         50 (99.9% of memory free)
```

variable name	storage type	display format	value label	variable label
family_id	int	%8.0g		Family ID number
child_id	byte	%8.0g		Child ID number
x1	byte	%8.0g		
x2	int	%8.0g		

```
Sorted by:  family_id
. list
```

	family~d	child_id	x1	x2
1.	1025	3	11	320
2.	1025	1	12	300
3.	1025	4	10	275
4.	1026	2	13	280
5.	1027	5	15	210

```
. use http://www.stata-press.com/data/r10/parent
(Data on Parents)

. describe

Contains data from http://www.stata-press.com/data/r10/parent.dta
  obs:             6                          Data on Parents
  vars:            4                          11 Dec 2006 03:06
  size:          108 (99.9% of memory free)
```

variable name	storage type	display format	value label	variable label
family_id	int	%8.0g		Family ID number
parent_id	float	%9.0g		Parent ID number
x1	float	%9.0g		
x3	float	%9.0g		

```
Sorted by:

. list, sep(0)
```

	family~d	parent~d	x1	x3
1.	1030	10	39	600
2.	1025	11	20	643
3.	1025	12	27	721
4.	1026	13	30	760
5.	1026	14	26	668
6.	1030	15	32	684

We want to join the information for the parents and their children. The data on parents are in memory, and the data on children are posted at http://www.stata-press.com. child.dta has been sorted by family_id, but parent.dta has not, so first we sort the parent data on family_id:

```
. sort family_id

. joinby family_id using http://www.stata-press.com/data/r10/child

. describe

Contains data
  obs:             8                          Data on Parents
  vars:            6
  size:          168 (99.4% of memory free)
```

variable name	storage type	display format	value label	variable label
family_id	int	%8.0g		Family ID number
parent_id	float	%9.0g		Parent ID number
x1	float	%9.0g		
x3	float	%9.0g		
child_id	byte	%8.0g		Child ID number
x2	int	%8.0g		

```
Sorted by:
      Note:  dataset has changed since last saved
```

```
. list, sepby(family_id) abbrev(12)
```

	family_id	parent_id	x1	x3	child_id	x2
1.	1025	12	27	721	3	320
2.	1025	11	20	643	3	320
3.	1025	11	20	643	1	300
4.	1025	12	27	721	1	300
5.	1025	12	27	721	4	275
6.	1025	11	20	643	4	275
7.	1026	14	26	668	2	280
8.	1026	13	30	760	2	280

1. family_id of 1027, which appears only in child.dta, and family_id of 1030, which appears only in parent.dta, are not in the combined dataset. Observations for which the matching variables are not in both datasets are omitted.

2. The x1 variable is in both datasets. Values for this variable in the joined dataset are the values from parent.dta—the dataset in memory when we issued the joinby command. If we had child.dta in memory and parent.dta on disk when we requested joinby, the values for x1 would have been those from child.dta. Values from the dataset in memory take precedence over the dataset on disk.

◁

Methods and Formulas

joinby is implemented as an ado-file.

Acknowledgment

joinby was written by Jeroen Weesie, Department of Sociology, Utrecht University, The Netherlands.

Also See

[D] **save** — Save datasets

[D] **append** — Append datasets

[D] **cross** — Form every pairwise combination of two datasets

[D] **fillin** — Rectangularize dataset

[D] **merge** — Merge datasets

[U] **22 Combining datasets**

Title

> **label** — Manipulate labels

Syntax

Label dataset

> <u>la</u>bel <u>d</u>ata [*"label"*]

Label variable

> <u>la</u>bel <u>var</u>iable *varname* [*"label"*]

Define value label

> <u>la</u>bel <u>de</u>fine *lblname* # *"label"* [# *"label"* ...] [, <u>a</u>dd modify nofix]

Assign value label to variable

> <u>la</u>bel <u>val</u>ues *varname* [*lblname*] [, nofix]

List names of value labels

> <u>la</u>bel <u>dir</u>

List names and contents of value labels

> <u>la</u>bel <u>l</u>ist [*lblname* [*lblname* ...]]

Drop value labels

> <u>la</u>bel drop { *lblname* [*lblname* ...] | _all }

Save value labels in do-file

> <u>la</u>bel save [*lblname* [*lblname* ...]] using *filename* [, replace]

where # is an integer or an extended missing value (.a, .b, ..., .z).

Description

label data attaches a label (up to 80 characters) to the dataset in memory. Dataset labels are displayed when you use the dataset and when you describe it. If no label is specified, any existing label is removed.

label variable attaches a label (up to 80 characters) to a variable. If no label is specified, any existing variable label is removed.

339

label define defines a list of up to 65,536 (1,000 for Small Stata) associations of integers and text called value labels. Value labels are attached to variables by label values.

label values attaches a value label to a variable. If no value label is specified, any existing value label is detached from that variable. The value label, however, is not deleted. Value labels may be up to 32,000 characters.

label dir lists the names of value labels stored in memory.

label list lists the names and contents of value labels stored in memory.

label drop eliminates value labels.

label save saves value labels in a do-file. This is particularly useful for value labels that are not attached to a variable because these labels are not saved with the data.

See [D] **label language** for information on the label language command.

Options

add allows you to add # ↔ *label* correspondences to *lblname*. If add is not specified, you may create only new *lblnames*. If add is specified, you may create new *lblnames* or add new entries to existing *lblnames*.

modify allows you to modify or delete existing # ↔ *label* correspondences and add new correspondences. Specifying modify implies add, even if you do not type the add option.

nofix prevents display formats from being widened according to the maximum length of the value label. Consider label values myvar mylab, and say that myvar has a %9.0g display format right now. Say that the maximum length of the strings in mylab is 12 characters. label values would change the format of myvar from %9.0g to %12.0g. nofix prevents this.

nofix is also allowed with label define, but it is relevant only when you are modifying an existing value label. Without the nofix option, label define finds all the variables that use this value label and considers widening their display formats. nofix prevents this.

replace allows *filename* to be replaced if it already exists.

Remarks

See [U] **12.6 Dataset, variable, and value labels** for a complete description of labels. This entry deals only with details not covered there.

label dir lists the names of all defined value labels. label list displays the contents of a value label.

▷ Example 1

Although describe shows the names of the value labels, those value labels may not exist. Stata does not consider it an error to label the values of a variable with a nonexistent label. When this occurs, Stata still shows the association on describe but otherwise acts as if the variable's values are unlabeled. This way, you can associate a value label name with a variable before creating the corresponding label. Similarly, you can define labels that you have not yet used.

```
. use http://www.stata-press.com/data/r10/hbp4

. describe

Contains data from http://www.stata-press.com/data/r10/hbp4.dta
  obs:        1,130
  vars:           7                          22 Jan 2007 11:12
  size:      23,730 (97.7% of memory free)

              storage  display    value
variable name   type   format     label    variable label

id             str10   %10s                 Record identification number
city           byte    %8.0g
year           int     %8.0g
age_grp        byte    %8.0g
race           byte    %8.0g
hbp            byte    %8.0g
female         byte    %8.0g      sexlbl

Sorted by:
```

The dataset is using the value label sexlbl. Let's define the value label yesno:

```
. label define yesno 0 "no" 1 "yes"
```

label dir shows you the labels that you have actually defined:

```
. label dir
yesno
sexlbl
```

We have two value labels stored in memory: yesno and sexlbl.

We can display the contents of a value label with the label list command:

```
. label list yesno
yesno:
           0 no
           1 yes
```

The value label yesno labels the values 0 as no and 1 as yes.

If you do not specify the name of the value label on the label list command, Stata lists all the value labels:

```
. label list
yesno:
           0 no
           1 yes
sexlbl:
           0 male
           1 female
```

◁

❑ Technical Note

Because Stata can have more value labels stored in memory than are actually used in the dataset, you may wonder what happens when you save the dataset. Stata stores only those value labels actually associated with variables.

When you use a dataset, Stata eliminates all the value labels stored in memory before loading the dataset.

❑

You can add new codings to an existing value label by using the `add` option with the `label define` command. You can modify existing codings by using the `modify` option.

▷ Example 2

The label `yesno` codes 0 as `no` and 1 as `yes`. You might wish later to add a third coding: 2 as `maybe`. Typing `label define` with no options results in an error:

```
. label define yesno 2 maybe
label yesno already defined
r(110);
```

If you do not specify the `add` or `modify` options, `label define` can be used only to create *new* value labels. The `add` option lets you add codings to an existing label:

```
. label define yesno 2 maybe, add
. label list yesno
yesno:
           0 no
           1 yes
           2 maybe
```

Perhaps you have accidentally mislabeled a value. For instance, 2 may not mean "maybe" but may instead mean "don't know". `add` does not allow you to change an existing label:

```
. label define yesno 2 "don't know", add
invalid attempt to modify label
r(180);
```

Instead, you specify the `modify` option:

```
. label define yesno 2 "don't know", modify
. label list yesno
yesno:
           0 no
           1 yes
           2 don't know
```

In this way, Stata attempts to protect you from yourself. If you type `label define` with no options, you can only create a new value label—you cannot accidentally change an existing one. If you specify the `add` option, you can add new labels to a label, but you cannot accidentally change any existing label. If you specify the `modify` option, which you may not abbreviate, you can change any existing label.

You can even use the `modify` option to eliminate existing labels. To do this, you map the numeric code to a *null string*, that is, `""`:

```
. label define yesno 2 "", modify
. label list yesno
yesno:
           0 no
           1 yes
```
◁

You can eliminate entire value labels by using the `label drop` command.

▷ Example 3

We currently have two value labels stored in memory—sexlbl and yesno—as shown by the label dir command:

```
. label dir
yesno
sexlbl
```

The dataset that we have in memory uses only one of the labels—sexlbl. describe reports that yesno is not being used:

```
. describe
Contains data from http://www.stata-press.com/data/r10/hbp4.dta
  obs:         1,130
  vars:            7                          22 Jan 2007 11:12
  size:       23,730 (97.7% of memory free)
```

variable name	storage type	display format	value label	variable label
id	str10	%10s		Record identification number
city	byte	%8.0g		
year	int	%8.0g		
age_grp	byte	%8.0g		
race	byte	%8.0g		
hbp	byte	%8.0g		
female	byte	%8.0g	sexlbl	

```
Sorted by:
```

We can eliminate the yesno label by typing label drop yesno:

```
. label drop yesno
. label dir
sexlbl
```

We could eliminate *all* the value labels in memory by typing

```
. label drop _all
. label dir
. _
```

The value label sexlbl, which no longer exists, was associated with the variable female. Even after dropping the value label, sexlbl is still associated with the variable:

(Continued on next page)

```
. describe
Contains data from http://www.stata-press.com/data/r10/hbp4.dta
  obs:          1,130
  vars:             7                          22 Jan 2007 11:12
  size:        23,730 (97.7% of memory free)

              storage  display   value
variable name   type   format    label     variable label

id            str10    %10s                 Record identification number
city          byte     %8.0g
year          int      %8.0g
age_grp       byte     %8.0g
race          byte     %8.0g
hbp           byte     %8.0g
female        byte     %8.0g     sexlbl

Sorted by:
```

Stata does not mind if a nonexistent value label is associated with a variable. When Stata uses such a variable, it simply acts as if the variable is not labeled:

```
. list in 1/4

            id    city   year   age_grp   race   hbp   female

 1.  8008238923      1   1993         2      2     0        1
 2.  8007143470      1   1992         5      .     0        .
 3.  8000468015      1   1988         4      2     0        0
 4.  8006167153      1   1991         4      2     0        0
```

◁

The `label save` command creates a *do-file* containing `label define` commands for each label you specify. If you do not specify the *lblnames*, all value labels are stored in the file. If you do not specify the extension for *filename*, .do is assumed.

▷ Example 4

Labels are automatically stored with your dataset when you `save` it. Conversely, the `use` command drops all labels before loading the new dataset. You may occasionally wish to move a value label from one dataset to another. The `label save` command allows you to do this.

For example, assume that we currently have the value label `yesno` in memory:

```
. label list yesno
yesno:
           1 yes
           2 no
           3 maybe
```

We have a dataset stored on disk called `survey.dta` to which we wish to add this value label. We might `use survey` and then retype the `label define yesno` command. Retyping the label would not be too tedious here but if the value label in memory mapped, say, the 50 states of the union, retyping it would be irksome. `label save` provides an alternative:

```
. label save yesno using ynfile
file ynfile.do saved
```

Typing `label save yesno using ynfile` caused Stata to create a do-file called `ynfile.do` containing the definition of the `yesno` label.

To see the contents of the file, we can use the Stata `type` command:

```
. type ynfile.do
label define yesno 1 '"yes"', modify
label define yesno 2 '"no"', modify
label define yesno 3 '"maybe"', modify
```

We can now use our new dataset, `survey.dta`:

```
. use survey
(Household survey data)
. label dir

. _
```

Using the new dataset causes Stata to eliminate all value labels stored in memory. The label `yesno` is now gone. Since we saved it in the file `ynfile.do`, however, we can get it back by typing either `do ynfile` or `run ynfile`. If we type `do`, we will see the commands in the file execute. If we type `run`, the file will execute silently:

```
. run ynfile
. label dir
yesno
```

The label is now restored just as if we had typed it from the keyboard.

◁

❏ Technical Note

You can also use the `label save` command to more easily edit value labels. You can save a label in a file, leave Stata and use your word processor or editor to edit the label, and then return to Stata. Using `do` or `run`, you can load the edited values.

❏

Saved Results

`label list` saves the following in `r()`:

Scalars

`r(k)`	number of mapped values, including missings
`r(min)`	minimum nonmissing value label
`r(max)`	maximum nonmissing value label
`r(hasemiss)`	1 if extended missing values labeled, 0 otherwise

`label dir` saves the following in `r()`:

Macros

`r(names)`	names of value labels

References

Gleason, J. R. 1998. dm56: A labels editor for Windows and Macintosh. *Stata Technical Bulletin* 43: 3–6. Reprinted in *Stata Technical Bulletin Reprints*, vol. 8, pp. 5–10.

——. 1999. dm56.1: Update to labedit. *Stata Technical Bulletin* 51: 2. Reprinted in *Stata Technical Bulletin Reprints*, vol. 9, p. 15.

Weesie, J. 1997. dm47: Verifying value label mappings. *Stata Technical Bulletin* 37: 7–8. Reprinted in *Stata Technical Bulletin Reprints*, vol. 7, pp. 39–40.

——. 2005a. Value label utilities: labeldup and labelrename. *Stata Journal* 5: 154–161.

——. 2005b. Multilingual datasets. *Stata Journal* 5: 162–187.

Also See

[D] **label language** — Labels for variables and values in multiple languages

[D] **labelbook** — Label utilities

[D] **encode** — Encode string into numeric and vice versa

[U] **12.6 Dataset, variable, and value labels**

Title

---

label language — Labels for variables and values in multiple languages

---

Syntax

List defined languages

    <u>la</u>bel <u>lang</u>uage

Change labels to specified language name

    <u>la</u>bel <u>lang</u>uage *languagename*

Create new set of labels with specified language name

    <u>la</u>bel <u>lang</u>uage *languagename*, new [copy]

Rename current label set

    <u>la</u>bel <u>lang</u>uage *languagename*, <u>ren</u>ame

Delete specified label set

    <u>la</u>bel <u>lang</u>uage *languagename*, delete

Description

    label language lets you create and use datasets that contain different sets of data, variable, and value labels. A dataset might contain one set in English, another in German, and a third in Spanish. A dataset may contain up to 100 sets of labels.

    We will write about the different sets as if they reflect different spoken languages, but you need not use the multiple sets in this way. You could create a dataset with one set of long labels and another set of shorter ones.

    One set of labels is in use at any instant, but a dataset may contain multiple sets. You can choose among the sets by typing

        . label language *languagename*

    When other Stata commands produce output (such as describe and tabulate), they use the currently set language. When you define or modify the labels by using the other label commands (see [D] **label**), you modify the current set.

    label language (without arguments)

        lists the available languages and the name of the current one. The current language refers to the labels you will see if you used, say, describe or tabulate. The available languages refer to the names of the other sets of previously created labels. For instance, you might currently be using the labels in en (English), but labels in de (German) and es (Spanish) may also be available.

`label language` *languagename*

changes the labels to those of the specified language. For instance, if `label language` revealed that `en`, `de`, and `es` were available, typing `label language de` would change the current language to `de`.

`label language` *languagename*`, new`

allows you to create a new set of labels and collectively name them *languagename*. You may name the set as you please, as long as the name does not exceed 24 characters. If the labels correspond to spoken languages, we recommend that you use the language's ISO 639-1 two-letter code, such as `en` for English, `de` for German, and `es` for Spanish. A list of codes for popular languages is listed below. For a complete list, see http://lcweb.loc.gov/standards/iso639-2/iso639jac.html.

`label language` *languagename*`, rename`

changes the name of the label set currently in use. If the label set in use were named `default` and you now wanted to change that to `en`, you could type `label language en, rename`.

Our choice of the name `default` in the example was not accidental. If you have not yet used `label language` to create a new language, the dataset will have one language, named `default`.

`label language` *languagename*`, delete`

deletes the specified label set. If *languagename* is also the current language, one of the other available languages becomes the current language.

Option

`copy` is used with `label language, new` and copies the labels from the current language to the new language.

Remarks

Remarks are presented under the following headings:

> *Creating labels in the first language*
> *Creating labels in the second and subsequent languages*
> *Creating labels from a clean slate*
> *Creating labels from a previously existing language*
> *Switching languages*
> *Changing the name of a language*
> *Deleting a language*
> *Appendix: Selected ISO 639-1 two-letter codes*

Creating labels in the first language

You can begin by ignoring the `label language` command. You create the data, variable, and value labels just as you would ordinarily; see [D] **label**.

```
. label data "1978 Automobile Data"
. label variable foreign "Car type"
. label values foreign origin
. label define origin  0 "Domestic"  1 "Foreign"
```

At some point—at the beginning, the middle, or the end—rename the language appropriately. For instance, if the labels you defined were in English, type

```
. label language en, rename
```

`label language, rename` simply changes the name of the currently set language. You may change the name as often as you wish.

Creating labels in the second and subsequent languages

After creating the first language, you can create a new language by typing

```
. label language newlanguagename, new
```

or by typing the two commands

```
. label language existinglanguagename
. label language newlanguagename, new copy
```

In the first case, you start with a clean slate: no data, variable, or value labels are defined. In the second case, you start with the labels from *existinglanguagename*, and you can make the changes from there.

Creating labels from a clean slate

To create new labels in the language named `de`, type

```
. label language de, new
```

If you were now to type `describe`, you would find that there are no data, variable, or value labels. You can define new labels in the usual way:

```
. label data "1978 Automobil Daten"
. label variable foreign "Art Auto"
. label values foreign origin_de
. label define origin_de  0 "Innen"  1 "Ausländisch"
```


Creating labels from a previously existing language

It is sometimes easier to start with the labels from a previously existing language, which you can then translate:

```
. label language en
. label language de, new copy
```

If you were now to type `describe`, you would see the English-language labels, even though the new language is named `de`. You can then work to translate the labels:

```
. label data "1978 Automobil Daten"
. label variable foreign "Art Auto"
```

Typing `describe`, you might also discover that the variable `foreign` has the value label `origin`. Do not change the contents of the value label. Instead, create a new value label:

```
. label define origin_de  0 "Innen"  1 "Ausländisch"
. label values foreign origin_de
```

Creating value labels with the `copy` option is no different from creating them from a clean slate, except that you start with an existing set of labels from another language. Using `describe` can make it easier to translate them.

Switching languages

You can discover the names of the previously defined languages by typing

```
. label language
```

You can switch to a previously defined language—say to `en`—by typing

```
. label language en
```

Changing the name of a language

To change the name of a previously defined language, make it the current language, and then specify the `rename` option:

```
. label language de
. label language German, rename
```

You may rename a language as often as you wish:

```
. label language de, rename
```

Deleting a language

To delete a previously defined language, such as `de`, type

```
. label language de, delete
```

The `delete` option deletes the specified language and, if the language was also the currently set language, resets the current language to one of the other languages or to `default` if there are none.

Appendix: Selected ISO 639-1 two-letter codes

You may name languages as you please. You may name German labels `Deutsch`, `German`, `Aleman`, or whatever else appeals to you. For consistency across datasets, if the language you are creating is a spoken language, we suggest that you use the ISO 639-1 two-letter codes. Some of them are listed below, and the full list can be found at http://lcweb.loc.gov/standards/iso639-2/iso639jac.html.

Two-letter code	English name of language
ar	Arabic
cs	Czech
cy	Welsh
de	German
el	Greek
en	English
es	Spanish; Castillian
fa	Persian
fi	Finnish
fr	French
ga	Irish
is	Icelandic
he	Hebrew
hi	Hindi
it	Italian
ja	Japanese
kl	Greenlandic; Kalaallisut
lt	Lithuanian
lv	Latvian
nl	Dutch; Flemish
no	Norwegian
pl	Polish
pt	Portuguese
ro	Romanian
ru	Russian
sk	Slovak
sr	Serbian
sv	Swedish
tr	Turkish
uk	Ukrainian
uz	Uzbek
zh	Chinese

Saved Results

label language without arguments saves the following in r():

Scalars

r(k)   number of languages defined

Macros

r(languages)   list of languages, listed one after the other

r(language)   name of current language

Methods and Formulas

This section is included for programmers who wish to access or extend the services label language provides.

Language sets are implemented using [P] **char**. The names of the languages and the name of the current language are stored in

$$\_dta\big[\_lang\_list\big] \quad \text{list of defined languages}$$
$$\_dta\big[\_lang\_c\big] \quad \text{currently set language}$$

If these characteristics are undefined, results are as if each contained the word "default". Do not change the contents of the above two macros except by using label language.

For each language *languagename* except the current language, data, variable, and value labels are stored in

$$\_dta\big[\_lang\_v\_languagename\big] \quad \text{data label}$$

$$varname\big[\_lang\_v\_languagename\big] \quad \text{variable label}$$
$$varname\big[\_lang\_l\_languagename\big] \quad \text{value label name}$$

Reference

Weesie, J. 2005. Multilingual datasets. *Stata Journal* 5: 162–187.

Also See

[D] **label** — Manipulate labels

[D] **codebook** — Describe data contents

[D] **labelbook** — Label utilities

Title

> **labelbook** — Label utilities

Syntax

Produce a codebook describing value labels

> labelbook [*lblname-list*] [, *labelbook_options*]

Prefix numeric values to value labels

> numlabel [*lblname-list*] , { <u>a</u>dd | <u>r</u>emove } [*numlabel_options*]

Make dataset containing value-label information

> uselabel [*lblname-list*] [using *filename*] [, clear <u>v</u>ar]

labelbook_options	description
<u>a</u>lpha	alphabetize label mappings
<u>l</u>ength(#)	check if value labels are unique to length #; default is length(12)
<u>l</u>ist(#)	list maximum of # mappings; default is list(32000)
<u>p</u>roblems	describe potential problems in a summary report
<u>d</u>etail	do not suppress detailed report on variables or value labels

numlabel_options	description
* <u>a</u>dd	prefix numeric values to value labels
* <u>r</u>emove	remove numeric values from value labels
<u>m</u>ask(*str*)	mask for formatting numeric labels; default mask is "#. "
force	force adding or removing of numeric labels
<u>d</u>etail	provide details about value labels, where some labels are prefixed with numbers and others are not

* Either add or remove must be specified.

Description

labelbook displays information for the value labels specified or, if no labels are specified, all the labels in the data.

For multilingual datasets (see [D] **label language**), labelbook lists the variables to which value labels are attached in all defined languages.

numlabel prefixes numeric values to value labels. For example, a value mapping of 2 -> "catholic" will be changed to 2 -> "2. catholic". See option mask for the different formats. Stata commands that display the value labels also show the associated numeric values. Prefixes are removed with the remove option.

uselabel is a programmer's command that reads the value-label information from the currently loaded dataset or from an optionally specified filename.

353

uselabel creates a dataset in memory that contains only that value-label information. The new dataset has four variables named label, lname, value, and trunc; is sorted by lname value; and has 1 observation per mapping. Value labels can be longer than the maximum string length in Stata; see help limits. The new variable trunc contains 1 if the value label is truncated to fit in a string variable in the dataset created by uselabel.

uselabel complements label, save, which produces an ASCII file of the variable labels in a format that allows easy editing of the value-label texts.

Specifying no list or _all is equivalent to specifying all value labels. Value-label names may not be abbreviated or specified with wildcards.

Options for labelbook

alpha specifies that the list of value-label mappings be sorted alphabetically on label. The default is to sort the list on value.

length(#) specifies the minimum length that labelbook checks to determine whether shortened value labels are still unique. It defaults to 12, the width used by most Stata commands. labelbook also reports whether value labels are unique at their full length.

list(#) specifies the maximum number of value-label mappings to be listed. If a value label defines more mappings, a random subset of # mappings is displayed. By default, labelbook displays all mappings. list(0) suppresses the listing of the value-label definitions.

problems specifies that a summary report be produced describing potential problems that were diagnosed:

1. Value label has gaps in mapped values (e.g., values 0 and 2 are labeled, while 1 is not)

2. Value label strings contain leading or trailing blanks

3. Value label contains duplicate labels, i.e., there are different values that map into the same string

4. Value label contains duplicate labels at length 12

5. Value label contains numeric → numeric mappings

6. Value label contains numeric → null string mappings

7. Value label is not used by variables

detail may be specified only with problems. It specifies that the detailed report on the variables or value labels not be suppressed.

Options for numlabel

add specifies that numeric values be prefixed to value labels. Value labels that are already numlabeled (using the same mask) are not modified.

remove specifies that numeric values be removed from the value labels. If you added numeric values by using a nondefault mask, you must specify the same mask to remove them. Value labels that are not numlabeled or are numlabeled using a different mask are not modified.

mask(str) specifies a mask for formatting the numeric labels. In the mask, # is replaced by the numeric label. The default mask is "#. " so that numeric value 3 is shown as "3. ". Spaces are relevant. For the mask "[#]", numeric value 3 would be shown as "[3]".

force specifies that adding or removing numeric labels be performed, even if some value labels are numlabeled using the mask and others are not. Here only labels that are not numlabeled will be modified.

detail specifies that details be provided about the value labels that are sometimes, but not always, numlabeled using the mask.

Options for uselabel

clear permits the dataset to be created, even if the dataset already in memory has changed since it was last saved.

var specifies that the varlists using value label *vl* be returned in r(*vl*).

Remarks

Remarks are presented under the following headings:

> *labelbook*
> *Diagnosing problems*
> *numlabel*
> *uselabel*

labelbook

labelbook produces a detailed report of the value labels in your data. You can restrict the report to a list of labels, meaning that no abbreviations or wildcards will be allowed. labelbook is a companion command to [D] **codebook**, which describes the data, focusing on the variables.

For multilingual datasets (see [D] **label language**), labelbook lists the variables to which value labels are attached in any of the languages.

▷ Example 1

We request a labelbook report for value labels in a large dataset on the internal organization of households. We restrict output to three value labels: agree5 (used for five-point Likert-style items), divlabor (division of labor between husband and wife), and noyes for simple no-or-yes questions.

(Continued on next page)

```
. use http://www.stata-press.com/data/r10/labelbook1

. labelbook agree5 divlabor noyes
```

---

value label agree5

---

values		**labels**	
range:	[1,5]	string length:	[8,11]
N:	5	unique at full length:	yes
gaps:	no	unique at length 12:	yes
missing .*:	0	null string:	no
		leading/trailing blanks:	no
		numeric -> numeric:	no

definition
```
            1    -- disagree
            2    - disagree
            3    indifferent
            4    + agree
            5    ++ agree
    variables:   rs056 rs057 rs058 rs059 rs060 rs061 rs062 rs063 rs064 rs065
                 rs066 rs067 rs068 rs069 rs070 rs071 rs072 rs073 rs074 rs075
                 rs076 rs077 rs078 rs079 rs080 rs081
```

---

value label divlabor

---

values		**labels**	
range:	[1,7]	string length:	[7,16]
N:	7	unique at full length:	yes
gaps:	no	unique at length 12:	yes
missing .*:	0	null string:	no
		leading/trailing blanks:	yes
		numeric -> numeric:	no

definition
```
            1    wife only
            2    wife >> husband
            3    wife > husband
            4    equally
            5    husband > wife
            6    husband >> wife
            7    husband only
    variables:   hm01_a hm01_b hm01_c hm01_d hm01_e hn19 hn21 hn25_a hn25_b
                 hn25_c hn25_d hn25_e hn27_a hn27_b hn27_c hn27_d hn27_e hn31
                 hn36 hn38 hn42 hn46_a hn46_b hn46_c hn46_d hn46_e ho01_a ho01_b
                 ho01_c ho01_d ho01_e
```

```
value label noyes

        values                              labels
        range:  [1,2]              string length:  [2,16]
            N:  4            unique at full length:  yes
         gaps:  yes           unique at length 12:  yes
   missing .*:  2                      null string:  no
                           leading/trailing blanks:  no
                                 numeric -> numeric:  no

definition
            1   no
            2   yes
           .a   not applicable
           .b   ambiguous answer
    variables:  hb12 hd01_a hd01_b hd03 hd04_a hd04_b he03_a he03_b hlat hn09_b
                hn24_a hn34 hn49 hu05_a hu06_1c hu06_2c hx07_a hx08 hlat2
                hfinish rh02 rj10_01 rk16_a rk16_b rl01 rl03 rl08_a rl08_b
                rl09_a rs047 rs048 rs049 rs050 rs051 rs052 rs053 rs054 rs093
                rs095 rs096 rs098
```

The report is largely self-explanatory. Extended missing values are denoted by ".*". In the definition of the mappings, the leading 12 characters of longer value labels are underlined to make it easier to check that the value labels still make sense after truncation. The following example emphasizes this feature. The option `alpha` specifies that the value-label mappings be sorted in alphabetical order by the label strings rather than by the mapped values.

```
. use http://www.stata-press.com/data/r10/labelbook2

. labelbook sports, alpha
```

```
value label sports

        values                              labels
        range:  [1,5]              string length:  [16,23]
            N:  4            unique at full length:  yes
         gaps:  yes           unique at length 12:  no
   missing .*:  0                      null string:  no
                           leading/trailing blanks:  no
                                 numeric -> numeric:  no

definition
            5   college baseball
            4   college basketball
            2   professional baseball
            1   professional basketball
    variables:  active passive
```

The report includes information about potential problems in the data. These are discussed in greater detail in the next section.

◁

Diagnosing problems

`labelbook` can diagnose a series of potential problems in the value-label mappings. `labelbook` produces warning messages for a series of problems:

1. Gaps in the labeled values (e.g., values 0 and 2 are labeled, whereas 1 is not) may occur when value labels of the intermediate values have not been defined.

2. Leading or trailing blanks in the value labels may distort Stata output.

3. Stata allows you to define blank labels, i.e., the mapping of a number to the empty string. Below we give you an example of the unexpected output that may result. Blank labels are most often the result of a mistaken value-label definition, for instance, the expansion of a nonexisting macro in the definition of a value label.

4. Stata does not require that the labels within each value label consist of *unique* strings, i.e., that different values be mapped into different strings. For instance, you might accidentally define the value label `gender` as

```
label define gender 1 female 2 female
```

You will probably catch most of the problems, but in more complicated value labels, it is easy to miss the error. `labelbook` finds such problems and displays a warning.

5. Stata allows long value labels (32,000 characters), so labels can be long. However, some commands may need to display truncated value labels, typically at length 12. Consequently, even if the value labels are unique, the truncated value labels may not be, which can cause problems. `labelbook` warns you for value labels that are not unique at length 12.

6. Stata allows value labels that can be interpreted as numbers. This is sometimes useful, but it can cause highly misleading output. Think about tabulating a variable for which the associated value label incorrectly maps 1 into "2", 2 into "3", and 3 into "1". `labelbook` looks for such problematic labels and warns you if they are found.

7. In Stata, value labels are defined as separate objects that can be associated with more than one variable:

```
label define labname # str # str ....
label value varname1 labname
label value varname2 labname
...
```

If you forget to associate a variable label with a variable, Stata considers the label unused and drops its definition. `labelbook` reports unused value labels so that you may fix the problem.

The related command `codebook` reports on two other potential problems concerning value labels:

a. A variable is value labeled, but some values of the variable are not labeled. You may have forgotten to define a mapping for some values, or you generated a variable incorrectly; e.g., your variable `sex` has an unlabeled value 3, and you are not working in experimental genetics!

b. A variable has been associated with an undefined value label.

`labelbook` can also be invoked with the option `problems`, specifying that only a report on potential problems be displayed without the standard detailed description of the value labels.

❏ Technical Note

The following two examples demonstrate some features of value labels that may be difficult to understand. In the first example, we `encode` a string variable with blank strings of various sizes; that is, we turn a string variable into a value-labeled numeric variable. Then we tabulate the generated variable.

```
. clear all
. set obs 5
obs was 0, now 5
. generate str10 horror = substr("        ", 1, _n)
. encode horror, gen(Ihorror)
. tabulate horror
```

horror	Freq.	Percent	Cum.
	1	20.00	20.00
	1	20.00	40.00
	1	20.00	60.00
	1	20.00	80.00
	1	20.00	100.00
Total	5	100.00	

It may look as if you have discovered a bug in Stata because there are no value labels in the first column of the table. This happened because we encoded a variable with only blank strings, so the associated value label maps integers into blank strings.

```
. label list Ihorror
Ihorror:
           1
           2
           3
           4
           5
```

In the first column of the table, `tabulate` displayed the value-label texts, just as it should. Since these texts are all blank, the first column is empty. As illustrated below, `labelbook` would have warned you about this odd value label.

Our second example illustrates what could go wrong with numeric values stored as string values. We want to turn this into a numeric variable, but we incorrectly `encode` the variable rather than using the appropriate command, `destring`.

```
. generate str10 horror2 = string(_n+1)
. encode horror2, gen(Ihorror2)
. tabulate Ihorror2
```

Ihorror2	Freq.	Percent	Cum.
2	1	20.00	20.00
3	1	20.00	40.00
4	1	20.00	60.00
5	1	20.00	80.00
6	1	20.00	100.00
Total	5	100.00	

```
. tabulate Ihorror2, nolabel
```

Ihorror2	Freq.	Percent	Cum.
1	1	20.00	20.00
2	1	20.00	40.00
3	1	20.00	60.00
4	1	20.00	80.00
5	1	20.00	100.00
Total	5	100.00	

```
. label list Ihorror2
Ihorror2:
            1 2
            2 3
            3 4
            4 5
            5 6
```

❑

labelbook skips the detailed descriptions of the value labels and reports only the potential problems in the value labels if the problems option is specified. This report would have alerted you to the problems with the value labels we just described.

```
. use http://www.stata-press.com/data/r10/data_in_trouble, clear
. labelbook, problem
   Potential problems in dataset    http://www.stata-press.com/data/r10/
> data_in_trouble.dta
                potential problem    value labels
   ─────────────────────────────────────────────────
            numeric -> numeric    Ihorror2
      leading or trailing blanks    Ihorror
          not used by variables    unused
   ─────────────────────────────────────────────────
```

Running labelbook, problems and codebook, problems on new data might catch a series of annoying problems.

numlabel

The command numlabel allows you to prefix numeric codes to value labels. The reason you might want to do this is best seen in an example using the automobile data. First, we create a value label for the variable rep78 (repair record in 1978),

```
. use http://www.stata-press.com/data/r10/auto
(1978 Automobile Data)
. label define repair 1 "very poor" 2 "poor" 3 "medium" 4 good 5 "very good"
. label values rep78 repair
```

and tabulate it.

```
. tabulate rep78
    Repair │
Record 1978 │      Freq.      Percent        Cum.
────────────┼───────────────────────────────────────
  very poor │          2         2.90        2.90
       poor │          8        11.59       14.49
     medium │         30        43.48       57.97
       good │         18        26.09       84.06
  very good │         11        15.94      100.00
────────────┼───────────────────────────────────────
      Total │         69       100.00
```

Suppose that we want to recode the variable by joining the categories *poor* and *very poor*. To do this, we need the numerical codes of the categories, not the value labels. However, Stata does not display both the numeric codes and the value labels. We could redisplay the table with the nolabel option. The numlabel provides a simple alternative: it modifies the value labels so that they also contain the numeric codes.

```
. numlabel, add

. tabulate rep78
```

Repair Record 1978	Freq.	Percent	Cum.
1. very poor	2	2.90	2.90
2. poor	8	11.59	14.49
3. medium	30	43.48	57.97
4. good	18	26.09	84.06
5. very good	11	15.94	100.00
Total	69	100.00	

If you do not like the way the numeric codes are formatted, you can use `numlabel` to change the formatting. First, we remove the numeric codes again.

```
. numlabel repair, remove
```

In this example, we specified the name of the label. If we had not typed it, `numlabel` would have removed the codes from all the value labels. We can include the numeric codes while specifying a mask:

```
. numlabel, add mask("[#]   ")

. tabulate rep78
```

Repair Record 1978	Freq.	Percent	Cum.
[1] very poor	2	2.90	2.90
[2] poor	8	11.59	14.49
[3] medium	30	43.48	57.97
[4] good	18	26.09	84.06
[5] very good	11	15.94	100.00
Total	69	100.00	

`numlabel` prefixes rather than postfixes the value labels with numeric codes. Because value labels can be fairly long (up to 80 characters), Stata usually displays only the first 12 characters.

(Continued on next page)

uselabel

uselabel is of interest primarily to programmers. Here we briefly illustrate it with the auto dataset.

▷ Example 2

```
. use http://www.stata-press.com/data/r10/auto
(1978 Automobile Data)
. uselabel
. describe
Contains data
  obs:               2
  vars:              4
  size:             40 (99.9% of memory free)
```

variable name	storage type	display format	value label	variable label
lname	str6	%9s		
value	byte	%9.0g		
label	str8	%9s		
trunc	byte	%9.0g		

```
Sorted by: lname  value
    Note:  dataset has changed since last saved
. list
```

	lname	value	label	trunc
1.	origin	0	Domestic	0
2.	origin	1	Foreign	0

uselabel created a dataset containing the labels and values for the value label origin.

The maximum length of the text associated with a value label is 32,000 characters, whereas the maximum length of a string variable in a Stata dataset is 244. uselabel uses only the first 244 characters of the label. Variable trunc will record a 1 if the text was truncated for this reason.

◁

Saved Results

labelbook saves the following in r():

Macros

r(names)	*lblname-list*
r(gaps)	gaps in mapped values
r(blanks)	leading or trailing blanks
r(null)	name of value label containing null strings
r(nuniq)	duplicate labels
r(nuniq_sh)	duplicate labels at length 12
r(ntruniq)	duplicate labels at maximum string length
r(notused)	not used by any of the variables
r(numeric)	name of value label containing mappings to numbers

uselabel saves the following in r():

Macros
    r(*lblname*)   list of variables that use value label *lblname*
                        (only when var option is specified)

Methods and Formulas

labelbook, numlabel, and uselabel are implemented as ado-files.

Acknowledgments

labelbook and numlabel were written by Jeroen Weesie, Department of Sociology, Utrecht University. A command similar to numlabel was written by J. M. Lauritsen (2001).

References

Lauritsen, J. M. 2001. dm84: Adding numerical codes to value labels. *Stata Technical Bulletin* 59: 6–7. Reprinted in *Stata Technical Bulletin Reprints*, vol. 10, pp. 35–37.

Weesie, J. 1997. dm47: Verifying value label mappings. *Stata Technical Bulletin* 37: 7–8. Reprinted in *Stata Technical Bulletin Reprints*, vol. 7, pp. 39–40.

Also See

[D] **codebook** — Describe data contents

[D] **describe** — Describe data in memory or in file

[D] **encode** — Encode string into numeric and vice versa

[D] **label** — Manipulate labels

[U] **12.6 Dataset, variable, and value labels**

[U] **15 Printing and preserving output**

Title

list — List values of variables

Syntax

$\underline{\text{list}}$ $\lceil$ *varlist* $\rceil$ $\lceil$ *if* $\rceil$ $\lceil$ *in* $\rceil$ $\lceil$, *options* $\rceil$

$\underline{\text{fl}}\text{ist}$ is equivalent to list with the fast option.

options	description
Main	
compress	compress width of columns in both table and display formats
nocompress	use display format of each variable
fast	synonym for nocompress; no delay in output of large datasets
abbreviate(#)	abbreviate variable names to # characters; default is ab(8)
string(#)	truncate string variables to # characters; default is string(10)
noobs	do not list observation numbers
Options	
table	force table format
display	force display format
header	display variable header once; default is table mode
noheader	suppress variable header
header(#)	display variable header every # lines
clean	force table format with no divider or separator lines
divider	draw divider lines between columns
separator(#)	draw a separator line every # lines; default is separator(5)
sepby(*varlist*)	draw a separator line whenever *varlist* values change
nolabel	display numeric codes rather than label values
Summary	
mean$\lceil$(*varlist*)$\rceil$	add line reporting the mean for the (specified) variables
sum$\lceil$(*varlist*)$\rceil$	add line reporting the sum for the (specified) variables
N$\lceil$(*varlist*)$\rceil$	add line reporting the number of nonmissing values for the (specified) variables
labvar(*varname*)	substitute Mean, Sum, or N for value of *varname* in last row of table
Advanced	
constant$\lceil$(*varlist*)$\rceil$	separate and list variables that are constant only once
notrim	suppress string trimming
absolute	display overall observation numbers when using by *varlist*:
nodotz	display numerical values equal to .z as field of blanks
subvarname	substitute characteristic for variable name in header
linesize(#)	columns per line; default is linesize(79)

varlist may contain time-series operators; see [U] **11.4.3 Time-series varlists**.
by is allowed with list; see [D] **by**.

364

Description

list displays the values of variables. If no *varlist* is specified, the values of all the variables are displayed. Also see browse in [D] **edit**.

Options

> Main

compress and nocompress change the width of the columns in both table and display formats. By default, list examines the data and allocates the needed width to each variable. For instance, a variable might be a string with a %18s format, and yet the longest string will be only 12 characters long. Or, a numeric variable might have %9.0g format, and yet, given the values actually present, the widest number needs only four columns.

nocompress prevents list from examining the data. Widths will be set according to the display format of each variable. Output generally looks better when nocompress is not specified, but for very large datasets (say, 1,000,000 observations or more), nocompress can speed up the execution of list.

compress allows list to engage in a little more compression than it otherwise would by telling list to abbreviate variable names to fewer than eight characters.

fast is a synonym for nocompress. fast may be of interest to those with very large datasets who wish to see output appear without delay.

abbreviate(#) is an alternative to compress that allows you to specify the minimum abbreviation of variable names to be considered. For example, you could specify abbreviate(16) if you never wanted variables abbreviated to less than 16 characters.

string(#) specifies that when string variables are listed, they be truncated to # characters in the output. Any value that is truncated will be appended with "..." to indicate the truncation. string() is useful for displaying just a part of long strings.

noobs suppresses the listing of the observation numbers.

> Options

table and display determine the style of output. By default, list determines whether to use table or display on the basis of the width of your screen and the linesize() option, if you specify it.

table forces table format. Forcing table format when list would have chosen otherwise generally produces impossible-to-read output because of the linewraps. However, if you are logging output in SMCL format and plan to print the output on wide paper later, specifying table can be a reasonable thing to do.

display forces display format.

header, noheader, and header(#) specify how the variable header is to be displayed.

header is the default in table mode and displays the variable header once, at the top of the table.

noheader suppresses the header altogether.

header(#) redisplays the variable header every # observations. For example, header(10) would display a new header every 10 observations.

The default in display mode is to display the variable names interweaved with the data:

	make	price	mpg	rep78	headroom	trunk	weight	length
1.	AMC Concord	4,099	22	3	2.5	11	2,930	186

	turn	displa~t	gear_r~o	foreign
	40	121	3.58	Domestic

However, if you specify `header`, the header is displayed once, at the top of the table:

make	price	mpg	rep78	headroom	trunk	weight	length
turn		displa~t		gear_r~o		foreign	

	make/AMC Concord	price	mpg	rep78	headroom	trunk	weight	length
1.	AMC Concord	4,099	22	3	2.5	11	2,930	186
	40		121		3.58		Domestic	

`clean` is a better alternative to `table` when you want to force table format and your goal is to produce more readable output on the screen. `clean` implies `table`, and it removes all dividing and separating lines, which is what makes wrapped table output nearly impossible to read.

`divider`, `separator(#)`, and `sepby(varlist)` specify how dividers and separator lines should be displayed. These three options affect only table format.

> `divider` specifies that divider lines be drawn between columns. The default is `nodivider`.

> `separator(#)` and `sepby(varlist)` indicate when separator lines should be drawn between rows.

> `separator(#)` specifies how often separator lines should be drawn between rows. The default is `separator(5)`, meaning every 5 observations. You may specify `separator(0)` to suppress separators altogether.

> `sepby(varlist)` specifies that a separator line be drawn whenever any of the variables in `sepby(varlist)` change their values; up to 10 variables may be specified. You need not make sure the data were sorted on `sepby(varlist)` before issuing the `list` command. The variables in `sepby(varlist)` also need not be among the variables being listed.

`nolabel` specifies that numeric codes be displayed rather than the label values.

⌐ Summary ⌐

`mean`, `sum`, `N`, `mean(varlist)`, `sum(varlist)`, and `N(varlist)` all specify that lines be added to the output reporting the mean, sum, or number of nonmissing values for the (specified) variables. If you do not specify the variables, all numeric variables in the *varlist* following `list` are used.

`labvar(varname)` is for use with $\text{mean}[()]$, $\text{sum}[()]$, and $\text{N}[()]$. `list` displays Mean, Sum, or N where the observation number would usually appear to indicate the end of the table—where a row represents the calculated mean, sum, or number of observations.

> `labvar(varname)` changes that. Instead, Mean, Sum, or N is displayed where the value for *varname* would be displayed. For instance, you might type

```
. list group costs profits, sum(costs profits) labvar(group)
```

	group	costs	profits
1.	1	47	5
2.	2	123	10
3.	3	22	2
	Sum	192	17

and then also specify the `noobs` option to suppress the observation numbers.

> Advanced

constant and constant(*varlist*) specify that variables that do not vary observation by observation be separated out and listed only once.

  constant specifies that `list` determine for itself which variables are constant.

  constant(*varlist*) allows you to specify which of the constant variables you want listed separately. `list` verifies that the variables you specify really are constant and issues an error message if they are not.

  constant and constant() respect if *exp* and in *range*. If you type

```
. list if group==3
```

  variable x might be constant in the selected observations, even though the variable varies in the entire dataset.

notrim affects how string variables are listed. The default is to trim strings at the width implied by the widest possible column given your screen width (or `linesize()`, if you specified that). `notrim` specifies that strings not be trimmed. `notrim` implies `clean` (see above) and, in fact, is equivalent to the `clean` option, so specifying either makes no difference.

absolute affects output only when `list` is prefixed with by *varlist*:. Observation numbers are displayed, but the overall observation numbers are used rather than the observation numbers within each by group. For example, if the first group had 4 observations and the second had 2, by default, the observations would be numbered 1, 2, 3, 4 and 1, 2. If `absolute` is specified, the observations will be numbered 1, 2, 3, 4 and 5, 6.

nodotz is a programmer's option that specifies that numerical values equal to .z be listed as a field of blanks rather than as .z.

subvarname is a programmer's option. If a variable has the characteristic *var*[varname] set, then the contents of that characteristic will be used in place of the variable's name in the headers.

linesize(*#*) specifies the width of the page to be used for determining whether table or display format should be used and for formatting the resulting table. Specifying a value of `linesize()` that is wider than your screen width can produce truly ugly output on the screen, but that output can nevertheless be useful if you are logging output and plan to print the log later on a wide printer.

(Continued on next page)

Syntax for clist

<u>cl</u>ist [*varlist*] [*if*] [*in*] [*clist_options*]

clist_options	description
[<u>no</u>]display	format into display or tabular nodisplay format
<u>noh</u>eader	omit variable or observation number header information
<u>nol</u>abel	display numeric codes; default displays label values
<u>noo</u>bs	suppress printing of observation numbers
<u>do</u>ublespace	insert a blank line between each observation when in nodisplay mode; has no effect in display mode

by is allowed with clist. See [D] **by**.

varlist may contain time-series operators; see [U] **11.4.3 Time-series varlists**.

Description of clist

clist is similar to list, clean; clist is the list command that appeared in Stata before Stata 8, options and all. list continues to be the preferred command. clist is provided for those instances when the old style of output is desired.

Options for clist

[no]display forces the format into display or tabular (nodisplay) format. If you do not specify one of these two options, Stata chooses the one it believes would be most readable.

noheader omits variable or observation number header information. The blank line and variable names at the top of the listing are omitted when in nodisplay mode. The observation number header and one blank line are omitted for each observation when in display mode.

nolabel specifies that numeric codes be displayed rather than label values.

noobs suppresses printing of the observation numbers.

doublespace produces a blank line between each observation in the listing when in nodisplay mode; it has no effect in display mode.

Remarks

list, typed by itself, lists all the observations and variables in the dataset. If you specify *varlist*, only those variables are listed. Specifying one or both of in *range* and if *exp* limits the observations listed.

list respects line size. That is, if you resize the Results window (in windowed versions of Stata) before running list, it will take advantage of the available horizontal space. Stata for Unix(console) users can instead use the set linesize command to take advantage of this feature.

list may not display all the large strings. You have two choices: 1) you can specify the clean option, which makes a different, less attractive listing, or 2) you can increase line size, as discussed above.

▷ Example 1

list has two output formats, known as table and display. The table format is suitable for listing a few variables, whereas the display format is suitable for listing an unlimited number of variables. Stata chooses automatically between those two formats:

```
. use http://www.stata-press.com/data/r10/auto
(1978 Automobile Data)
. list in 1/2
```

1.	make AMC Concord	price 4,099	mpg 22	rep78 3	headroom 2.5	trunk 11	weight 2,930	length 186
	turn 40	displa~t 121			gear_r~o 3.58		foreign Domestic	

2.	make AMC Pacer	price 4,749	mpg 17	rep78 3	headroom 3.0	trunk 11	weight 3,350	length 173
	turn 40	displa~t 258			gear_r~o 2.53		foreign Domestic	

```
. list make mpg weight displ rep78 in 1/5
```

	make	mpg	weight	displa~t	rep78
1.	AMC Concord	22	2,930	121	3
2.	AMC Pacer	17	3,350	258	3
3.	AMC Spirit	22	2,640	121	.
4.	Buick Century	20	3,250	196	3
5.	Buick Electra	15	4,080	350	4

The first case is an example of display format; the second is an example of table format. The table format is more readable and takes less space, but it is effective only if the variables can fit on one line across the screen. Stata chose to list all 12 variables in display format, but when the *varlist* was restricted to five variables, Stata chose table format.

If you are dissatisfied with Stata's choice, you can decide for yourself. You can specify the display option to force display format and the nodisplay option to force table format.

◁

❑ Technical Note

If you have long string variables in your data—say, str75 or longer—by default, list displays only the first 70 or so characters of each; the exact number is determined by the width of your Results window. The first 70 or so characters will be shown followed by "...". If you need to see the entire contents of the string, you can

1. specify the clean option, which makes a different (and uglier) style of list, or

2. make your Results window wider [Stata for Unix(console) users: increase set linesize].

❑

❑ Technical Note

Among the things that determine the widths of the columns, the variable names play a role. Left to itself, `list` will never abbreviate variable names to fewer than eight characters. You can use the `compress` option to abbreviate variable names to fewer characters than that.

❑

❑ Technical Note

When Stata lists a string variable in table output format, the variable is displayed right-justified by default.

When Stata lists a string variable in display output format, it decides whether to display the variable right-justified or left-justified according to the display format for the string variable; see [U] **12.5 Formats: controlling how data are displayed**. In our previous example, make has a display format of %-18s.

```
. describe make

              storage  display   value
variable name  type    format    label    variable label

make           str18   %-18s               Make and Model
```

The negative sign in the %-18s instructs Stata to left-justify this variable. If the display format had been %18s, Stata would have right-justified the variable.

Note that `foreign` appears to be a string variable, but if we `describe` it, we see that it is not:

```
. describe foreign

              storage  display   value
variable name  type    format    label    variable label

foreign        byte    %8.0g     origin    Car type
```

`foreign` is stored as a byte, but it has an associated value label named origin; see [U] **12.6.3 Value labels**. Stata decides whether to right-justify or left-justify a numeric variable with an associated value label by using the same rule used for string variables: it looks at the display format of the variable. Here the display format of %8.0g tells Stata to right-justify the variable. If the display format had been %-8.0g, Stata would have left-justified this variable.

❑

❑ Technical Note

You can `list` the variables in any order. When you specify the *varlist*, `list` displays the variables in the order you specify. You may also include variables more than once in the *varlist*.

❑

▷ Example 2

Sometimes you may wish to suppress the observation numbers. You do this by specifying the `noobs` option:

```
. list make mpg weight displ foreign in 46/55, noobs
```

make	mpg	weight	displa~t	foreign
Plym. Volare	18	3,330	225	Domestic
Pont. Catalina	18	3,700	231	Domestic
Pont. Firebird	18	3,470	231	Domestic
Pont. Grand Prix	19	3,210	231	Domestic
Pont. Le Mans	19	3,200	231	Domestic
Pont. Phoenix	19	3,420	231	Domestic
Pont. Sunbird	24	2,690	151	Domestic
Audi 5000	17	2,830	131	Foreign
Audi Fox	23	2,070	97	Foreign
BMW 320i	25	2,650	121	Foreign

After seeing the table, we decide that we want to separate the "Domestic" observations from the "Foreign" observations, so we specify sepby(foreign).

```
. list make mpg weight displ foreign in 46/55, noobs sepby(foreign)
```

make	mpg	weight	displa~t	foreign
Plym. Volare	18	3,330	225	Domestic
Pont. Catalina	18	3,700	231	Domestic
Pont. Firebird	18	3,470	231	Domestic
Pont. Grand Prix	19	3,210	231	Domestic
Pont. Le Mans	19	3,200	231	Domestic
Pont. Phoenix	19	3,420	231	Domestic
Pont. Sunbird	24	2,690	151	Domestic
Audi 5000	17	2,830	131	Foreign
Audi Fox	23	2,070	97	Foreign
BMW 320i	25	2,650	121	Foreign

◁

▷ Example 3

We want to add vertical lines in the table to separate the variables, so we specify the divider option. We also want to draw a horizontal line after every 2 observations, so we specify separator(2).

```
. list make mpg weight displ foreign in 46/55, divider separator(2)
```

	make	mpg	weight	displa~t	foreign
46.	Plym. Volare	18	3,330	225	Domestic
47.	Pont. Catalina	18	3,700	231	Domestic
48.	Pont. Firebird	18	3,470	231	Domestic
49.	Pont. Grand Prix	19	3,210	231	Domestic
50.	Pont. Le Mans	19	3,200	231	Domestic
51.	Pont. Phoenix	19	3,420	231	Domestic
52.	Pont. Sunbird	24	2,690	151	Domestic
53.	Audi 5000	17	2,830	131	Foreign
54.	Audi Fox	23	2,070	97	Foreign
55.	BMW 320i	25	2,650	121	Foreign

After seeing the table, we decide that we do not want to abbreviate displacement, so we specify abbreviate(12).

. list make mpg weight displ foreign in 46/55, divider sep(2) abbreviate(12)

	make	mpg	weight	displacement	foreign
46.	Plym. Volare	18	3,330	225	Domestic
47.	Pont. Catalina	18	3,700	231	Domestic
48.	Pont. Firebird	18	3,470	231	Domestic
49.	Pont. Grand Prix	19	3,210	231	Domestic
50.	Pont. Le Mans	19	3,200	231	Domestic
51.	Pont. Phoenix	19	3,420	231	Domestic
52.	Pont. Sunbird	24	2,690	151	Domestic
53.	Audi 5000	17	2,830	131	Foreign
54.	Audi Fox	23	2,070	97	Foreign
55.	BMW 320i	25	2,650	121	Foreign

◁

❏ Technical Note

You can suppress the use of value labels by specifying the nolabel option. For instance, the variable foreign in the examples above really contains numeric codes, with 0 meaning Domestic and 1 meaning Foreign. When we list the variable, however, we see the corresponding value labels rather than the underlying numeric code:

. list foreign in 51/55

	foreign
51.	Domestic
52.	Domestic
53.	Foreign
54.	Foreign
55.	Foreign

Specifying the nolabel option displays the underlying numeric codes:

. list foreign in 51/55, nolabel

	foreign
51.	0
52.	0
53.	1
54.	1
55.	1

❏

References

Harrison, D. A. 2006. Stata tip 34: Tabulation by listing. *Stata Journal* 6: 425–427.

Lauritsen, J. M. 2001. dm85: listjl: List one variable in a condensed form. *Stata Technical Bulletin* 59: 7–8. Reprinted in *Stata Technical Bulletin Reprints*, vol. 10, pp. 37–38.

Riley, A. R. 1993. dm15: Interactively list values of variables. *Stata Technical Bulletin* 16: 2–6. Reprinted in *Stata Technical Bulletin Reprints*, vol. 3, pp. 37–41.

Royston, P., and P. Sasieni. 1994. dm16: Compact listing of a single variable. *Stata Technical Bulletin* 17: 7–8. Reprinted in *Stata Technical Bulletin Reprints*, vol. 3, pp. 41–43.

Weesie, J. 1999. dm68: Display of variables in blocks. *Stata Technical Bulletin* 50: 3–4. Reprinted in *Stata Technical Bulletin Reprints*, vol. 9, pp. 27–29.

Also See

[D] **edit** — Edit and list data with Data Editor

[P] **display** — Display strings and values of scalar expressions

[P] **tabdisp** — Display tables

Title

> **lookfor** — Search for string in variable names and labels

Syntax

lookfor *string* [*string* [...]]

Description

lookfor helps you find variables by searching for *string* among all variable names and labels. If multiple *strings* are specified, lookfor will search for each of them separately. You may search for a phrase by enclosing *string* in double quotes.

Remarks

▷ Example 1

lookfor finds variables by searching for *string*, ignoring case, among the variable names and labels.

```
. use http://www.stata-press.com/data/r10/nlswork
(National Longitudinal Survey. Young Women 14-26 years of age in 1968)
. lookfor code
```

variable name	storage type	display format	value label	variable label
idcode	int	%8.0g		NLS ID
ind_code	byte	%8.0g		industry of employment
occ_code	byte	%8.0g		occupation

Three variable names contain the word code.

```
. lookfor married
```

variable name	storage type	display format	value label	variable label
msp	byte	%8.0g		1 if married, spouse present
nev_mar	byte	%8.0g		1 if never married

Two variable labels contain the word married.

```
. lookfor gnp
```

variable name	storage type	display format	value label	variable label
ln_wage	float	%9.0g		ln(wage/GNP deflator)

lookfor ignores case, so lookfor gnp found GNP in a variable label.

◁

▷ Example 2

If multiple strings are specified, all variable names or labels containing any of the strings are listed.

```
. lookfor code married

                storage  display   value
variable name   type     format    label     variable label

idcode          int      %8.0g               NLS ID
msp             byte     %8.0g               1 if married, spouse present
nev_mar         byte     %8.0g               1 if never married
ind_code        byte     %8.0g               industry of employment
occ_code        byte     %8.0g               occupation
```
◁

To search for a phrase, enclose *string* in double quotes.

```
. lookfor "never married"

                storage  display   value
variable name   type     format    label     variable label

nev_mar         byte     %8.0g               1 if never married
```

Saved Results

lookfor saves the following in r():

Macros
    r(varlist)   the varlist of found variables

Methods and Formulas

lookfor is implemented as an ado-file.

Also See

[D] **describe** — Describe data in memory or in file

Title

> **memory** — Memory size considerations

Syntax

Change amount of memory allocated to Stata

> set m<u>em</u>ory #[b|k|m|g] [, <u>perma</u>nently]

Report on Stata's memory usage

> memory

Display memory settings

> <u>q</u>uery memory

Use virtual memory more efficiently

> set <u>vir</u>tual {on|off}

Set maximum number of variables in Stata/MP and Stata/SE

> set maxvar # [, <u>perma</u>nently]

where # is specified in bytes, kilobytes, megabytes, or gigabytes (b, k, m, and g may be typed in uppercase), with the default being **k**,

and where $2{,}048 \leq \# \leq 32{,}767$ in set maxvar.

Description

You can set the size of memory only if you are using Stata/MP, Stata/SE, or Stata/IC. For Small Stata, the amount of memory used is fixed.

set memory allows you to increase or decrease the amount of memory allocated to Stata by the operating system while Stata is running.

memory reports Stata's memory usage. memory is available on all Stata/MP, Stata/SE, and Stata/IC, regardless of platform. query memory for Stata/MP and Stata/SE reports more information.

query memory displays the current memory settings; see [R] **query**.

set virtual specifies that Stata arrange data in memory to use virtual memory more efficiently. You do not need to set virtual on to use virtual memory. set virtual is off by default and is available only in Stata/MP, Stata/SE, and Stata/IC.

set maxvar allows you to set the maximum number of variables in Stata/MP and Stata/SE; Stata/IC and Small Stata do not allow you to change the maximum number of variables.

Option for set memory

permanently specifies that, in addition to making the change right now, the memory setting be remembered and become the default setting when you invoke Stata.

Option for set maxvar

permanently specifies that, in addition to making the change right now, the new limit be remembered and become the default setting when you invoke Stata.

Remarks

Remarks are presented under the following headings:

> *Resetting the amount of memory*
> *Obtaining the memory report and how Stata uses memory*
> *Using virtual memory*
> *Resetting the maximum number of variables in Stata/MP and Stata/SE*

Resetting the amount of memory

If you use Stata/MP, Stata/SE, or Stata/IC, you can change the amount of memory Stata has allocated while Stata is running:

```
. set memory 4m
no; data in memory would be lost
r(4);
```

You can change the amount of memory, but only when there are no data in memory:

```
. drop _all
. set memory 4m
(4096k)
```

You can increase the amount of memory,

```
. set memory 32m
(32768k)
```

or decrease it:

```
. set memory 1m
(1024k)
```

If you ask for more memory than your operating system can provide, the following error is returned:

```
. set memory 800m
op. sys. refuses to provide memory
r(909);
```

The number you type can be specified in megabytes or kilobytes. A number followed by m indicates megabytes. Numbers with k (or nothing) indicate kilobytes.

```
. set memory 4000k
(4000k)
. set memory 1000
(1000k)
```

❏ Technical Note

(This note is relevant only if you use Stata/MP, Stata/SE, or Stata/IC for Unix.) Operating systems vary in how they check out and return memory. Typically, operating systems handle returned memory in one of three ways:

1. The instant memory is returned, it is marked as returned and it is available for other programs to check out.

2. When memory is returned, it is put into a special bin, and, 5 or 10 minutes later, it will be marked as returned for other programs to check out. In the meantime, you could check it out again if you want, but no other program can.

3. When memory is returned, it is put into the special bin and never moved from there. You can use the memory again, but no other program can ever use that memory.

Windows follows policy 1. The various types of Unix differ on the policies they follow, and this has implications.

Let's imagine that you are pushing your Unix computer to its limits and have allocated lots of memory to Stata. You suddenly want to jump out of Stata and do something in Unix, so you use Stata's `shell` command to obtain a new shell:

```
. shell
op. sys. refused to start new process
r(702);
```

This can happen if there is no free memory. This reminds you that Stata has all the memory, but since you no longer need it, you return most of it:

```
. set memory 4m
(4096k)
```

Now you try the `shell` command again. What happens?

1. If your system follows policy 1, `shell` works.

2. If your system follows policy 2, `shell` does not work but will work 5 or 10 minutes from now.

3. If your system follows policy 3, `shell` does not work.

The result hinges on whether the operating system really takes back the memory Stata returns and when it returns. If your operating system follows policy 3, you must `exit` and restart Stata. If your operating system follows policy 2 and you are in a hurry, you can also `exit` and restart.

❏

Obtaining the memory report and how Stata uses memory

Type memory, and Stata will give you a memory report. We just started Stata/SE:

```
. memory
```

	bytes	
Details of **set memory** usage		
overhead (pointers)	0	0.00%
data	0	0.00%
data + overhead	0	0.00%
free	10,485,752	100.00%
Total allocated	10,485,752	100.00%
Other memory usage		
set maxvar usage	1,816,666	
set matsize usage	1,315,200	
programs, saved results, etc.	205	
Total	3,132,071	
Grand total	13,617,823	

Stata/IC differs slightly. Below we just started Stata/IC:

```
. memory
```

	bytes	
Details of **set memory** usage		
overhead (pointers)	0	0.00%
data	0	0.00%
data + overhead	0	0.00%
free	1,048,568	100.00%
Total allocated	1,048,568	100.00%
Other memory usage		
system overhead	745,090	
set matsize usage	16,320	
programs, saved results, etc.	599	
Total	762,009	
Grand total	1,810,577	

If you perform this experiment on your computer, you will probably see different numbers. Here is our memory report after we load the automobile dataset that comes with Stata using Stata/SE:

(Continued on next page)

```
. use http://www.stata-press.com/data/r10/auto
(1978 Automobile Data)

. memory
```

	bytes	
Details of **set memory** usage		
overhead (pointers)	296	0.00%
data	3,182	0.03%
data + overhead	3,478	0.03%
free	10,482,274	99.97%
Total allocated	10,485,752	100.00%
Other memory usage		
set maxvar usage	2,001,666	
set matsize usage	1,315,200	
programs, saved results, etc.	4,757	
Total	3,321,623	
Grand total	13,807,375	

Total memory refers to the total amount of memory Stata has allocated to its data areas—the number that can be specified at startup time or reset by set memory.

Overhead, data, and data + overhead refer to the amount of memory necessary to hold the dataset currently in memory. Let's start with the middle number.

The total amount of memory necessary to hold the automobile dataset is 3,182 bytes, and you could work this out for yourself from a describe, detail. The automobile dataset has 74 observations, and each observation requires 43 bytes (called the width), and $74 \times 43 = 3,182$.

The pointer overhead associated with this dataset is 296 bytes. Stata needs something called a pointer to keep track of where each observation is stored in memory. On this computer, pointers are 4 bytes—but that varies—and the dataset has 74 observations, so $4 \times 74 = 296$.

Data + overhead is just the sum of the two numbers: $296 + 3,182 = 3,478$ is the total amount of memory that Stata needs to store and keep track of these data.

Programs, saved results, etc., make up the total amount of memory Stata has used to store just what it says: Stata's programs (ado-files), macros, matrices, value labels, and all sorts of other things. This is sometimes referred to as Stata's dynamic memory. The report shows 4,757 bytes at this time, but this number often changes.

Here is a memory report from another session using Stata/IC in which we have loaded a dataset with 69,515 observations on 23 variables, and we are in the midst of analyzing it with xtgee:

```
. memory
                                                     bytes

Details of set memory usage
      overhead (pointers)                          278,060        5.30%
      data                                        2,850,115       54.36%

      data + overhead                             3,128,175       59.67%
      free                                        2,114,697       40.33%

      Total allocated                             5,242,872      100.00%

Other memory usage
      system overhead                               677,289
      set matsize usage                              16,320
      programs, saved results, etc.                     989

      Total                                         694,598

Grand total                                       5,937,470
```

Using virtual memory

Virtual memory refers to the ability to use more memory than is physically present on your computer. This is a feature provided by the operating system, not Stata, and is one that you may find yourself sometimes using.

Virtual memory is slow. You will be unhappy if you need to use virtual memory daily. On the other hand, virtual memory can get you out of a bind, and that is the right way to use it with Stata.

You do not need to set virtual on for Stata to use virtual memory. set virtual on merely possibly makes Stata run a little faster when the operating system is paging a lot. set virtual on will not make Stata run fast, just faster.

Virtual memory is most efficient (which is not to say efficient) when the program being executed exhibits something called locality of reference. This is the idea that if the program accesses one location in memory, later memory references will be to a location near that. If you set virtual on, Stata's memory management routines will go to extra work to arrange things so that the idea is true more often. Hence, Stata will run a little faster. If Stata is not using virtual memory, setting virtual on will make Stata run a little slower because Stata will be going to extra work for no good reason.

You set virtual on by typing the command

```
. set virtual on
```

You can check whether virtual is on or off by using query:

(Continued on next page)

```
. query
```

Memory settings (also see **query memory** for a more complete report)		
set memory	10M	
set maxvar	5000	
set matsize	400	

Output settings		
set more	on	
set rmsg	off	
set dp	period	may be period or comma
set linesize	79	characters
set pagesize	22	lines
set logtype	smcl	may be smcl or text

Interface settings		
set dockable	off	
set dockingguides	on	
set floatresults	off	
set floatwindows	off	
set locksplitters	off	
set pinnable	off	
set persistfv	off	
set persistvtopic	off	
set doublebuffer	on	
set linegap	1	pixels
set scrollbufsize	32000	characters
set varlabelpos	32	column number
set reventries	5000	lines
set maxdb	50	dialog boxes

Graphics settings		
set graphics	on	
set autotabgraphs	off	
set scheme	s2color	
set printcolor	automatic	may be automatic, asis, gs1, gs2, gs3
set copycolor	automatic	may be automatic, asis, gs1, gs2, gs3

Efficiency settings		
set adosize	500	kilobytes
set virtual	off	

(*output omitted*)

virtual is reported under **Efficiency settings**. To set virtual on, type

```
. set virtual on
```

Resetting the maximum number of variables in Stata/MP and Stata/SE

If you use Stata/MP or Stata/SE, you can change the maximum number of variables while Stata is running:

```
. set maxvar 2200
```

Current memory allocation

settable	current value	description	memory usage (1M = 1024k)
set maxvar	2200	max. variables allowed	0.841M
set memory	10M	max. data space	10.000M
set matsize	400	max. RHS vars in models	1.254M
			12.095M

Stata/MP and Stata/SE allow you to change the maximum number of variables allowed, but only when there are no data in memory.

```
. drop _all
. set maxvar 3000
```

Current memory allocation

settable	current value	description	memory usage (1M = 1024k)
set maxvar	3000	max. variables allowed	1.146M
set memory	10M	max. data space	10.000M
set matsize	400	max. RHS vars in models	1.254M
			12.400M

The minimum number of variables allowed is 2,048; the maximum number is 32,767. If you specify a number smaller or larger than these, an error is returned:

```
. set maxvar 2000
# must be between 2048 and 32767
r(198);
```

Saved Results

memory saves the following in r():

Scalars

r(k)	number of variables
r(width)	width of dataset
r(N_cur)	maximum observations (current partition)
r(N_curmax)	max. max. observations (current partition)
r(k_cur)	maximum variables (current partition)
r(w_cur)	maximum width (current partition)
r(M_data)	total memory available to data (bytes)
r(size_ptr)	size of memory pointer (bytes)
r(matsize)	matsize
r(adosize)	adosize
r(M_oh)	total memory for set maxvar (fixed if Stata/IC or Small Stata)
r(M_matsize)	total memory for set matsize
r(M_dm)	total memory for programs, saved results, etc.

There are four saved results that refer to the current partition. At any instant, Stata has partitioned the memory into observations and variables. The characteristics of the partition can change at any time, including right in the middle of a command, so the first four numbers do not reflect any real constraint, but they do reflect efficiency. If something should occur that violates any of those limits, Stata will have to work silently to reform the partition, which it can do reasonably efficiently and with no disk accesses. When Stata receives a request that violates the current partition's limits, it considers discarding memory copies of ado-files that have not been used recently. Ado-files are loaded automatically as needed, so how long they are kept in memory is only an efficiency issue. Stata considers reducing the memory requirement as an alternative to repartitioning.

The output produced by `memory` can be calculated from the saved results by

$$\text{total memory} = \texttt{r(M\_data)}$$

$$\text{overhead (pointers)} = \_N \times \texttt{r(size\_ptr)}$$

$$\text{data} = \_N \times \texttt{r(width)}$$

$$\text{programs, saved results, etc.} = \texttt{r(M\_dm)}$$

Reference

Sasieni, P. 1997. ip20: Checking for sufficient memory to add variables. *Stata Technical Bulletin* 40: 13. Reprinted in *Stata Technical Bulletin Reprints*, vol. 7, p. 86.

Also See

[R] **query** — Display system parameters

[P] **creturn** — Return c-class values

[R] **matsize** — Set the maximum number of variables in a model

[U] **6 Setting the size of memory**

Title

> **merge** — Merge datasets

Syntax

<u>merge</u> [*varlist*] using *filename* [*filename* ...] [, *options*]

options	description
Options	
keep(*varlist*)	keep only the specified variables from data in *filename*
_merge(*newvar*)	*newvar* marks source of resulting observation; default is _merge
<u>nol</u>abel	do not copy value label definitions from *filename*
<u>nonotes</u>	do not copy notes from *filename*
update	replace missing data in memory with data from *filename*
replace	replace nonmissing data in memory with data from *filename*
<u>nokeep</u>	drop obs. in using dataset that do not match
<u>nosummary</u>	drop summary variables when multiple *filenames* are specified
* <u>unique</u>	match variables uniquely identify observations in both data in memory and in *filename*
* <u>uniqmaster</u>	match variables uniquely identify observations in memory
* <u>uniqusing</u>	match variables uniquely identify observations in *filename*
* <u>sort</u>	sort master and using datasets by match variables before merge; sort implies unique

* unique, uniqmaster, uniqusing, and sort require *varlist* (the match variables) be specified.

Description

merge joins corresponding observations from the dataset currently in memory (called the *master* dataset) with those from Stata-format datasets stored as *filename* (called the *using* datasets) into single observations. If *filename* is specified without an extension, .dta is assumed.

merge can perform both one-to-one and match merges.

Options

> Options

keep(*varlist*) specifies the variables to be kept from the using dataset. If keep() is not specified, all variables are kept.

The *varlist* in keep(*varlist*) differs from standard Stata varlists in two ways: variable names in *varlist* may not be abbreviated, except by the use of wildcard characters; and you may not refer to a range of variables, such as price-weight.

_merge(*newvar*) specifies the name of the variable to be created that will mark the source of the resulting observation. The default is _merge(_merge); that is, if you do not specify this option, the new variable will be named _merge. See *The two kinds of merges* below for details.

nolabel prevents Stata from copying the value label definitions from the using dataset into the result. Even if you do not specify this option, label definitions from the using dataset do not replace label definitions in the master dataset.

nonotes prevents notes in the using dataset from being incorporated into the result. The default is to incorporate notes from the using dataset that do not already appear in the master dataset.

update specifies that the values from the using dataset be retained in cases where the master dataset contains missing. By default, the master dataset is held inviolate—values from the master dataset are retained when variables are found in both datasets.

replace, allowed with update only, specifies that even when the master dataset contains nonmissing values, they are to be replaced with corresponding values from the using dataset when the corresponding values are not equal. A nonmissing value, however, will never be replaced with a missing value.

nokeep causes merge to ignore observations in the using dataset that have no corresponding observation in the master. The default is to add these observations to the merged result and mark such observations with _merge = 2.

nosummary causes merge to drop the summary variables created when multiple using datasets are specified. The default is to create _merge1 recording results from merging the first disk dataset, _merge2 recording results from merging the second disk dataset, and so on. _merge1, _merge2, ..., contain 1 if an observation was found in the respective disk dataset and 0 otherwise.

Whether or not nosummary is specified, overall status variable _merge is created.

unique, uniqmaster, and uniqusing specify that the match variables in a match-merge uniquely identify the observations. (See *Match-merge* below.)

unique specifies that the match variables uniquely identify the observations in the master dataset and in the using dataset. For most match-merges, you should specify unique. merge does nothing differently when you specify the option, unless the assumption you are making is false, in which case an error message is issued and the data are not merged.

uniqmaster specifies that the match variables uniquely identify the observations in memory, in the master dataset, but not necessarily the ones in the using dataset.

uniqusing specifies that the match variables uniquely identify the observations in the using dataset, but not necessarily the ones in the master data.

unique is thus equivalent to specifying uniqmaster and uniqusing.

Things are more complicated when multiple using datasets are specified. unique still means unique in all datasets, and uniqusing still means unique in each of the using datasets, just as you would expect, but uniqmaster takes on a whole new meaning: uniqmaster means unique in the master and in all using datasets except the last! It asserts that the match variables uniquely identify observations in the master at each step, meaning that when the master is merged with the first using dataset, then when the (new) master (equal to original plus first using) is merged with the second using dataset, and so on. In summary, uniqmaster is simply not useful when multiple using datasets are specified.

If none of the three unique options are specified, observations in neither the master nor the using datasets are required to be unique, although they could be. If they are not unique, records that have the same values of the match variables are joined by observation until all the records on one side or the other are matched; after that, the final record on the shorter side is duplicated over and over again to match with the remaining records needing to be matched on the longer side.

sort specifies that the master and using datasets be sorted by the match variables, before the datasets are merged, if they are not already sorted by them. Match variables are required with `sort`. `sort` implies `unique`.

Remarks

Remarks are presented under the following headings:

> The two kinds of merges
> One-to-one merge
> Match-merge
> Match-merging with multiple using datasets
> Updating data

Distinguish carefully between merging and appending datasets and the corresponding Stata commands `merge` and `append`. Appending refers to adding new observations on existing variables. If you think of a dataset as a rectangle with observations going down and variables going across, appending increases the dataset's length. Merging adds new variables to existing observations, increasing the dataset's width. See [U] **22 Combining datasets** for more information.

Say that you have a dataset in which each observation records the characteristics of a particular automobile, such as the car's price and weight. If you have two such datasets, one for domestic and another for imported cars, and you wish to combine them, see [D] **append**.

On the other hand, if you have two datasets on the same types of cars, one recording price and the other weight, mileage, etc., and you wish to combine them, continue reading; `merge` does this.

Another command, `joinby`, forms all pairwise combinations of observations within group. Say that you have one dataset on mothers and fathers and another on their children. If you wish to combine them to match each parent with every one of their children (each child is matched with both parents) so that a two-parent, three-child family results in $2 \times 3 = 6$ observations, see [D] **joinby**.

The two kinds of merges

`merge` joins the observations stored in memory with the observations stored in *filename*. The disk dataset must be a Stata-format dataset; that is, it must have been created with the `save` command.

Stata performs two kinds of merges. If no *varlist* is specified, Stata performs a *one-to-one* merge, meaning that the first observation of one dataset is joined with the first observation of the other dataset, the second observation is joined with the second, and so on. If a *varlist* is specified, however, Stata uses those variables to perform a *match* merge, in which observations are joined only if the values of the variables in the specified *varlist* match.

Regardless of the style of merge being performed, `merge` always adds a new variable called (by default) _merge to the dataset. This variable takes on the values 1, 2, or 3 to mark the source of the resulting observation. The coding is as follows:

1.   The observation occurred only in the master dataset.

2.   The observation occurred only in the using dataset.

3.   The observation is the result of joining an observation from the master dataset with one from the using dataset.

When you use the `update` option, this coding is extended to include

4. The same as 3, except that missing values in the master dataset were updated with values from the using dataset.

5. The same as 3, except that some values in the master dataset disagree with values in the using dataset.

One-to-one merge

In a one-to-one merge, the first observation in the master dataset is joined with the first observation in the using dataset, the second observation is joined with the second, and so on. If variables with the same name occur in both the master and the using datasets, the joined observation retains those variables' *original* values—the values of the variables in the master dataset. When the master and using datasets contain different numbers of observations, missing values are joined with the remaining observations from the longer dataset.

▷ Example 1

We have two datasets that we wish to merge into one dataset. The first dataset, `odd.dta`, contains the first five positive odd numbers. The second dataset, `even1.dta`, contains the fifth through eighth positive even numbers. (Our example is admittedly not realistic, but it does illustrate the concept.) The datasets are

```
. use http://www.stata-press.com/data/r10/odd
(First five odd numbers)
. list
```

	number	odd
1.	1	1
2.	2	3
3.	3	5
4.	4	7
5.	5	9

```
. use http://www.stata-press.com/data/r10/even1
(5th through 8th even numbers)
. list
```

	number	even
1.	5	10
2.	6	12
3.	7	14
4.	8	16

We will join these two datasets using a one-to-one merge. Since `even1.dta` is already in memory (we just used it above), we type `merge using odd`. The result is

```
. merge using http://www.stata-press.com/data/r10/odd
number was byte now float
```

```
. list
```

	number	even	odd	_merge
1.	5	10	1	3
2.	6	12	3	3
3.	7	14	5	3
4.	8	16	7	3
5.	5	.	9	2

Notice the new variable _merge. Every time Stata merges two datasets, it creates this variable and assigns a value of 1, 2, or 3 to each observation. The value 1 indicates that the resulting observation occurred only in the master dataset, 2 indicates that the observation occurred only in the using dataset, and 3 indicates that the observation occurred in both datasets and is thus the result of joining an observation from the master dataset with an observation from the using dataset.

Here the first 4 observations are marked by _merge with a value of 3, and the last observation by _merge with a value of 2. The first 4 observations are the result of joining observations from the two datasets, and the last observation is a result of adding a new observation from the using dataset. These values reflect that the original dataset in memory had 4 observations, and the odd dataset stored on disk had 5 observations. The new observation is from the odd dataset exclusively: number is 5, odd is 9, and even has been filled in with *missing*.

number takes on the values 5–8 for the first 4 observations. Those are the values of number from the original dataset in memory—the even dataset—and conflict with the value of number stored in the first 4 observations of the odd dataset. number in that dataset took on the values 1–4, and those values were lost during the merge process. When Stata joins observations and there is a conflict between the value of a variable in memory and the value stored in the using dataset, Stata retains the value stored in memory by default.

When the command merge using odd was issued, Stata responded with "number was byte now float". Let's describe the datasets in this example:

```
. describe using http://www.stata-press.com/data/r10/odd
```

Contains data			First five odd numbers
obs:	5		9 Jan 2007 08:50
vars:	2		
size:	60		

variable name	storage type	display format	value label	variable label
number	float	%9.0g		
odd	float	%9.0g		Odd numbers

Sorted by:

```
. describe using http://www.stata-press.com/data/r10/even1
```

Contains data			5th through 8th even numbers
obs:	4		8 Jan 2007 10:02
vars:	2		
size:	36		

variable name	storage type	display format	value label	variable label
number	byte	%8.0g		
even	float	%9.0g		Even numbers

Sorted by:

number is stored as a `float` in `odd.dta` but as a `byte` in `even1.dta`; see [U] **12.2.2 Numeric storage types**. When we `merge` two datasets, Stata engages in automatic variable promotion; that is, if there are conflicts in numeric storage types, the more precise storage type will be used. The resulting dataset, therefore, will have `number` stored as a `float`, and Stata told us this when it said "number was byte now float".

◁

Match-merge

In a match-merge, observations are joined if the values of the variables in the *varlist* are the same. Since the values must be the same, obviously the variables in the *varlist* must appear in both the master and the using datasets.

A match-merge proceeds by taking an observation from the master dataset and one from the using dataset and comparing the values of the variables in the *varlist*. If the *varlist* values match, the observations are joined. If the *varlist* values do not match, the observation from the *earlier* dataset (the dataset whose *varlist* value comes first in the sort order) is joined with a pseudoobservation from the *later* dataset. All the variables in the pseudoobservation contain missing values. The actual observation from the later dataset is retained and compared with the next observation in the earlier dataset, and the process repeats.

▷ Example 2

The result is easier to understand than the explanation. Let's return to the dataset used in example 1 and `merge` the two datasets on variable `number`. We first use the `even` dataset and then type `merge number using odd`:

```
. use http://www.stata-press.com/data/r10/even1, clear
(5th through 8th even numbers)
. merge number using http://www.stata-press.com/data/r10/odd
master data not sorted
r(5);
```

Instead of merging the datasets, Stata reports the error message "master data not sorted". Match-merges require that the data be sorted in the order of the *varlist*, which here means ascending order of `number`. In the previous example, the data are in such an order, so the message is more than a little confusing. Before Stata can merge two datasets, however, both datasets must be sorted, and Stata must *know* that the data are sorted.

The basis of Stata's knowledge is the internal information that it keeps on the sort order, which Stata reveals whenever we `describe` the dataset:

```
. describe
Contains data from http://www.stata-press.com/data/r10/even1.dta
  obs:           4                          5th through 8th even numbers
  vars:          2                          8 Jan 2007 10:02
  size:         36 (99.9% of memory free)

              storage  display    value
variable name   type   format     label      variable label

number          byte   %8.0g
even            float  %9.0g                  Even numbers

Sorted by:
```

The last line of the description shows that the data are "Sorted by:" nothing. As far as Stata can tell, the data are not sorted. Similarly, `odd.dta` is sorted in ascending order of `number`, but as before, typing `describe` shows that Stata does not know this.

If the datasets are not sorted on the merge variable, we need to use the `sort` option with `merge` to tell Stata to sort both datasets before doing the merge:

```
. merge number using http://www.stata-press.com/data/r10/odd, sort
(First five odd numbers)
number was byte now float
. list
```

	number	even	odd	_merge
1.	5	10	9	3
2.	6	12	.	1
3.	7	14	.	1
4.	8	16	.	1
5.	1	.	1	2
6.	2	.	3	2
7.	3	.	5	2
8.	4	.	7	2

It worked! Let's understand what happened. Even though Stata sorted both datasets by `number` since we specified the `sort` option, we immediately discern that the result is no longer in ascending order of `number`. It will be easier to understand what happened if we re-`sort` the data and then `list` the data again:

```
. sort number
. list
```

	number	even	odd	_merge
1.	1	.	1	2
2.	2	.	3	2
3.	3	.	5	2
4.	4	.	7	2
5.	5	10	9	3
6.	6	12	.	1
7.	7	14	.	1
8.	8	16	.	1

`number` now goes from 1 to 8, with no repeated values and no values left out of the sequence. The odd dataset defined observations for `number` between 1 and 5, whereas the even dataset defined observations between 5 and 8. Thus the variable `odd` is defined for `number` variables equal to 1–5, and `even` is defined for `number` variables equal to 5–8.

For instance, in the first observation, `number` is 1, `even` is *missing*, and `odd` is 1. The value of _merge, 2, indicates that this observation came from the using dataset—`odd.dta`. In the last observation, `number` is 8, `even` is 16, and `odd` is *missing*. The value of _merge, 1, indicates that this observation came from the master dataset—`even1.dta`.

For the fifth observation, `number` is 5, `even` is 10, and `odd` is 9. Both `even` and `odd` are defined since both the even and the odd datasets had information for `number` with a value of 5. The value of _merge, 3, also tells us that both datasets contributed to the formation of the observation.   ◁

▷ Example 3

Although the previous example demonstrated, in glorious detail, how the match-merging process works, it was not a practical example of how you will ordinarily use it. Here is a more realistic application.

We have two datasets containing information on automobiles. The identifying variable in each dataset is make, a string variable containing the manufacturer and the model. An *identifying* variable is one that is unique for every observation in the dataset. Values for make—for instance, Honda Accord—are sufficient for identifying each observation.

One dataset, autotech.dta, also contains mpg, weight, and length. The other dataset, auto-cost.dta, contains price and rep78, the 1978 repair record.

```
. describe using http://www.stata-press.com/data/r10/autotech
Contains data                              1978 Automobile Data
  obs:           74                        8 Jan 2007 11:19
  vars:           4
  size:       2,072
```

variable name	storage type	display format	value label	variable label
make	str18	%18s		Make and Model
mpg	int	%8.0g		Mileage (mpg)
weight	int	%8.0g		Weight (lbs.)
length	int	%8.0g		Length (in.)

```
Sorted by:  make
. describe using http://www.stata-press.com/data/r10/autocost
Contains data                              1978 Automobile Data
  obs:           74                        8 Jan 2007 08:20
  vars:           3
  size:       1,924
```

variable name	storage type	display format	value label	variable label
make	str18	%18s		Make and Model
price	int	%8.0g		Price
rep78	int	%8.0g		Repair Record 1978

```
Sorted by:  make
```

We want to merge these two datasets into one dataset:

```
. use http://www.stata-press.com/data/r10/autotech, clear
(1978 Automobile Data)
. merge make using http://www.stata-press.com/data/r10/autocost
```

Let's now examine the result:

```
. describe
Contains data from http://www.stata-press.com/data/r10/autotech.dta
  obs:            74                           1978 Automobile Data
  vars:            7                           8 Jan 2007 11:19
  size:         2,442 (99.8% of memory free)

              storage  display     value
variable name   type   format      label      variable label

make            str18  %18s                    Make and Model
mpg             int    %8.0g                   Mileage (mpg)
weight          int    %8.0g                   Weight (lbs.)
length          int    %8.0g                   Length (in.)
price           int    %8.0g                   Price
rep78           int    %8.0g                   Repair Record 1978
_merge          byte   %8.0g

Sorted by:
     Note:  dataset has changed since last saved
```

We now have one dataset containing all the information from the two original datasets—or at least it appears that we do. We need to verify the result. We think that we entered data for the same cars in each dataset, so every variable should be defined for every car. It is possible that we made a mistake and accidentally left some cars out of one or the other dataset. We can reassure ourselves of our infallibility by tabulating _merge:

```
. tabulate _merge
   _merge |      Freq.     Percent        Cum.

        3 |         74      100.00      100.00

    Total |         74      100.00
```

We see that _merge is 3 for every observation in the dataset. We made no mistake—for every observation in autocost.dta, there is an observation in autotech.dta, and vice versa.

Now pretend that we have another dataset containing more information on these automobiles, automore.dta, which we want to merge as well. We use the sort option, since after a merge the sort order may have changed:

```
. merge make using http://www.stata-press.com/data/r10/automore, sort
_merge already defined
r(110);
```

Stata refused to merge the new dataset, complaining instead that _merge is already defined. Every time that Stata merges datasets, it wants to create a variable called _merge (or *varname* if the _merge(*varname*) option was specified). Here there is an _merge variable left over from the last time we merged. We have three choices: We can rename the _merge variable, we can drop it, or we can specify a different variable name with the _merge() option. Here _merge contains no useful information—we have already verified that the previous merge went as expected—so we drop it and try again:

```
. drop _merge
. merge make using http://www.stata-press.com/data/r10/automore, sort
```

Stata performed our request; whatever new variables were contained in `automore.dta` should be contained in our single master dataset. After a match-merge, we should *always* tabulate _merge to verify that the expected actually happened, as we do below:

```
. tabulate _merge
```

_merge	Freq.	Percent	Cum.
1	1	1.33	1.33
2	1	1.33	2.67
3	73	97.33	100.00
Total	75	100.00	

Surprise! Here something strange did happen. Some 73 of the observations merged as we anticipated. However, the new dataset `automore.dta` added one new car to the dataset (identified by _merge equal to 2) and failed to define new variables for another car in our original dataset (identified by _merge equal to 1). Most likely, we have a mistake in `automore.dta`. We probably misidentified one car so that, to Stata, it appeared as data on a new car, resulting in one new observation and missing data on another.

If this happened to us, we could figure out why it happened. We could type `list make if _merge==1` to learn the identity of the car that did not appear in `automore.dta`, and we could type `list make if _merge==2` to learn the identity of the car that `automore.dta` added to our data. ◁

❑ Technical Note

Always tabulate _merge, no matter how sure you are that you have no errors. It takes only a second and can save you hours of grief. Along the same lines, one-to-one merges are a bad idea. In the example above, we could have performed all the merges as one-to-one merges and saved a little typing. Let's examine what would have happened.

We first merged `autotech.dta` with `autocost.dta` by typing `merge make using autocost`. We could perform a one-to-one merge by typing `merge using autocost`. The result would be the same; the datasets line up and are in the same sort order, so sequentially matching the observations from the two datasets would have resulted in a perfectly matched dataset.

In the second case, we merged the data in memory with `automore.dta` by typing `merge make using automore`. A one-to-one merge would have led to disaster, and we would never have known it! If we had typed `merge using automore`, Stata would have sequentially, and blindly, joined observations. Since there are the same number of observations in each dataset, everything would have appeared to merge perfectly.

We speculated in example 3 that we had an error in `automore.dta`. Remember that `automore.dta` included data on one new car and lacked data on an existing car. Even if there is no error, things have gone awry. No matter what, the data in memory and `automore.dta` do not match. For instance, assume that this new car is the first observation of `automore.dta` and that it is some (perhaps mistaken) model of Ford. Assume that the first observation of the data in memory is on a Chevrolet. Stata could and would silently join data on the Chevrolet with data on the Ford, and thereafter, data on a Volvo with data on a Saab, and even data on a Volkswagen with data on a Cadillac, and you would never know.

Every dataset should carry a variable or a set of variables that *uniquely* identifies each observation, and you should always use those variables when merging data. Ignore this advice at your own peril.

❑

❏ Technical Note

You may need to merge two datasets, knowing that there will be mismatches. Say that you have an analysis dataset on patients from the cancer ward of a particular hospital, and you have just received another dataset containing their demographic information. Actually, this other dataset contains not just their demographic information but also the demographic information on every patient in the hospital during the year. You could

```
. merge patid using demog
. drop if _merge==2
```

or

```
. merge patid using demog, nokeep
```

The `nokeep` option tells `merge` not to store observations from the using data that do not appear in the master. There is an advantage in this. When we merged and dropped, we stored the irrelevant observations and then discarded them, so the data in memory temporarily grew. When we merge with the `nokeep` option, the data never grow beyond what is absolutely necessary.

❏

❏ Technical Note

In our automobile example, we had one identifying variable. Sometimes there will be multiple identifying variables that, taken together, are unique for every observation.

Let's imagine that, rather than having one variable called `make`, we had two variables: `manuf` and `model`. `manuf` contains the manufacturer, and `model` contains the model. Rather than having one variable recording, say, "Honda Accord", we have two variables, one recording "Honda" and another recording "Accord". Stata can deal with this type of data. We could go back through our previous example and substitute `manuf model` everywhere we see `make`. For instance, rather than typing `merge make using autocost`, we would have typed `merge manuf model using autocost`.

Now let's make one more change in our assumptions. Let's assume that `manuf` and `model` are not string variables but are instead numerically coded variables. Perhaps the number 15 stands for Honda in the `manuf` variable, and the number 2 stands for Accord in the `model` variable. We do not have to remember our numeric codes because we have smartly created value labels telling Stata what number stands for what string of characters. We now go back to the step when we merged `autotech.dta` with `autocost.dta`:

```
. use autotech, clear
(1978 Automobile Data)
. merge manuf model using autocost
(label manuf already defined)
(label model already defined)
```

Stata makes two minor comments but otherwise carries out our request. It tells us that the labels `manuf` and `model` are already defined. The messages refer to the *value labels* named `manuf` and `model`.

Both datasets contain value label definitions that turn the numeric codes for manufacturer and model into words. When Stata merged the two datasets, it already had one set of definitions in memory (obtained when we typed `use autotech`) and thus ignored the second set of definitions contained in `autocost.dta`. Stata felt obliged to mention the second set of definitions while otherwise ignoring them, since they *might* contain different codings. Here we know that they are the same since we created them. (*Hint:* You should never give the same name to value labels containing different codings.)

❏

▷ Example 4

We might want the match-merge to take place only if the match variable(s) uniquely identify the observations in both datasets. We can make this condition a requirement for `merge` by specifying the `unique` option. Suppose that we wanted to merge `autotech.dta` and `autocost.dta`, with `make` as the match variable. Now suppose that `autotech.dta` has 2 observations in which the value of `make` is "Dodge Colt". If we specify the `unique` option, `merge` will refuse to execute the merge, and Stata will display a message telling us why.

```
. use http://www.stata-press.com/data/r10/autotech, clear
(1978 Automobile Data)
. generate fweight = 2 if make == "Dodge Colt"
(73 missing values generated)
. expand fweight
(73 missing counts ignored; observations not deleted)
(1 observation created)
. merge make using http://www.stata-press.com/data/r10/autocost, unique sort
variable make does not uniquely identify observations in the master data
r(459);
```

To find out which observations have duplicate values in the match variables, we use the `duplicates` command (see [D] **duplicates**).

```
. duplicates list make

Duplicates in terms of make
```

obs:	make
27	Dodge Colt
28	Dodge Colt

```
. drop in 28
(1 observation deleted)
```

Once we have resolved the duplicates problem, we can successfully `merge` the two datasets.

```
. merge make using http://www.stata-press.com/data/r10/autocost, unique sort
```

If your only criterion is that the match variable be unique in the master dataset or using dataset, use `uniqmaster` or `uniqusing`, respectively, rather than `unique`.

◁

In a match-merge, the master and using datasets may have multiple observations with the same *varlist* value. These multiple observations are joined sequentially, as in a one-to-one merge. If the datasets have an unequal number of observations with the same *varlist* value, the last such observation in the *shorter* dataset is replicated until the number of observations is equal.

▷ Example 5

The process of replicating the observation from the shorter dataset is known as *spreading* and can be put to practical use. Suppose that we have two datasets. `dollars.dta` contains the dollar sales and costs of our firm, by region, for the last year:

```
. use http://www.stata-press.com/data/r10/dollars, clear
(Regional Sales & Costs)
. list
```

	region	sales	cost
1.	N Cntrl	419,472	227,677
2.	NE	360,523	138,097
3.	South	532,399	330,499
4.	West	310,565	165,348

`sforce.dta` contains the names of the individuals in our sales force along with the region in which they operate:

```
. use http://www.stata-press.com/data/r10/sforce
(Sales Force)
. list
```

	region	name
1.	N Cntrl	Krantz
2.	N Cntrl	Phipps
3.	N Cntrl	Willis
4.	NE	Ecklund
5.	NE	Franks
6.	South	Anderson
7.	South	Dubnoff
8.	South	Lee
9.	South	McNiel
10.	West	Charles
11.	West	Cobb
12.	West	Grant

We now wish to merge these two datasets by `region`, spreading the sales and cost information across all observations for which it is relevant; that is, we want to add the variables `sales` and `costs` to the sales force data. The variable `sales` will assume the value $360,523 for the first 2 observations, $419,472 for the next 3 observations, and so on.

(*Continued on next page*)

```
. merge region using http://www.stata-press.com/data/r10/dollars
variable region does not uniquely identify observations in the master data
(label region already defined)
. list
```

	region	name	sales	cost	_merge
1.	N Cntrl	Krantz	419,472	227,677	3
2.	N Cntrl	Phipps	419,472	227,677	3
3.	N Cntrl	Willis	419,472	227,677	3
4.	NE	Ecklund	360,523	138,097	3
5.	NE	Franks	360,523	138,097	3
6.	South	Anderson	532,399	330,499	3
7.	South	Dubnoff	532,399	330,499	3
8.	South	Lee	532,399	330,499	3
9.	South	McNiel	532,399	330,499	3
10.	West	Charles	310,565	165,348	3
11.	West	Cobb	310,565	165,348	3
12.	West	Grant	310,565	165,348	3

Even though there are 12 observations in the sales force data and only 4 observations in the sales and cost data, all the records merged. The dollars.dta contained 1 observation for the NE region. The sforce.dta contained two observations for the same region. Thus the single observation in dollars.dta was matched to both the observations in sforce.dta. In technical jargon, the single record in dollars.dta was replicated, or *spread*, across the observations in sforce.dta.

◁

Match-merging with multiple using datasets

▷ Example 6

To demonstrate merging with multiple using files, we will use the even dataset as the master and some new using datasets odd3 and letter. The odd3 dataset is just the odd2 dataset with an added entry for the sixth odd number.

```
. use http://www.stata-press.com/data/r10/odd3, clear
(First six odd numbers)
. list
```

	number	odd
1.	1	1
2.	2	3
3.	3	5
4.	4	7
5.	5	9
6.	6	11

The letter dataset contains a collection of letters from the alphabet, sorted by number.

```
. use http://www.stata-press.com/data/r10/letter
(Some letters from the alphabet)
. list
```

	number	letter
1.	3	c
2.	4	d
3.	5	e
4.	8	h
5.	9	i

We will first use the even dataset and then merge with both the odd3 dataset and the letter dataset as the using datasets.

```
. use http://www.stata-press.com/data/r10/even
(6th through 8th even numbers)
. list
```

	number	even
1.	6	12
2.	7	14
3.	8	16

```
. merge number using http://www.stata-press.com/data/r10/odd3
> http://www.stata-press.com/data/r10/letter
number was byte now float
. list
```

	number	even	odd	_merge1	letter	_merge2	_merge
1.	1	.	1	1		0	2
2.	2	.	3	1		0	2
3.	3	.	5	1	c	1	3
4.	4	.	7	1	d	1	3
5.	5	.	9	1	e	1	3
6.	6	12	11	1		0	3
7.	7	14	.	0		0	1
8.	8	16	.	0	h	1	3
9.	9	.	.	0	i	1	2

We have three new variables, _merge1, _merge2, and _merge. _merge is the standard result variable that we have discussed before: 1 means that the observation came from the master, 2 means that it came from the using, and 3 means that it came from both. Here, however, we have two using datasets, and _merge1 and _merge2 clarify what is meant by "from the using" in _merge.

_merge1 contains 1 if the first using dataset (odd.dta in this case) contributed to the result, and it contains 0 otherwise. _merge2 works the same way, but for the second using dataset (letter.dta here).

◁

Updating data

merge with the update option varies merge's actions when an observation in the master dataset is matched with an observation in the using dataset. Without the update option, merge leaves the values in the master dataset alone and adds the data for the new variables. With the update option, merge adds the new variables, but it also replaces missing values in the master observation with corresponding values from the using. Missing values are numeric missing (., .a, ..., .z) and empty strings ("").

The values for _merge are extended:

_merge	Meaning
1	observations from master data
2	observations from using data
3	observations from both, master agrees with using
4	observations from both, missing in master updated
5	observations from both, master disagrees with using

For _merge = 5, the master values are retained unless replace is specified, in which case the master values are updated as if they had been missing.

Suppose that dataset 1 contains variables id, a, and b, and that dataset 2 contains id, a, and x. You merge the two datasets by id, dataset 1 being the master dataset in memory and dataset 2 being the using dataset on disk. Consider two observations that match, and call the values from the first dataset id_1, etc., and those from the second id_2, etc. The resulting dataset will have variables id, a, b, x, and _merge. merge's typical logic is as follows:

1. The fact that the observations match means $id_1 = id_2$. Set $id = id_1$.

2. Variable a occurs in both datasets. Ignore a_2, and set $a = a_1$.

3. Variable b occurs in only dataset 1. Set $b = b_1$.

4. Variable x occurs in only dataset 2. Set $x = x_2$.

5. Set _merge = 3.

With update, the logic is modified:

1. (unchanged.) Since the observations match, $id_1 = id_2$. Set $id = id_1$.

2. Variable a occurs in both datasets:

   a. If $a_1 = a_2$, set $a = a_1$, and set _merge = 3.

   b. If a_1 contains missing and a_2 is nonmissing, set $a = a_2$ and set _merge = 4, indicating that an update was made.

   c. If a_2 contains missing, set $a = a_1$ and set _merge = 3, indicating no update.

   d. If $a_1 \neq a_2$ and both contain nonmissing, set $a = a_1$, or, if replace was specified, $a = a_2$. Regardless, set _merge = 5, indicating a disagreement.

Rules 3 and 4 remain unchanged.

▷ Example 7

In original.dta, we have data on some cars that include the make, price, and mileage rating. In updates.dta, we have some updated data on these cars, along with a new variable recording engine displacement. The data contain

```
. use http://www.stata-press.com/data/r10/original, clear
(original data)

. list
```

	make	price	mpg
1.	Chev. Chevette	3,299	29
2.	Chev. Malibu	4,504	.
3.	Datsun 510	5,079	24
4.	Merc. XR-7	6,303	.
5.	Olds Cutlass	4,733	19
6.	Renault Le Car	3,895	26
7.	VW Dasher	7,140	23

```
. use http://www.stata-press.com/data/r10/updates
(updates, mpg and displacement)

. list
```

	make	mpg	displa~t
1.	Chev. Chevette	.	231
2.	Chev. Malibu	22	200
3.	Datsun 510	24	119
4.	Merc. XR-7	14	302
5.	Olds Cutlass	19	231
6.	Renault Le Car	25	79
7.	VW Dasher	23	97

By updating our data, we obtain

```
. use http://www.stata-press.com/data/r10/original
(original data)

. merge make using http://www.stata-press.com/data/r10/updates, update

. list
```

	make	price	mpg	displa~t	_merge
1.	Chev. Chevette	3,299	29	231	3
2.	Chev. Malibu	4,504	22	200	4
3.	Datsun 510	5,079	24	119	3
4.	Merc. XR-7	6,303	14	302	4
5.	Olds Cutlass	4,733	19	231	3
6.	Renault Le Car	3,895	26	79	5
7.	VW Dasher	7,140	23	97	3

All observations merged because all have _merge $\geq$ 3. The observations having _merge $= 3$ have mpg just as it was recorded in the original dataset. In observation 1, mpg is 29 because the updated dataset had mpg $= .$; in observation 3, mpg remains 24 because the updated dataset also stated that mpg is 24.

The observations having _merge $= 4$ have had their mpg data updated. The mpg variable was missing in observations 2 and 4, and new values were obtained from the update data.

The observation having _merge = 5 has its mpg as it was recorded in the original dataset, as do the _merge = 3 observations, but there is an important difference. There is a disagreement about the value of mpg; the original has a value of 26, and the updated has 25. Had we specified the replace option, mpg would now contain the updated 25, but the observation would still be marked _merge = 5. replace affects only the value that is retained in the case of disagreement.

◁

Methods and Formulas

merge is implemented as an ado-file.

References

Nash, J. D. 1994. dm19: Merging raw data and dictionary files. *Stata Technical Bulletin* 20: 3–5. Reprinted in *Stata Technical Bulletin Reprints*, vol. 4, pp. 22–25.

Weesie, J. 2000. dm75: Safe and easy matched merging. *Stata Technical Bulletin* 53: 6–17. Reprinted in *Stata Technical Bulletin Reprints*, vol. 9, pp. 62–77.

Also See

[D] **save** — Save datasets

[D] **sort** — Sort data

[D] **append** — Append datasets

[D] **cross** — Form every pairwise combination of two datasets

[D] **joinby** — Form all pairwise combinations within groups

[U] **22 Combining datasets**

Title

missing values — Quick reference for missing values

Description

This entry provides a quick reference for Stata's missing values.

Remarks

Stata has 27 numeric missing values:

., the default, which is called the *system missing value* or `sysmiss`

and

.a, .b, .c, ..., .z, which are called the *extended missing values.*

Numeric missing values are represented by large positive values. The ordering is

$$\text{all nonmissing numbers} < . < .a < .b < \cdots < .z$$

Thus the expression

$$\text{age} > 60$$

is true if variable `age` is greater than 60 or missing.

To exclude missing values, ask whether the value is less than '.'.

```
. list if age > 60 & age < .
```

To specify missing values, ask whether the value is greater than or equal to '.'. For instance,

```
. list if age >=.
```

Stata has one string missing value, which is denoted by "" (blank).

Also See

[U] **12.2.1 Missing values**

Title

mkdir — Create directory

Syntax

mkdir *directoryname* $\left[\, , \underline{\text{public}}\right]$

Double quotes may be used to enclose *directoryname*, and the quotes must be used if *directoryname* contains embedded spaces.

Description

mkdir creates a new directory (folder).

Option

public specifies that *directoryname* be readable by everyone; otherwise, the directory will be created according to the default permissions of your operating system.

Remarks

Examples:

Windows

. mkdir myproj
. mkdir c:\projects\myproj
. mkdir "c:\My Projects\Project 1"

Macintosh and Unix

. mkdir myproj
. mkdir ~/projects/myproj

Also See

[D] **cd** — Change directory

[D] **copy** — Copy file from disk or URL

[D] **dir** — Display filenames

[D] **erase** — Erase a disk file

[D] **shell** — Temporarily invoke operating system

[D] **type** — Display contents of a file

[U] **11.6 File-naming conventions**

Title

> **mvencode** — Change missing values to numeric values and vice versa

Syntax

Change missing values to numeric values

> mvencode *varlist* $\left[\,if\,\right]$ $\left[\,in\,\right]$, mv(*#* | *mvc* = *#* $\left[\,\backslash\,mvc = \#\ldots\,\right]$ $\left[\,\backslash$ else = #$\right]$) $\left[\,\underline{o}verride\,\right]$

Change numeric values to missing values

> mvdecode *varlist* $\left[\,if\,\right]$ $\left[\,in\,\right]$, mv(*numlist* | *numlist=mvc* $\left[\,\backslash\,numlist = mvc\ \ldots\,\right]$)

where *mvc* is one of . | .a | .b | ... | .z.

Description

mvencode changes missing values in the specified *varlist* to numeric values.

mvdecode changes occurrences of a *numlist* in the specified *varlist* to a missing-value code.

Missing-value codes may be sysmiss (.) and the extended missing-value codes .a, .b, ..., .z.

String variables in *varlist* are ignored.

Options

> **Main**

mv(*#* | *mvc* = *#* $\left[\,\backslash\,mvc = \#\ldots\,\right]$ $\left[\,\backslash$ else = #$\right]$) is required and specifies the numeric values to which the missing values are to be changed.

mv(#) specifies that all types of missing values be changed to *#*.

mv(*mvc*=*#*) specifies that occurrences of missing-value code *mvc* be changed to *#*. Multiple transformation rules may be specified, separated by a backward slash (\). The list may be terminated by the special rule else=*#*, specifying that all types of missing values not yet transformed be set to *#*.

Examples: mv(9), mv(.=99\.a=98\.b=97), mv(.=99\ else=98)

mv(*numlist* | *numlist=mvc* $\left[\,\backslash\,numlist = mvc\ \ldots\,\right]$) is required and specifies the numeric values that are to be changed to missing values.

mv(*numlist*=*mvc*) specifies that the values in *numlist* be changed to missing-value code *mvc*. Multiple transformation rules may be specified, separated by a backward slash (\). See [P] **numlist** for the syntax of a numlist.

Examples: mv(9), mv(99=.\98=.a\97=.b), mv(99=.\ 100/999=.a)

override specifies that the protection provided by mvencode be overridden. Without this option, mvencode refuses to make the requested change if any of the numeric values are already used in the data.

405

Remarks

You may occasionally read data in which missing (e.g., a respondent failed to answer a survey question, or the data were not collected) is coded with a special numeric value. Popular codings are 9, 99, −9, −99, and the like. If missing were encoded as −99,

```
. mvdecode _all, mv(-99)
```

would translate the special code to the Stata missing value ".". Use this command cautiously because, even if −99 were not a special code, all −99s in the data would be changed to missing.

Sometimes different codes are used to represent different reasons for missing values. For instance, 98 may be used for "refused to answer" and 99 for "not applicable". Extended missing values (.a, .b, and so on) may be used to code these differences.

```
. mvdecode _all, mv(98=.a\ 99=.b)
```

Conversely, you might need to export data to software that does not understand that "." indicates a missing value, so you might code missing with a special numeric value. To change all missings to −99, you could type

```
. mvencode _all, mv(-99)
```

To change extended missing values back to numeric values, type

```
. mvencode _all, mv(.a=98\ .b=99)
```

This would leave sysmiss and all other extended missing values unchanged. To encode in addition sysmiss . to 999 and all other extended missing values to 97, you might type

```
. mvencode _all, mv(.=999\ .a=98\ .b=99\ else=97)
```

mvencode will automatically recast variables upward, if necessary, so even if a variable is stored as a byte, its missing values can be recoded to, say, 999. Also mvencode refuses to make the change if # (−99 here) is already used in the data, so you can be certain that your coding is unique. You can override this feature by including the override option.

Be aware of another potential problem with encoding and decoding missing values: value labels are not automatically adapted to the changed codings. You have to do this yourself. For example, the value label divlabor maps the value 99 to the string "not applicable". You used mvdecode to recode 99 to .a for all variables that are associated with this label. To fix the value label, clear the mapping for 99 and define it again for .a.

```
. label define divlabor 99 "", modify
. label define divlabor .a "not applicable", add
```

▷ Example 1

Our automobile dataset contains 74 observations and 12 variables. Let's first attempt to translate the missing values in the data to 1:

```
. use http://www.stata-press.com/data/r10/auto
(1978 Automobile Data)

. mvencode _all, mv(1)
        make: string variable ignored
       rep78: already 1 in     2 observations
     foreign: already 1 in    22 observations
no action taken
r(9);
```

Our attempt failed. mvencode first informed us that make is a string variable—this is not a problem but is reported merely for our information. String variables are ignored by mvencode. It next informed us that rep78 was already coded 1 in 2 observations and that foreign was already coded 1 in 22 observations. Thus 1 would be a poor choice for encoding missing values because, after encoding, we could not tell a real 1 from a coded missing value 1.

We could force mvencode to encode the data with 1, anyway, by typing mvencode _all, mv(1) override. That would be appropriate if the 1s in our data already represented missing data. They do not, however, so we code missing as 999:

```
. mvencode _all, mv(999)
        make: string variable ignored
       rep78: 5 missing values
```

This worked, and we are informed that the only changes necessary were to 5 observations of rep78.

◁

▷ Example 2

Let us now pretend that we just read in the automobile data from some raw dataset in which all the missing values were coded 999. We can convert the 999s to real missings by typing

```
. mvdecode _all, mv(999)
        make: string variable ignored
       rep78: 5 missing values
```

We are informed that make is a string variable, so it was ignored, and that rep78 contained 5 observations with 999. Those observations have now been changed to contain missing.

◁

Methods and Formulas

mvencode and mvdecode are implemented as ado-files.

Acknowledgment

These versions of mvencode and mvdecode were written by Jeroen Weesie, Department of Sociology, Utrecht University, The Netherlands.

Also See

[D] **generate** — Create or change contents of variable

[D] **recode** — Recode categorical variables

Title

notes — Place notes in data

Syntax

Attach notes to dataset

notes [*varname*] : *text*

List all notes

notes

List specific notes

notes [list] *evarlist* [in #[/#]]

Drop notes

notes drop *evarlist* [in #[/#]]

where *evarlist* is a *varlist* that may also contain the word _dta and # is a number or the letter 1.

If *text* includes the letters TS surrounded by blanks, the TS is removed, and a time stamp is substituted in its place.

Description

notes attaches notes to the dataset in memory. These notes become a part of the dataset and are saved when the dataset is saved and retrieved when the dataset is used; see [D] **save** and [D] **use**. notes can be attached generically to the dataset or specifically to a variable within the dataset.

Remarks

Remarks are presented under the following headings:

> Attaching and listing notes
> Selectively listing notes
> Deleting notes
> Warnings

Attaching and listing notes

A note is nothing formal; it is merely a string of text reminding you to do something, cautioning you against something, or saying anything else you might feel like jotting down. People who work with real data invariably end up with paper notes plastered around their terminal saying things like, "Send the new sales data to Bob", "Check the income variable in salary95; I don't believe it", or "The gender dummy was significant!" It would be better if these notes were attached to the dataset.

Adding a note to your dataset requires typing note or notes (they are synonyms), a colon (:), and whatever you want to remember. The note is added to the dataset currently in memory.

```
. note:  Send copy to Bob once verified.
```

408

You can replay your notes by typing `notes` (or `note`) by itself.

```
. notes
_dta:
   1.  Send copy to Bob once verified.
```

Once you resave your data, you can replay the note in the future, too. You add more notes just as you did the first:

```
. note: Mary wants a copy, too.

. notes
_dta:
   1.  Send copy to Bob once verified.
   2.  Mary wants a copy, too.
```

You can place time stamps on your notes by placing the word TS (in capitals) in the text of your note:

```
. note: TS merged updates from JJ&F

. notes
_dta:
   1.  Send copy to Bob once verified.
   2.  Mary wants a copy, too.
   3.  19 Apr 2007 15:38 merged updates from JJ&F
```

Notes may contain SMCL directives:

```
. use http://www.stata-press.com/data/r10/auto
(1978 Automobile Data)

. note: check reason for missing values in {cmd:rep78}

. notes
_dta:
   1.  from Consumer Reports with permission
   2.  check reason for missing values in rep78
```

The notes we have added so far are attached to the dataset generically, which is why Stata prefixes them with _dta when it lists them. You can attach notes to variables:

```
. note mpg: is the 44 a mistake?  Ask Bob.

. note mpg: what about the two missing values?

. notes
_dta:
   1.  Send copy to Bob once verified.
   2.  Mary wants a copy, too.
   3.  19 Apr 2007 15:38 merged updates from JJ&F
mpg:
   1.  is the 44 a mistake? Ask Bob.
   2.  what about the two missing values?
```

Up to 9,999 generic notes can be attached to _dta, and another 9,999 notes can be attached to each variable.

(Continued on next page)

Selectively listing notes

Typing notes by itself lists all the notes. In full syntax, notes is equivalent to typing notes _all in 1/1. Here are some variations:

notes _dta	list all generic notes
notes mpg	list all notes for variable mpg
notes _dta mpg	list all generic notes and mpg notes
notes _dta in 3	list generic note 3
notes _dta in 3/5	list generic notes 3–5
notes mpg in 3/5	list mpg notes 3–5
notes _dta in 3/l	list generic notes 3 through last

Deleting notes

notes drop works much like listing notes, except that typing notes drop by itself does not delete all notes; you must type notes drop _all. Here are some variations:

notes drop _dta	delete all generic notes
notes drop _dta in 3	delete generic note 3
notes drop _dta in 3/5	delete generic notes 3–5
notes drop _dta in 3/l	delete generic notes 3 through last
notes drop mpg in 4	delete mpg note 4

Warnings

1. Notes are stored with the data, and as with other updates you make to the data, the additions and deletions are not permanent until you save the data; see [D] **save**.

2. The maximum length of one note is 67,784 characters for Stata/MP, Stata/SE, and Stata/IC; it is 8,681 characters for Small Stata.

Methods and Formulas

notes is implemented as an ado-file.

Reference

Gleason, J. R. 1998. dm57: A notes editor for Windows and Macintosh. *Stata Technical Bulletin* 43: 6–9. Reprinted in *Stata Technical Bulletin Reprints*, vol. 8, pp. 10–13.

Also See

[D] **describe** — Describe data in memory or in file

[D] **save** — Save datasets

[D] **codebook** — Describe data contents

[U] **12.8 Characteristics**

Title

> **obs** — Increase the number of observations in a dataset

Syntax

set <u>obs</u> #

Description

set obs changes the number of observations in the current dataset. # must be at least as large as the current number of observations. If there are variables in memory, the values of all new observations are set to missing.

Remarks

▷ Example 1

set obs can be useful for creating artificial datasets. For instance, if we wanted to graph the function $y = x^2$ over the range 1–100, we could type

```
. drop _all
. set obs 100
obs was 0, now 100
. generate x = _n
. generate y = x^2
. scatter y x
(graph not shown)
```

◁

▷ Example 2

If we want to add an extra data point in a program, we could type

```
. local np1 = _N + 1
. set obs 'np1'
```

or

```
. set obs '=_N + 1'
```

◁

Also See

[D] **describe** — Describe data in memory or in file

Title

odbc — Load, write, or view data from ODBC sources

Syntax

List ODBC sources to which Stata can connect

    odbc list

Retrieve available names from specified data source

    odbc query ["DataSourceName", connect_options]

List column names and types associated with specified table

    odbc describe ["TableName", dsn("DataSourceName") connect_options]

Read ODBC table into memory

    odbc load [extvarlist] [if] [in], { table("TableName") | exec("SqlStmt") }
       [load_options connect_options]

Write data to an ODBC table

    odbc insert [varlist], table("TableName") dsn("DataSourceName")
       [insert_options connect_options]

Allow SQL statements to be issued directly to ODBC data source

    odbc exec("SqlStmt"), dsn("DataSourceName") [connect_options]

Batch job alternative to odbc exec

    odbc sqlfile("filename"), dsn("DataSourceName") [loud connect_options]

Specify ODBC driver manager (Unix only)

    set odbcmgr { iodbc | unixodbc } [, permanently]

where

   DataSourceName is the name of the ODBC source (database, spreadsheet, etc.)

   TableName is the name of a table within the ODBC data source

   SqlStmt is a SQL SELECT statement

   filename is pure SQL commands separated by semicolons

412

and where *extvarlist* contains
> *sqlvarname*
> *varname* = *sqlvarname*

connect_options	description
<u>u</u>ser(*UserID*)	user ID of user establishing connection
<u>p</u>assword(*Password*)	password of user establishing connection
<u>d</u>ialog(noprompt)	do not display ODBC connection-information dialog, and do not prompt user for connection information
<u>d</u>ialog(prompt)	display ODBC connection-information dialog
<u>d</u>ialog(complete)	display ODBC connection-information dialog only if there is not enough information
<u>d</u>ialog(required)	display ODBC connection-information dialog only if there is not enough mandatory information provided
* dsn("*DataSourceName*")	name of data source

* dsn("*DataSourceName*") is not allowed with odbc query. It is required with odbc insert, odbc exec, and odbc sqlfile.

load_options	description
* <u>t</u>able("*TableName*")	name of table stored in data source
* <u>e</u>xec("*SqlStmt*")	SQL SELECT statement to generate a table to be read into Stata
<u>cl</u>ear	load dataset even if there is one in memory
<u>no</u>quote	alter Stata's internal use of SQL commands; seldom used
<u>l</u>owercase	read variable names as lowercase
<u>s</u>qlshow	show all SQL commands issued
<u>a</u>llstring	read all variables as strings
<u>d</u>atestring	read date-formatted variables as strings

* Either table("*TableName*") or exec("*SqlStmt*") must be specified with odbc load.

insert_options	description
* <u>t</u>able("*TableName*")	name of table stored in data source
<u>cr</u>eate	create a simple ODBC table
<u>o</u>verwrite	clear data in ODBC table before data in memory is written to the table
<u>in</u>sert	default mode of operation for the odbc insert command
<u>q</u>uoted	quote all values with single quotes as they are inserted in ODBC table
<u>s</u>qlshow	show all SQL commands issued
<u>a</u>s("*varlist*")	ODBC variables on the data source that correspond to the variables in Stata's memory

* table("*TableName*") is required for odbc insert.

Description

odbc allows you to load, write, and view data from ODBC sources into Stata. ODBC, an acronym for Open DataBase Connectivity, is a standardized set of function calls for accessing data stored in both relational and nonrelational database-management systems. By default on Unix platforms iODBC is the ODBC driver manager Stata uses, but you can use unixODBC by using the command `set odbcmgr unixodbc`.

ODBC's architecture consists of four major components (or layers): the client interface, the ODBC driver manager, the ODBC drivers, and the data sources. Stata provides odbc as the client interface. The system is illustrated as follows:

odbc `list` produces a list of ODBC data source names to which Stata can connect.

odbc `query` retrieves a list of table names available from a specified data source's system catalog.

odbc `describe` lists column names and types associated with a specified table.

odbc `load` reads an ODBC table into memory. You can load an ODBC table specified in the `table()` option or load an ODBC table generated by a SQL SELECT statement specified in the `exec()` option. In both cases, you can choose which columns and rows of the ODBC table to read by specifying *extvarlist* and `if`, `in` conditions. *extvarlist* specifies the columns to be read and allows you to rename variables. For example,

```
. odbc load id=EmployeeID LastName, table(Employees) dsn(Northwind)
```

reads two columns, `EmployeeID` and `LastName`, from the `Employees` table of the `Northwind` data source. It will also rename variable `EmployeeID` to `id`.

odbc `insert` writes data from memory to an ODBC table. The data can be appended to an existing table, replace an existing table, or be placed in a newly created ODBC table.

odbc `exec` allows for most SQL statements to be issued directly to any ODBC data source. Statements that produce output, such as SELECT, have their output neatly displayed. By using Stata's ado language, you can also generate SQL commands on the fly to do positional updates or whatever the situation requires.

odbc `sqlfile` provides a "batch job" alternative to the odbc exec command. A file is specified that contains any number of any length SQL commands. Every SQL command in this file should be delimited by a semicolon and must be constructed as pure SQL. Stata macros and ado-language syntax are not permitted. The advantage in using this command, as opposed to odbc exec, is that only one connection is established for multiple SQL statements. A similar sequence of SQL commands used via odbc exec would require constructing an ado-file that issued a command and, thus, a connection for every SQL command. Another slight difference is that any output that might be generated from a SQL command is suppressed by default. A `loud` option is provided to toggle output back on.

`set odbcmgr iodbc` specifies that the ODBC driver manager is iODBC (the default). `set odbcmgr unixodbc` specifies that the ODBC driver manager is unixODBC.

Options

user(*UserID*) specifies the user ID of the user attempting to establish the connection to the data source. By default, Stata assumes that the user ID is the same as the one specified in the previous odbc command or is empty if user() has never been specified in the current session of Stata.

password(*Password*) specifies the password of the user attempting to establish the connection to the data source. By default, Stata assumes that the password is the same as the one previously specified or is empty if the password has not been used during the current session of Stata. Typically, the password() option will not be specified apart from the user() option.

dialog(noprompt | prompt | complete | required) specifies the mode the ODBC Driver Manager uses to display the ODBC connection-information dialog to prompt for more connection information.

noprompt is the default value. The ODBC connection-information dialog is not displayed, and you are not prompted for connection information. If there is not enough information to establish a connection to the specified data source, an error is returned.

prompt causes the ODBC connection-information dialog to be displayed.

complete causes the ODBC connection-information dialog to be displayed only if there is not enough information, even if the information is not mandatory.

required causes the ODBC connection-information dialog to be displayed only if there is not enough mandatory information provided to establish a connection to the specified data source. You are prompted only for mandatory information; controls for information that is not required to connect to the specified data source are disabled.

dsn("*DataSourceName*") specifies the name of a data source, as listed by the odbc list command. If a name contains spaces, it must be enclosed in double quotes. By default, Stata assumes that the data source name is the same as the one specified in the previous odbc command.

table("*TableName*") specifies the name of an ODBC table stored in a specified data source's system catalog, as listed by the odbc query command. If a table name contains spaces, it must be enclosed in double quotes. Either the table() option or the exec() option—but not both—is required with the odbc load command.

exec("*SqlStmt*") allows you to issue a SQL SELECT statement to generate a table to be read into Stata. An error message is returned if the SELECT statement is an invalid SQL statement. The statement must be enclosed in double quotes. Either the table() option or the exec() option—but not both—is required with the odbc load command.

clear permits the data to be loaded, even if there is a dataset already in memory, and even if that dataset has changed since the data were last saved.

noquote is a rarely used option that alters Stata's internal use of SQL commands, specifically those relating to quoted table names, to better accommodate various drivers. This option has been particularly helpful for DB2 drivers.

lowercase causes all the variable names to be read as lowercase.

sqlshow is a useful option for showing all SQL commands issued to the ODBC data source from the odbc insert or odbc load command. This can help you debug any issues related to inserting or loading.

allstring causes all variables to be read as string data types.

datestring causes all date- and time-formatted variables to be read as string data types.

create specifies that a simple ODBC table be created on the specified data source and populated with the data in memory. Column data types are approximated based on the existing format in Stata's memory.

overwrite allows data to be cleared from an ODBC table before the data in memory are written to the table. All data from the ODBC table are erased, not just the data from the variable columns that will be replaced.

insert appends data to an existing ODBC table and is the default mode of operation for the odbc insert command.

quoted is useful for ODBC data sources that require all inserted values to be quoted. This option specifies that all values be quoted with single quotes as they are inserted into an ODBC table.

as("*varlist*") allows you to specify the ODBC variables on the data source that correspond to the variables in Stata's memory. If this option is specified, the number of variables must equal the number of variables being inserted, even if some names are identical.

loud specifies that output be displayed for SQL commands.

permanently (set odbcmgr only) specifies that, in addition to making the change right now, the setting be remembered and become the default setting when you invoke Stata.

Remarks

When possible, the examples in this manual entry are developed using the Northwind sample database that is automatically installed with Microsoft Access. If you do not have Access, you can still use odbc, but you will need to consult the documentation for your other ODBC sources to determine how to set them up.

Remarks are presented under the following headings:

> *Setting up the data sources*
> *Listing ODBC data-source names*
> *Listing available table names from a specified data source's system catalog*
> *Describing a specified table*
> *Loading data from ODBC sources*

Setting up the data sources

Before using Stata's ODBC commands, you must register your ODBC database with the *ODBC Data Source Administrator*. This process varies depending on platform, but the following example shows the steps necessary for Windows.

Using Windows XP or Vista, follow these steps to create an ODBC User Data Source for the Northwind sample database:

1. From the *Start Menu*, select the *Control Panel*.

2. In the *Control Panel* window, click *Administrative Tools*.

3. In the *Administrative Tools* window, double-click *Data Sources (ODBC)*. Vista users will have to click *Classic View* on the left side of the Control Panel window before *Administrative Tools* is visible.

4. In the *Data Sources (ODBC)* window,

    a. click the *User DSN* tab;

    b. click *Add ...*;

c. choose *Microsoft Access Driver (*.mdb)* driver on the *Create New Data Source* panel; and

b. click *Finish*.

5. In the *ODBC Microsoft Access Setup* panel, enter "Northwind" in the Data Source Name field and click *Select ....*

6. Locate the `Northwind.mdb` database in your computer.

7. Click *OK* to finish creating the data source.

❏ Technical Note

In earlier versions of Windows, the exact location of the *Data Source (ODBC)* panel varies, but it is always somewhere within the *Control Panel.*

❏

Listing ODBC data-source names

`odbc list` is used to produce a list of data source names to which Stata can connect. For a specific data source name to be shown in the list, the data source has to be registered with the *ODBC Data Source Administrator.* See *Setting up the data sources* for information on how to do this.

▷ Example 1

```
. odbc list
```

Data Source Name	Driver
Visual FoxPro Database	Microsoft Visual FoxPro Driver
Visual FoxPro Tables	Microsoft Visual FoxPro Driver
dBase Files - Word	Microsoft dBase VFP Driver (*.dbf)
FoxPro Files - Word	Microsoft FoxPro VFP Driver (*.dbf)
MS Access Database	Microsoft Access Driver (*.mdb)
Northwind	Microsoft Access Driver (*.mdb)
dBASE Files	Microsoft dBase Driver (*.dbf)
DeluxeCD	Microsoft Access Driver (*.mdb)
Excel Files	Microsoft Excel Driver (*.xls)
ECDCMusic	Microsoft Access Driver (*.mdb)

In the above list, `Northwind` is one of the sample Microsoft Access databases that Access installs by default.

◁

(Continued on next page)

Listing available table names from a specified data source's system catalog

odbc query is used to list table names available from a specified data source.

▷ Example 2

```
. odbc query "Northwind"
DataSource: Northwind
Path     : C:\Program Files\Microsoft Office\Office\Samples\Northwind

Categories
Customers
Employees
Order Details
Orders
Products
Shippers
Suppliers
```

◁

❏ Technical Note

To query a *Microsoft Excel* data source, you must define a database as a named range within Excel. Multiple name ranges can exist within an Excel file, and each one is treated as a separate table.

To define a named range within Excel, highlight the entire range, including all columns of interest; from the Excel menu, select *Insert*, select *Name*, click *Define*, enter the desired name, and save the file.

You can also describe a worksheet associated with an Excel file without defining a named range for the worksheet. To do so, you must specify the name of the worksheet in the odbc describe command followed by a dollar sign ($).

❏

Describing a specified table

odbc describe is used to list column (variable) names and their SQL data types that are associated with a specified table.

▷ Example 3

Here we specify that we want to list all variables in the Employees table of the Northwind data source.

```
. odbc describe "Employees", dsn("Northwind")
```

DataSource: Northwind (query)
Table: Employees (load)

Variable Name	Variable Type
EmployeeID	COUNTER
LastName	VARCHAR
FirstName	VARCHAR
Title	VARCHAR
TitleOfCourtesy	VARCHAR
BirthDate	DATETIME
HireDate	DATETIME
Address	VARCHAR
City	VARCHAR
Region	VARCHAR
PostalCode	VARCHAR
Country	VARCHAR
HomePhone	VARCHAR
Extension	VARCHAR
Photo	LONGBINARY
Notes	LONGCHAR
ReportsTo	INTEGER

◁

Loading data from ODBC sources

odbc load is used to load an ODBC table into memory.

To load an ODBC table listed in the odbc query output, specify the table name in the table() option and the data-source name in the dsn() option.

▷ Example 4

We want to load the Employees table from the Northwind data source.

(Continued on next page)

```
. clear

. odbc load, table("Employees") dsn("Northwind")
note: Photo is of a type not supported in Stata; skipped

. describe

Contains data
  obs:             9
  vars:           16
  size:        3,222 (99.6% of memory free)
```

| | storage | display | value | |
variable name	type	format	label	variable label
EmployeeID	long	%12.0g		
LastName	str20	%20s		
FirstName	str10	%10s		
Title	str30	%30s		
TitleOfCourtesy	str25	%25s		
BirthDate	double	%td		
HireDate	double	%td		
Address	str60	%60s		
City	str15	%15s		
Region	str15	%15s		
PostalCode	str10	%10s		
Country	str15	%15s		
HomePhone	str24	%24s		
Extension	str4	%9s		
Notes	str80	%80s		
ReportsTo	long	%12.0g		

```
Sorted by:
     Note:  dataset has changed since last saved
```

❏

❏ Technical Note

When Stata loads the ODBC table, data are converted from SQL data types to Stata data types. Stata does not support all SQL data types. If the column cannot be read because of incompatible data types, Stata will issue a note and skip a column. The following table lists the supported SQL data types and their corresponding Stata data types:

SQL data type	Stata data type
SQL_BIT	byte
SQL_TINYINT	
SQL_SMALLINT	int
SQL_INTEGER	long
SQL_BIGINT	
SQL_DECIMAL	double
SQL_NUMERIC	
SQL_REAL	float
SQL_FLOAT	double
SQL_DOUBLE	
SQL_CHAR	string
SQL_VARCHAR	
SQL_LONGVARCHAR	
SQL_WCHAR	
SQL_WVARCHAR	
SQL_WLONGVARCHAR	
SQL_TIME	
SQL_DATE	
SQL_TIMESTAMP	
SQL_TYPE_TIME	double
SQL_TYPE_DATE	
SQL_TYPE_TIMESTAMP	
SQL_BINARY	*not supported*
SQL_VARBINARY	*not supported*
SQL_LONGVARBINARY	*not supported*

❏

You can also load an ODBC table generated by a SQL SELECT statement specified in the exec() option.

▷ Example 5

Suppose that, from the Northwind data source, we want a list of all the customers who have placed orders. We might use the SQL SELECT statement

```
SELECT DISTINCT c.CustomerID, c.CompanyName
FROM Customers c
INNER JOIN Orders o
        ON c.CustomerID = o.CustomerID
```

To load the table into Stata, we use odbc load with the exec() option.

(Continued on next page)

```
. odbc load, exec("SELECT DISTINCT c.CustomerID, c.CompanyName FROM Customers c
> INNER JOIN Orders o ON c.CustomerID = o.CustomerID") dsn("Northwind") clear
. describe
Contains data
  obs:           89
  vars:           2
  size:       4,361 (99.5% of memory free)
```

variable name	storage type	display format	value label	variable label
CustomerID	str5	%9s		
CompanyName	str40	%40s		

```
Sorted by:
    Note:  dataset has changed since last saved
```

◁

The *extvarlist* is optional. It allows you to choose which columns (variables) are to be read and to rename variables when they are read.

▷ Example 6

Suppose that we want to load the EmployeeID column and the LastName column from the Employees table of the Northwind data source. Moreover, we want to rename EmployeeID as id and LastName as name.

```
. odbc load id=EmployeeID name=LastName, table("Employees") dsn("Northwind") clear
. describe
Contains data
  obs:            9
  vars:           2
  size:         252 (99.9% of memory free)
```

variable name	storage type	display format	value label	variable label
id	long	%12.0g		EmployeeID
name	str20	%20s		LastName

```
Sorted by:
    Note:  dataset has changed since last saved
```

◁

The if and in qualifiers allow you to choose which rows are to be read. You can also use a WHERE clause in the SQL SELECT statement to select the rows to be read.

▷ Example 7

Suppose that we want the information from the Order Details table, where Quantity is greater than 50. We can specify the if and in qualifiers

```
. odbc load if Quantity>50, table("Order Details") dsn("Northwind") clear
. summarize Quantity
```

Variable	Obs	Mean	Std. Dev.	Min	Max
Quantity	159	72.56604	18.38255	52	130

or we can issue the SQL SELECT statement directly.

```
. odbc load, exec("SELECT * FROM [Order Details] WHERE Quantity>50")
> dsn("Northwind") clear
. summarize Quantity
```

Variable	Obs	Mean	Std. Dev.	Min	Max
Quantity	159	72.56604	18.38255	52	130

◁

Also See

[R] **net** — Install and manage user-written additions from the Internet

[D] **fdasave** — Save and use datasets in FDA (SAS XPORT) format

[D] **infix (fixed format)** — Read ASCII (text) data in fixed format

[D] **insheet** — Read ASCII (text) data created by a spreadsheet

[D] **xmlsave** — Save and use datasets in XML format

[TS] **haver** — Load data from Haver Analytics database

Title

order — Reorder variables in dataset

Syntax

Move specified variables to front of dataset

    order *varlist*

Move one variable to specified position

    <u>mov</u>e *varname₁* *varname₂*

Alphabetize specified variables and move to front of dataset

    aorder [*varlist*]

Description

order changes the order of the variables in the current dataset. The variables specified in *varlist* are moved, in the specified order, to the front of the dataset.

move relocates $varname_1$ to the position of $varname_2$ and shifts the remaining variables, including $varname_2$, to make room.

aorder alphabetizes the variables specified in *varlist* and moves them to the front of the dataset. If no *varlist* is specified, _all is assumed.

Remarks

▷ Example 1

When using order, you must specify a *varlist*, but you do not need to specify all the variables in the dataset. For example, we want to move the make and mpg variables to the front of the auto dataset.

```
. use http://www.stata-press.com/data/r10/auto4
(1978 Automobile Data)

. describe

Contains data from http://www.stata-press.com/data/r10/auto4.dta
  obs:            74                          1978 Automobile Data
 vars:             6                          6 Apr 2007 00:20
 size:         2,368 (99.6% of memory free)
```

	storage	display	value	
variable name	type	format	label	variable label
price	int	%8.0gc		Price
weight	int	%8.0gc		Weight (lbs.)
mpg	int	%8.0g		Mileage (mpg)
make	str18	%-18s		Make and Model
length	int	%8.0g		Length (in.)
rep78	int	%8.0g		Repair Record 1978

```
Sorted by:

. order make mpg

. describe

Contains data from http://www.stata-press.com/data/r10/auto4.dta
  obs:            74                          1978 Automobile Data
 vars:             6                          6 Apr 2007 00:20
 size:         2,368 (99.6% of memory free)
```

	storage	display	value	
variable name	type	format	label	variable label
make	str18	%-18s		Make and Model
mpg	int	%8.0g		Mileage (mpg)
price	int	%8.0gc		Price
weight	int	%8.0gc		Weight (lbs.)
length	int	%8.0g		Length (in.)
rep78	int	%8.0g		Repair Record 1978

```
Sorted by:
```

We now want `length` to be the last variable in our dataset, so we could type `order make mpg price weight rep78 length`, but it would be easier to use `move`:

```
. move length rep78

. describe

Contains data from http://www.stata-press.com/data/r10/auto4.dta
  obs:            74                          1978 Automobile Data
 vars:             6                          6 Apr 2007 00:20
 size:         2,368 (99.6% of memory free)
```

	storage	display	value	
variable name	type	format	label	variable label
make	str18	%-18s		Make and Model
mpg	int	%8.0g		Mileage (mpg)
price	int	%8.0gc		Price
weight	int	%8.0gc		Weight (lbs.)
rep78	int	%8.0g		Repair Record 1978
length	int	%8.0g		Length (in.)

```
Sorted by:
```

We now change our mind and decide that we prefer that the variables be alphabetized.

```
. aorder

. describe
Contains data from http://www.stata-press.com/data/r10/auto4.dta
  obs:            74                          1978 Automobile Data
 vars:             6                          6 Apr 2007 00:20
 size:         2,368 (99.4% of memory free)
```

variable name	storage type	display format	value label	variable label
length	int	%8.0g		Length (in.)
make	str18	%-18s		Make and Model
mpg	int	%8.0g		Mileage (mpg)
price	int	%8.0gc		Price
rep78	int	%8.0g		Repair Record 1978
weight	int	%8.0gc		Weight (lbs.)

```
Sorted by:
```

◁

❑ Technical Note

If your data contain variables named year1, year2, ..., year19, year20, aorder will order them correctly, even though to most computer programs, year10 is alphabetically between year1 and year2.

❑

Methods and Formulas

aorder is implemented as an ado-file.

References

Gleason, J. R. 1997. dm51: Defining and recording variable orderings. *Stata Technical Bulletin* 40: 10–12. Reprinted in *Stata Technical Bulletin Reprints*, vol. 7, pp. 49–52.

Weesie, J. 1999. dm74: Changing the order of variables in a dataset. *Stata Technical Bulletin* 52: 8–9. Reprinted in *Stata Technical Bulletin Reprints*, vol. 9, pp. 61–62.

Also See

[D] **describe** — Describe data in memory or in file

[D] **edit** — Edit and list data with Data Editor

[D] **rename** — Rename variable

Title

> **outfile** — Write ASCII-format dataset

Syntax

<u>out</u>file [*varlist*] using *filename* [*if*] [*in*] [, *options*]

options	description
Main	
<u>dict</u>ionary	write the file in Stata's dictionary format
<u>nol</u>abel	output numeric values (not labels) of labeled variables; the default is to write labels in double quotes
<u>noq</u>uote	do not enclose strings in double quotes
<u>c</u>omma	write file in comma-separated (instead of space separated) format
<u>w</u>ide	force 1 observation per line (no matter how wide)
Advanced	
rjs	right-justify string variables; the default is to left-justify
fjs	left-justify if format width < 0; right-justify if format width > 0
runtogether	all on one line, no quotes, no space between, and ignore formats
<u>m</u>issing	retain missing values; use only with comma
† <u>r</u>eplace	overwrite the existing file

† replace is not shown in the dialog box.

Description

outfile writes data to a disk file in ASCII format, which can be read by other programs. The new file is *not* in Stata format; see [D] **save** for instructions on saving data for later use in Stata.

The data saved by outfile can be read back by infile; see [D] **infile**. If *filename* is specified without an extension, .raw is assumed unless the dictionary option is specified, in which case, .dct is assumed. If your *filename* contains embedded spaces, remember to enclose it in double quotes.

Options

> **Main**

dictionary writes the file in Stata's data dictionary format. See [D] **infile (fixed format)** for a description of dictionaries. comma, missing, and wide are not allowed with dictionary.

nolabel causes Stata to write the numeric values of labeled variables. The default is to write the labels enclosed in double quotes.

noquote prevents Stata from placing double quotes around the contents of strings, meaning string variables and value labels.

comma causes Stata to write the file in comma-separated-value format. In this format, values are separated by commas rather than by blanks. Missing values are written as two consecutive commas unless `missing` is specified.

wide causes Stata to write the data with 1 observation per line. The default is to split observations into lines of 80 characters or fewer, but strings longer than 80 characters are never split across lines.

⌐ Advanced ⌐

rjs and fjs affect how strings are justified; you probably do not want to specify either of these options. By default, `outfile` outputs strings left-justified in their field.

If rjs is specified, strings are output right-justified. rjs stands for "right-justified strings".

If fjs is specified, strings are output left- or right-justified according to the variable's format: left-justified if the format width is negative and right-justified if the format width is positive. fjs stands for "format-justified strings".

runtogether is a programmer's option that is valid only when all variables of the specified *varlist* are of type string. runtogether specifies that the variables be output in the order specified, without quotes, with no spaces between, and ignoring the display format attached to each variable. Each observation ends with a new line character.

missing, valid only with comma, specifies that missing values be retained. When comma is specified without missing, missing values are changed to null strings ("").

The following option is available with `outfile` but is not shown in the dialog box:

replace permits `outfile` to overwrite an existing dataset.

Remarks

`outfile` enables data to be sent to a disk file for processing by a non-Stata program. Each observation is written as one or more records that will not exceed 80 characters unless you specify the `wide` option. Each column, other than the first, is prefixed by two blanks.

`outfile` is careful to put the data in columns in case you want to read the data by using formatted input. String variables and value labels are output in left-justified fields by default. You can change this behavior by using the `rjs` or `fjs` options.

Numeric variables are output right-justified in the field width specified by their display format. A numeric variable with a display format of `%9.0g` will be right-justified in a nine-character field. Commas are not written in numeric variables, even if a comma format is used.

If you specify the `dictionary` option, the data are written in the same way, but preceding the data, `outfile` writes a data dictionary describing the contents of the file.

▷ Example 1

We have entered into Stata some data on seven employees in our firm. The data contain employee name, employee identification number, salary, and sex:

```
. list
```

	name	empno	salary	sex
1.	Carl Marks	57213	24,000	male
2.	Irene Adler	47229	27,000	female
3.	Adam Smith	57323	24,000	male
4.	David Wallis	57401	24,500	male
5.	Mary Rogers	57802	27,000	female
6.	Carolyn Frank	57805	24,000	female
7.	Robert Lawson	57824	22,500	male

The last variable in our data, sex, is really a numeric variable, but it has an associated value label.

If we now wish to use a program other than Stata with these data, we must somehow get the data over to that other program. The standard Stata-format dataset created by save will not do the job—it is written in a special format that only Stata understands. Most programs, however, understand ASCII datasets, such as those produced by a text editor. We can tell Stata to produce such a dataset by using outfile. Typing outfile using employee creates a dataset called employee.raw that contains all the data. We can use the Stata type command to review the resulting file:

```
. outfile using employee

. type employee.raw
"Carl Marks"        57213    24000   "male"
"Irene Adler"       47229    27000   "female"
"Adam Smith"        57323    24000   "male"
"David Wallis"      57401    24500   "male"
"Mary Rogers"       57802    27000   "female"
"Carolyn Frank"     57805    24000   "female"
"Robert Lawson"     57824    22500   "male"
```

We see that the file contains the four variables and that Stata has surrounded the string variables with double quotes.

◁

❏ Technical Note

The nolabel option prevents Stata from substituting value-label strings for the underlying numeric values; see [U] **12.6.3 Value labels**. The last variable in our data is really a numeric variable:

```
. outfile using employ2, nolabel

. type employ2.raw
"Carl Marks"        57213    24000    0
"Irene Adler"       47229    27000    1
"Adam Smith"        57323    24000    0
"David Wallis"      57401    24500    0
"Mary Rogers"       57802    27000    1
"Carolyn Frank"     57805    24000    1
"Robert Lawson"     57824    22500    0
```

❏

❑ Technical Note

If you do not want Stata to place double quotes around the contents of string variables, you can specify the noquote option:

```
. outfile using employ3, noquote
. type employ3.raw
Carl Marks          57213     24000  male
Irene Adler         47229     27000  female
Adam Smith          57323     24000  male
David Wallis        57401     24500  male
Mary Rogers         57802     27000  female
Carolyn Frank       57805     24000  female
Robert Lawson       57824     22500  male
```

❑

▷ Example 2

Stata never writes over an existing file unless explicitly told to do so. For instance, if the file employee.raw already exists and we attempt to overwrite it by typing outfile using employee, here is what would happen:

```
. outfile using employee
file employee.raw already exists
r(602);
```

We can tell Stata that it is okay to overwrite a file by specifying the replace option:

```
. outfile using employee, replace
```

◁

❑ Technical Note

Some programs prefer data to be separated by commas rather than by blanks. Stata produces such a dataset if you specify the comma option:

```
. outfile using employee, comma replace
. type employee.raw
"Carl Marks",57213,24000,"male"
"Irene Adler",47229,27000,"female"
"Adam Smith",57323,24000,"male"
"David Wallis",57401,24500,"male"
"Mary Rogers",57802,27000,"female"
"Carolyn Frank",57805,24000,"female"
"Robert Lawson",57824,22500,"male"
```

❑

▷ Example 3

Finally, outfile can create data dictionaries that infile can read. Dictionaries are perhaps the best way to organize your raw data. A dictionary describes your data so that you do not have to remember the order of the variables, the number of variables, the variable names, or anything else. The file in which you store your data becomes self-documenting so that you can understand the data in the future. See [D] **infile (fixed format)** for a full description of data dictionaries.

When you specify the `dictionary` option, Stata writes a `.dct` file:

```
. outfile using employee, dict replace

. type employee.dct
dictionary {
        str15   name                    '"Employee name"'
        float   empno                   '"Employee number"'
        float   salary                  '"Annual salary"'
        float   sex     :sexlbl         '"Sex"'
}
"Carl Marks"            57213       24000   "male"
"Irene Adler"           47229       27000   "female"
"Adam Smith"            57323       24000   "male"
"David Wallis"          57401       24500   "male"
"Mary Rogers"           57802       27000   "female"
"Carolyn Frank"         57805       24000   "female"
"Robert Lawson"         57824       22500   "male"
```

◁

Also See

[D] **infile** — Overview of reading data into Stata

[D] **outsheet** — Write spreadsheet-style dataset

[U] **21 Inputting data**

Title

outsheet — Write spreadsheet-style dataset

Syntax

outsheet [*varlist*] using *filename* [*if*] [*in*] [, *options*]

options	description
Main	
comma	output in comma-separated (instead of tab separated) format
delimiter("*char*")	use *char* as delimiter
nonames	do not write variable names on the first line
nolabel	output numeric values (not labels) of labeled variables
noquote	do not enclose strings in double quotes
† replace	overwrite existing *filename*

† replace is not shown in the dialog box.

If *filename* is specified without a suffix, .out is assumed.

If your *filename* contains embedded spaces, remember to enclose it in double quotes.

Description

outsheet, by default, writes data into a file in tab-separated format. outsheet also allows users to specify comma-separated format or any separation character that they prefer.

Options

⌐ Main ⌐

comma specifies comma-separated format rather than the default tab-separated format.

delimiter("*char*") allows you to specify other separation characters. For instance, if you want the values in the file to be separated by a semicolon, specify delimiter(";").

nonames specifies that variable names not be written in the first line of the file; the file is to contain data values only.

nolabel specifies that the numeric values of labeled variables be written into the file rather than the label associated with each value.

noquote specifies that string variables not be enclosed in double quotes.

The following option is available with outsheet but is not shown in the dialog box:

replace specifies that *filename* be replaced if it already exists.

Remarks

If you wish to move your data into another program, you can do any of the following:

1. Use an external data-transfer program; see [U] **21.4 Transfer programs**.

2. Cut and paste from Stata's Data Editor; see *Getting Started, Chapter 6.*

3. Use outsheet.

4. Use outfile; see [D] **outfile**.

outsheet is typically preferred to outfile for moving the data to a spreadsheet, and outfile is probably better for moving data to another statistical program.

If your goal is to send data to another Stata user, you could use outsheet or outfile, but it is easiest to send the .dta dataset. This will work even if you use Stata for Windows and your colleague uses Stata for Macintosh. All Statas can read each other's .dta files.

▷ Example 1

outsheet copies the data currently loaded in memory into the specified file. It is easy to use.

```
. use http://www.stata-press.com/data/r10/auto
(1978 Automobile Data)
. keep make price mpg rep78 foreign
. keep in 1/10
(64 observations deleted)
```

Let's write our shortened version of the auto dataset in tab-separated ASCII format to the file myauto.out:

```
. outsheet using myauto
. type myauto.out
make     price   mpg    rep78   foreign
"AMC Concord"  4099   22    3       "Domestic"
"AMC Pacer"    4749   17    3       "Domestic"
"AMC Spirit"   3799   22            "Domestic"
"Buick Century" 4816  20    3       "Domestic"
"Buick Electra" 7827  15    4       "Domestic"
"Buick LeSabre" 5788  18    3       "Domestic"
"Buick Opel"   4453   26            "Domestic"
"Buick Regal"  5189   20    3       "Domestic"
"Buick Riviera" 10372 16    3       "Domestic"
"Buick Skylark" 4082  19    3       "Domestic"
```

We remember that we are not copying our data to a spreadsheet, so we want to suppress the dataset names from the first line of the file.

```
. outsheet using myauto, nonames
file myauto.out already exists
r(602);
```

We can erase myauto.out (see [D] **erase**), specify the replace option, or use a different filename.

```
. outsheet using myauto, nonames replace
. type myauto.out
"AMC Concord"    4099   22    3        "Domestic"
"AMC Pacer"      4749   17    3        "Domestic"
"AMC Spirit"     3799   22             "Domestic"
"Buick Century"  4816   20    3        "Domestic"
"Buick Electra"  7827   15    4        "Domestic"
"Buick LeSabre"  5788   18    3        "Domestic"
"Buick Opel"     4453   26             "Domestic"
"Buick Regal"    5189   20    3        "Domestic"
"Buick Riviera" 10372   16    3        "Domestic"
"Buick Skylark"  4082   19    3        "Domestic"
```

◁

Also See

[D] **insheet** — Read ASCII (text) data created by a spreadsheet

[D] **outfile** — Write ASCII-format dataset

[U] **21 Inputting data**

Title

pctile — Create variable containing percentiles

Syntax

Create variable containing percentiles

> pctile $\big[$ *type* $\big]$ *newvar* = *exp* $\big[$ *if* $\big]$ $\big[$ *in* $\big]$ $\big[$ *weight* $\big]$ $\big[$, *pctile_options* $\big]$

Create variable containing quantile categories

> xtile *newvar* = *exp* $\big[$ *if* $\big]$ $\big[$ *in* $\big]$ $\big[$ *weight* $\big]$ $\big[$, *xtile_options* $\big]$

Compute percentiles and store them in r()

> _pctile *varname* $\big[$ *if* $\big]$ $\big[$ *in* $\big]$ $\big[$ *weight* $\big]$ $\big[$, _*pctile_options* $\big]$

pctile_options	description
Main	
nquantiles(*#*)	number of quantiles; default is nquantiles(2)
genp(*newvar$_p$*)	generate *newvar$_p$* variable containing percentages
altdef	use alternative formula for calculating percentiles

xtile_options	description
Main	
nquantiles(*#*)	number of quantiles; default is nquantiles(2)
cutpoints(*varname*)	use values of *varname* as cutpoints
altdef	use alternative formula for calculating percentiles

_*pctile_options*	description
nquantiles(*#*)	number of quantiles; default is nquantiles(2)
percentiles(*numlist*)	calculate percentiles corresponding to the specified percentages
altdef	use alternative formula for calculating percentiles

aweights, fweights, and pweights are allowed (see [U] **11.1.6 weight**), except when the altdef option is specified, in which case no weights are allowed.

Description

pctile creates a new variable containing the percentiles of *exp*, where the expression *exp* is typically just another variable.

xtile creates a new variable that categorizes *exp* by its quantiles. If the cutpoints(*varname*) option is specified, it categorizes *exp* using the values of *varname* as category cutpoints. For example, *varname* might contain percentiles of another variable, generated by pctile.

435

_pctile is a programmer's command that computes up to 1,000 percentiles and places the results in r(); see [U] **18.8 Accessing results calculated by other programs**. summarize, detail computes some percentiles (1, 5, 10, 25, 50, 75, 90, 95, and 99th); see [R] **summarize**.

Options

___Main___

nquantiles(#) specifies the number of quantiles. It computes percentiles corresponding to percentages $100\,k/m$ for $k = 1, 2, \ldots, m - 1$, where $m = \#$. For example, nquantiles(10) requests
that the 10th, 20th, ..., 90th percentiles be computed. The default is nquantiles(2); i.e., the
median is computed.

genp(*newvar$_p$*) (pctile only) specifies a new variable to be generated containing the percentages
corresponding to the percentiles.

altdef uses an alternative formula for calculating percentiles. The default method is to invert the
empirical distribution function by using averages $\{(x_i + x_{i+1})/2\}$, where the function is flat (the
default is the same method used by summarize; see [R] **summarize**). The alternative formula uses
an interpolation method. See *Methods and Formulas* at the end of this entry. Weights cannot be
used when altdef is specified.

cutpoints(*varname*) (xtile only) requests that xtile use the values of *varname*, rather than
quantiles, as cutpoints for the categories. All values of *varname* are used, regardless of any if or
in restriction; see the technical note in the xtile section below.

percentiles(*numlist*) (_pctile only) requests percentiles corresponding to the specified percentages. Percentiles are placed in r(r1), r(r2), ..., etc. For example, percentiles(10(20)90)
requests that the 10th, 30th, 50th, 70th, and 90th percentiles be computed and placed into r(r1),
r(r2), r(r3), r(r4), and r(r5). Up to 1,000 (inclusive) percentiles can be requested. See
[U] **11.1.8 numlist** for details about specifying a *numlist*.

Remarks

Remarks are presented under the following headings:

> *pctile*
> *xtile*
> *_pctile*

pctile

pctile creates a new variable containing percentiles. You specify the number of quantiles that
you want, and pctile computes the corresponding percentiles. Here we use Stata's auto dataset
and compute the deciles of mpg:

```
. use http://www.stata-press.com/data/r10/auto
(1978 Automobile Data)
. pctile pct = mpg, nq(10)
```

```
. list pct in 1/10
```

	pct
1.	14
2.	17
3.	18
4.	19
5.	20
6.	22
7.	24
8.	25
9.	29
10.	.

If we use the `genp()` option to generate another variable with the corresponding percentages, it is easier to distinguish between the percentiles.

```
. drop pct
. pctile pct = mpg, nq(10) genp(percent)
. list percent pct in 1/10
```

	percent	pct
1.	10	14
2.	20	17
3.	30	18
4.	40	19
5.	50	20
6.	60	22
7.	70	24
8.	80	25
9.	90	29
10.	.	.

`summarize, detail` calculates standard percentiles.

```
. summarize mpg, detail
```

Mileage (mpg)

	Percentiles	Smallest		
1%	12	12		
5%	14	12		
10%	14	14	Obs	74
25%	18	14	Sum of Wgt.	74
50%	20		Mean	21.2973
		Largest	Std. Dev.	5.785503
75%	25	34		
90%	29	35	Variance	33.47205
95%	34	35	Skewness	.9487176
99%	41	41	Kurtosis	3.975005

`summarize, detail` can calculate only these particular percentiles. The `pctile` and `_pctile` commands allow you to compute any percentile.

Weights can be used with pctile, xtile, and _pctile:

```
. drop pct percent
. pctile pct = mpg [w=weight], nq(10) genp(percent)
(analytic weights assumed)
. list percent pct in 1/10
```

	percent	pct
1.	10	14
2.	20	16
3.	30	17
4.	40	18
5.	50	19
6.	60	20
7.	70	22
8.	80	24
9.	90	28
10.	.	.

The result is the same, no matter which weight type you specify—aweight, fweight, or pweight.

xtile

xtile creates a categorical variable that contains categories corresponding to quantiles. We illustrate this with a simple example. Suppose that we have a variable bp containing blood pressure measurements:

```
. use http://www.stata-press.com/data/r10/bp1, clear
. list bp, sep(4)
```

	bp
1.	98
2.	100
3.	104
4.	110
5.	120
6.	120
7.	120
8.	120
9.	125
10.	130
11.	132

xtile can be used to create a variable quart that indicates the quartiles of bp.

```
. xtile quart = bp, nq(4)
. list bp quart, sepby(quart)
```

	bp	quart
1.	98	1
2.	100	1
3.	104	1
4.	110	2
5.	120	2
6.	120	2
7.	120	2
8.	120	2
9.	125	3
10.	130	4
11.	132	4

The categories created are

$$(-\infty, x_{[25]}], \quad (x_{[25]}, x_{[50]}], \quad (x_{[50]}, x_{[75]}], \quad (x_{[75]}, +\infty)$$

where $x_{[25]}$, $x_{[50]}$, and $x_{[75]}$ are, respectively, the 25th, 50th (median), and 75th percentiles of bp. We could use the pctile command to generate these percentiles:

```
. pctile pct = bp, nq(4) genp(percent)
. list bp quart percent pct, sepby(quart)
```

	bp	quart	percent	pct
1.	98	1	25	104
2.	100	1	50	120
3.	104	1	75	125
4.	110	2	.	.
5.	120	2	.	.
6.	120	2	.	.
7.	120	2	.	.
8.	120	2	.	.
9.	125	3	.	.
10.	130	4	.	.
11.	132	4	.	.

xtile can categorize a variable on the basis of any set of cutpoints, not just percentiles. Suppose that we wish to create the following categories for blood pressure:

$$(-\infty, 100], \quad (100, 110], \quad (110, 120], \quad (120, 130], \quad (130, +\infty)$$

To do this, we simply create a variable containing the cutpoints

```
. input class
          class
1. 100
2. 110
3. 120
4. 130
5. end
```

and then use xtile with the cutpoints() option.

```
. xtile category = bp, cutpoints(class)
. list bp class category, sepby(category)
```

	bp	class	category
1.	98	100	1
2.	100	110	1
3.	104	120	2
4.	110	130	2
5.	120	.	3
6.	120	.	3
7.	120	.	3
8.	120	.	3
9.	125	.	4
10.	130	.	4
11.	132	.	5

The cutpoints can, of course, come from anywhere. They can be the quantiles of another variable or the quantiles of a subgroup of the variable. Suppose that we had a variable case that indicated whether an observation represented a case (case = 1) or control (case = 0).

```
. use http://www.stata-press.com/data/r10/bp2, clear
. list in 1/11, sep(4)
```

	bp	case
1.	98	1
2.	100	1
3.	104	1
4.	110	1
5.	120	1
6.	120	1
7.	120	1
8.	120	1
9.	125	1
10.	130	1
11.	132	1

We can categorize the cases on the basis of the quantiles of the controls. To do this, we first generate a variable pct containing the percentiles of the controls' blood pressure data

```
. pctile pct = bp if case==0, nq(4)
. list pct in 1/4
```

	pct
1.	104
2.	117
3.	124
4.	.

and then use these percentiles as cutpoints to classify bp for all subjects.

```
. xtile category = bp, cutpoints(pct)
. gsort -case bp
. list bp case category in 1/11, sepby(category)
```

	bp	case	category
1.	98	1	1
2.	100	1	1
3.	104	1	1
4.	110	1	2
5.	120	1	3
6.	120	1	3
7.	120	1	3
8.	120	1	3
9.	125	1	4
10.	130	1	4
11.	132	1	4

❑ Technical Note

In the last example, if we wanted to categorize only cases, we could have issued the command

```
. xtile category = bp if case==1, cutpoints(pct)
```

Most Stata commands follow the logic that using an if *exp* is equivalent to dropping observations that do not satisfy the expression and running the command. This is not true of xtile when the cutpoints() option is used. (When the cutpoints() option is not used, the standard logic is true.) xtile uses all nonmissing values of the cutpoints() variable whether or not these values belong to observations that satisfy the if expression.

If you do not want to use all the values in the cutpoints() variable as cutpoints, simply set the ones that you do not need to missing. xtile does not care about the order of the values or whether they are separated by missing values.

❑

❑ Technical Note

Quantiles are not always unique. If we categorize our blood pressure data by quintiles rather than quartiles, we get

```
. use http://www.stata-press.com/data/r10/bp1, clear
. xtile quint = bp, nq(5)
. pctile pct = bp, nq(5) genp(percent)
```

```
. list bp quint pct percent, sepby(quint)
```

	bp	quint	pct	percent
1.	98	1	104	20
2.	100	1	120	40
3.	104	1	120	60
4.	110	2	125	80
5.	120	2	.	.
6.	120	2	.	.
7.	120	2	.	.
8.	120	2	.	.
9.	125	4	.	.
10.	130	5	.	.
11.	132	5	.	.

The 40th and 60th percentile are the same; they are both 120. When two (or more) percentiles are the same, they are given the lower category number.

❑

_pctile

_pctile is a programmer's command. It computes percentiles and stores them in r(); see [U] **18.8 Accessing results calculated by other programs**.

You can use _pctile to compute quantiles, just as you can with pctile:

```
. use http://www.stata-press.com/data/r10/auto, clear
(1978 Automobile Data)
. _pctile weight, nq(10)
. return list

scalars:
            r(r1)     =     2020
            r(r2)     =     2160
            r(r3)     =     2520
            r(r4)     =     2730
            r(r5)     =     3190
            r(r6)     =     3310
            r(r7)     =     3420
            r(r8)     =     3700
            r(r9)     =     4060
```

The percentiles() option (abbreviation p()) can be used to compute any percentile you wish:

```
. _pctile weight, p(10, 33.333, 45, 50, 55, 66.667, 90)
. return list

scalars:
            r(r1)     =     2020
            r(r2)     =     2640
            r(r3)     =     2830
            r(r4)     =     3190
            r(r5)     =     3250
            r(r6)     =     3400
            r(r7)     =     4060
```

_pctile, pctile, and xtile each have an option that uses an alternative definition of percentiles, based on an interpolation scheme; see *Methods and Formulas* below.

```
. _pctile weight, p(10, 33.333, 45, 50, 55, 66.667, 90) altdef
. return list

scalars:
        r(r1)          =    2005
        r(r2)          =    2639.985
        r(r3)          =    2830
        r(r4)          =    3190
        r(r5)          =    3252.5
        r(r6)          =    3400.005
        r(r7)          =    4060
```

The default formula inverts the empirical distribution function. The default formula is more commonly used, although some consider the "alternative" formula to be the standard definition. One drawback of the alternative formula is that it does not have an obvious generalization to noninteger weights.

❏ Technical Note

summarize, detail computes the 1st, 5th, 10th, 25th, 50th (median), 75th, 90th, 95th, and 99th percentiles. There is no real advantage in using _pctile to compute these percentiles. Both summarize, detail and _pctile use the same internal code. _pctile is slightly faster because summarize, detail computes a few extra things. The value of _pctile is its ability to compute percentiles other than these standard ones.

❏

Saved Results

pctile and _pctile save the following in r():

Scalars
   r(r#) value of #-requested percentile

Methods and Formulas

pctile and xtile are implemented as ado-files.

The default formula for percentiles is as follows: Let $x_{(j)}$ refer to the x in ascending order for $j = 1, 2, \ldots, n$. Let $w_{(j)}$ refer to the corresponding weights of $x_{(j)}$; if there are no weights, $w_{(j)} = 1$. Let $N = \sum_{j=1}^{n} w_{(j)}$.

To obtain the pth percentile, which we will denote as $x_{[p]}$, let $P = Np/100$, and let

$$W_{(i)} = \sum_{j=1}^{i} w_{(j)}$$

Find the first index i such that $W_{(i)} > P$. The pth percentile is then

$$x_{[p]} = \begin{cases} \dfrac{x_{(i-1)} + x_{(i)}}{2} & \text{if } W_{(i-1)} = P \\ x_{(i)} & \text{otherwise} \end{cases}$$

When the option `altdef` is specified, the following alternative definition is used. Here weights are not allowed.

Let i be the integer floor of $(n + 1)p/100$; i.e., i is the largest integer $i \leq (n + 1)p/100$. Let h be the remainder $h = (n + 1)p/100 - i$. The pth percentile is then

$$x_{[p]} = (1 - h)x_{(i)} + hx_{(i+1)}$$

where $x_{(0)}$ is taken to be $x_{(1)}$ and $x_{(n+1)}$ is taken to be $x_{(n)}$.

`xtile` produces the categories

$$(-\infty, x_{[p_1]}], \ (x_{[p_1]}, x_{[p_2]}], \ \dots, \ (x_{[p_{m-2}]}, x_{[p_{m-1}]}], \ (x_{[p_{m-1}]}, +\infty)$$

numbered, respectively, $1, 2, \dots, m$, based on the m quantiles given by the p_kth percentiles, where $p_k = 100\,k/m$ for $k = 1, 2, \dots, m - 1$.

If $x_{[p_{k-1}]} = x_{[p_k]}$, the kth category is empty. All elements $x = x_{[p_{k-1}]} = x_{[p_k]}$ are put in the $(k-1)$th category: $(x_{[p_{k-2}]}, x_{[p_{k-1}]}]$.

If `xtile` is used with the `cutpoints(`*varname*`)` option, the categories are

$$(-\infty, y_{(1)}], \ (y_{(1)}, y_{(2)}], \ \dots, \ (y_{(m-1)}, y_{(m)}], \ (y_{(m)}, +\infty)$$

and they are numbered, respectively, $1, 2, \dots, m + 1$, based on the m nonmissing values of *varname*: $y_{(1)}, y_{(2)}, \dots, y_{(m)}$.

Acknowledgment

`xtile` is based on a command originally posted on Statalist (see [U] **3.4 The Stata listserver**) by Philip Ryan of the University of Adelaide, Australia.

Also See

[R] **centile** — Report centile and confidence interval

[R] **summarize** — Summary statistics

[U] **18.8 Accessing results calculated by other programs**

Title

> **range** — Generate numerical range

Syntax

range *varname* #$_{\text{first}}$ #$_{\text{last}}$ $\left[\,\text{\#}_{\text{obs}}\,\right]$

Description

range generates a numerical range, which is useful for evaluating and graphing functions.

Remarks

range constructs the variable *varname*, taking on values #$_{\text{first}}$ to #$_{\text{last}}$, inclusive, over #$_{\text{obs}}$. If #$_{\text{obs}}$ is not specified, the number of observations in the current dataset is used.

range can be used to produce increasing sequences such as

```
. range x 0 12.56 100
```

or it can be used to produce decreasing sequences:

```
. range z 100 1
```

▷ Example 1

To graph $y = e^{-x/6}\sin(x)$ over the interval $[0, 12.56]$, we can type

```
. range x 0 12.56 100
obs was 0, now 100
. generate y = exp(-x/6)*sin(x)
. scatter y x, yline(0) ytitle(y = exp(-x/6) sin(x))
```

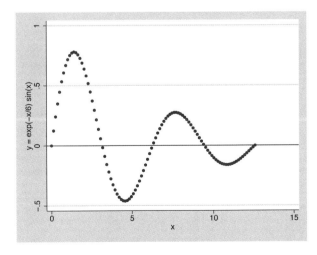

◁

▷ Example 2

Stata is not limited solely to graphing functions—it can draw parameterized curves as well. For instance, consider the curve given by the polar coordinate relation $r = 2\sin(2\theta)$. The conversion of polar coordinates to parameterized form is $(y, x) = (r \sin \theta, r \cos \theta)$, so we can type

```
. clear
. range theta 0 2*_pi 400
(obs was 100, now 400)
. generate r = 2*sin(2*theta)
. generate y = r*sin(theta)
. generate x = r*cos(theta)
. scatter y x, c(l) m(i) yline(0) xline(0) aspectratio(1)
```

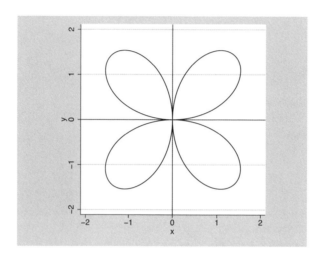

◁

Methods and Formulas

range is implemented as an ado-file.

Also See

[D] **egen** — Extensions to generate

[D] **obs** — Increase the number of observations in a dataset

Title

> **recast** — Change storage type of variable

Syntax

> recast *type varlist* [, force]

where *type* is byte, int, long, float, double, or str1, str2, ..., str244.

Description

> recast changes the storage type of the variables identified in *varlist* to *type*.

Option

> force makes recast unsafe by causing the variables to be given the new storage type even if that will cause a loss of precision, introduction of missing values, or, for string variables, the truncation of strings.

> force should be used with caution. force is for those instances where you have a variable saved as a double but would now be satisfied to have the variable stored as a float, even though that would lead to a slight rounding of its values.

Remarks

See [U] **12 Data** for a description of storage types. Also see [D] **compress** and [D] **destring** for alternatives to recast.

▷ Example 1

recast refuses to change a variable's type if that change is inappropriate for the values actually stored, so it is always safe to try:

```
. use http://www.stata-press.com/data/r10/auto
(1978 Automobile Data)
. describe headroom
```

variable name	storage type	display format	value label	variable label
headroom	float	%6.1f		Headroom (in.)

```
. recast int headroom
headroom:  37 values would be changed; not changed
```

Our attempt to change headroom from a float to an int was ignored—if the change had been made, 37 values would have changed. Here is an example where the type can be changed:

```
. describe mpg
```

variable name	storage type	display format	value label	variable label
mpg	int	%8.0g		Mileage (mpg)

```
. recast byte mpg

. describe mpg

              storage  display   value
variable name  type    format    label     variable label

mpg            byte    %8.0g                Mileage (mpg)
```

recast works with string variables as well as numeric variables, and it provides all the same protections:

```
. describe make

              storage  display   value
variable name  type    format    label     variable label

make           str18   %-18s                Make and Model
. recast str16 make
make:  2 values would be changed; not changed
```

recast can be used both to promote and to demote variables:

```
. recast str20 make

. describe make

              storage  display   value
variable name  type    format    label     variable label

make           str20   %-20s                Make and Model
```

◁

Methods and Formulas

recast is implemented as an ado-file.

Also See

[D] **compress** — Compress data in memory

[D] **destring** — Convert string variables to numeric variables and vice versa

[U] **12.2.2 Numeric storage types**

[U] **12.4.4 String storage types**

Title

> **recode** — Recode categorical variables

Syntax

Basic syntax

> recode *varlist* (*rule*) $\big[$(*rule*) ...$\big]$ $\big[$, <u>gen</u>erate(*newvar*)$\big]$

Full syntax

> recode *varlist* (*erule*) $\big[$(*erule*) ...$\big]$ $\big[$*if*$\big]$ $\big[$*in*$\big]$ $\big[$, *options*$\big]$

where the most common forms for *rule* are

rule	Example	Meaning
# = #	3 = 1	3 recoded to 1
# # = #	2 . = 9	2 and . recoded to 9
#/# = #	1/5 = 4	1 through 5 recoded to 4
<u>nonm</u>issing = #	nonmiss = 8	all other nonmissing to 8
<u>miss</u>ing = #	miss = 9	all other missings to 9

where *erule* has the form

> *element* $\big[$*element* ...$\big]$ = *el* $\big[$"*label*"$\big]$

> <u>nonm</u>issing = *el* $\big[$"*label*"$\big]$

> <u>miss</u>ing = *el* $\big[$"*label*"$\big]$

> else | * = *el* $\big[$"*label*"$\big]$

element has the form

> *el* | *el/el*

and *el* is

> # | min | max

The keyword rules missing, nonmissing, and else must be the last rules specified. else may not be combined with missing or nonmissing.

options	description
Options	
<u>g</u>enerate(*newvar*)	generate *newvar* containing transformed variables; default is to replace existing variables
<u>p</u>refix(*str*)	generate new variables with *str* prefix
<u>l</u>abel(*name*)	specify a name for the value label defined by the transformation rules
<u>copyr</u>est	copy out-of-sample values from original variables
<u>test</u>	test that rules are invoked and do not overlap

Description

recode changes the values of numeric variables according to the rules specified. Values that do not meet any of the conditions of the rules are left unchanged, unless an *otherwise* rule is specified.

A range *#1/#2* refers to *all* (real and integer) values between *#1* and *#2*, including the boundaries *#1* and *#2*. This interpretation of *#1/#2* differs from that in numlists.

min and max provide a convenient way to refer to the minimum and maximum for each variable in *varlist* and may be used in both the from-value and the to-value parts of the specification. Combined with if and in, the minimum and maximum are determined over the restricted dataset.

The keyword rules specify transformations for values not changed by the previous rules:

nonmissing	all nonmissing values not changed by the rules
missing	all missing values (., .a, .b, ..., .z) not changed by the rules
else	all nonmissing and missing values not changed by the rules
*	synonym for else

recode provides a convenient way to define value labels for the generated variables during the definition of the transformation, reducing the risk of inconsistencies between the definition and value labeling of variables. Value labels may be defined for integer values and for the extended missing values (.a, .b, ..., .z), but not for noninteger values or for sysmiss (.).

Although this is not shown in the syntax diagram, the parentheses around the *rules* and keyword clauses are optional if you transform only one variable and if you do not define value labels.

Options

┌─────────┐
│ Options │
└─────────┘

generate(*newvar*) specifies the names of the variables that will contain the transformed variables. into() is a synonym for generate(). Values outside the range implied by if or in are set to missing (.), unless the option copyrest is specified.

If generate() is not specified, the input variables are overwritten; values outside the if or in range are not modified. Overwriting variables is dangerous (you cannot undo changes, value labels may be wrong, etc.), so we strongly recommend specifying generate().

prefix(*str*) specifies that the recoded variables be returned in new variables formed by prefixing the names of the original variables with *str*.

label(*name*) specifies a name for the value label defined from the transformation rules. label() may be defined only with generate() (or its synonym, into()) and prefix(). If a variable is recoded, the label name defaults to *newvar* unless a label with that name already exists.

copyrest specifies that out-of-sample values be copied from the original variables. In line with other data management commands, recode defaults to setting *newvar* to missing (.) outside the observations selected by the if *exp* and in *range*.

test specifies that Stata test whether rules are ever invoked or that rules overlap; for example, (1/5=1) (3=2).

Remarks

Remarks are presented under the following headings:

Simple examples
Setting up value labels with recode
Referring to the minimum and maximum in rules
Recoding missing values
Recoding subsets of the data
Otherwise rules
Test for overlapping rules

Simple examples

Many users experienced with other statistical software use the `recode` command often, but easier and faster solutions in Stata are available. On the other hand, `recode` often provides simple ways to manipulate variables that are not easily accomplished otherwise. Therefore, we show other ways to perform a series of tasks with and without `recode`.

We want to change 1 to 2, leave all other values unchanged, and store the results in the new variable `nx`.

```
. recode x (1 = 2), gen(nx)
```

or

```
. gen nx = x
. replace nx = 2 if nx==1
```

or

```
. gen nx = cond(x==1,2,x)
```

We want to swap 1 and 2, saving them in `nx`.

```
. recode x (1 = 2) (2 = 1), gen(nx)
```

or

```
. gen nx = cond(x==1,2,cond(x==2,1,x))
```

We want to recode `item` by collapsing 1 and 2 into 1, 3 to 2, and 4 to 7 (boundaries included) into 3.

```
. recode item (1 2 = 1) (3 = 2) (4/7 = 3), gen(Ritem)
```

or

```
. gen Ritem = item
. replace Ritem = 1 if inlist(item,1,2)
. replace Ritem = 2 if item==3
. replace Ritem = 3 if inrange(item,4,7)
```

We want to change the "direction" of the $1, \ldots, 5$ valued variables `x1`, `x2`, `x3`, storing the transformed variables in `nx1`, `nx2`, and `nx3` (i.e., we form new variable names by prefixing old variable names with an "n").

```
. recode x1 x2 x3 (1=5) (2=4) (3=3) (4=2) (5=1), pre(n) test
```

or

```
. gen nx1 = 6-x1
. gen nx2 = 6-x2
```

```
. gen nx3 = 6-x3
. forvalues i = 1/3 {
        generate nx'i' = 6-x'i'
  }
```

In the categorical variable `religion`, we want to change 1, 3, and the real and integer numbers 3 through 5 into 6; we want to set 2, 8, and 10 to 3 and leave all other values unchanged.

```
. recode religion 1 3/5 = 6 2 8 10 = 3
```

or

```
. replace religion = 6 if religion==1 | inrange(religion,3,5)
. replace religion = 3 if inlist(religion,2,8,10)
```

This example illustrates two features of `recode` that were included for backward compatibility with previous versions of `recode` but that we do not recommend. First, we omitted the parentheses around the rules. This is allowed if you recode one variable and you do not plan to define value labels with `recode` (see below for an explanation of this feature). Personally, we find the syntax without parentheses hard to read, although we admit that we could have used blanks more sensibly. Because difficulties in reading may cause us to overlook errors, we recommend always including parentheses. Second, because we did not specify a `generate()` option, we overwrite the variable `religion`. This is often dangerous, especially for "original" variables in a dataset. We recommend that you always specify `generate()` unless you want to overwrite your data.

Setting up value labels with recode

The `recode` command is most often used to transform categorical variables, which are many times value labeled. When a value-labeled variable is overwritten by `recode`, it may well be that the value label is no longer appropriate. Consequently, output that is labeled using these value labels may be misleading or wrong.

When `recode` creates one or more new variables with a new classification, you may want to put value labels on these new variables. It is possible to do this in three steps:

(1) Create the new variables (`recode ... , gen()`).

(2) Define the value label (`label define ...`).

(3) Link the value label to the variables (`label value ...`).

Inconsistencies may emerge from mistakes between steps (1) and (2). Especially when you make a change to the recode (1), it is easy to forget to make a similar adjustment to the value label (2). Therefore, `recode` can perform steps (2) and (3) itself.

Consider recoding a series of items with values

$$1 = \text{strongly agree}$$
$$2 = \text{agree}$$
$$3 = \text{neutral}$$
$$4 = \text{disagree}$$
$$5 = \text{strongly disagree}$$

into three items

$$1 = \text{positive} \ (= \text{"strongly agree" or "agree"})$$
$$2 = \text{neutral}$$
$$3 = \text{negative} \ (= \text{"strongly disagree" or "disagree"})$$

This is accomplished by typing

```
. recode item* (1 2 = 1 positive) (3 = 2 neutral) (4 5 = 3 negative), pre(R)
> label(Item3)
```

which is much simpler and safer than

```
. recode item1-item7 (1 2 = 1) (3 = 2) (4 5 = 3), pre(R)
. label define Item3  1 positive 2 neutral 3 negative
. forvalues i = 1/7 {
        label value Ritem`i' Item3
  }
```

▷ Example 1

As another example, let us recode vote (voting intentions) for 12 political parties in the Dutch parliament into left, center, and right parties. We then tabulate the original and new variables so that we can check that everything came out correctly.

```
. use http://www.stata-press.com/data/r10/recodexmpl
. label list pparty
pparty:
            1 pvda
            2 cda
            3 d66
            4 vvd
            5 groenlinks
            6 sgp
            7 rpf
            8 gpv
            9 aov
           10 unie55
           11 sp
           12 cd
. recode polpref (1 5 11 = 1 left) (2 3 = 2 center) (4 6/10 12 = 3 right),
> gen(polpref3)
(2020 differences between polpref and polpref3)
. tabulate polpref polpref3
```

pol party choice if elections	RECODE of polpref (pol party choice if elections)			Total
	left	center	right	
pvda	622	0	0	622
cda	0	525	0	525
d66	0	634	0	634
vvd	0	0	930	930
groenlinks	199	0	0	199
sgp	0	0	54	54
rpf	0	0	63	63
gpv	0	0	30	30
aov	0	0	17	17
unie55	0	0	23	23
sp	45	0	0	45
cd	0	0	25	25
Total	866	1,159	1,142	3,167

◁

Referring to the minimum and maximum in rules

recode allows you to refer to the minimum and maximum of a variable in the transformation rules. The keywords min and max may be included as a from-value, as well as a to-value.

For example, we might divide age into age categories, storing in iage.

```
. recode age (0/9=1) (10/19=2) (20/29=3) (30/39=4) (40/49=5) (50/max=6),
> gen(iage)
```

or

```
. gen iage = 1 + irecode(age,9,19,29,39,49)
```

or

```
. gen iage = min(6, 1+int(age/10))
```

As another example, we could set all incomes less than 10,000 to 10,000 and those more than 200,000 to 200,000, storing the data in ninc.

```
. recode inc (min/10000 = 10000) (200000/max = 200000), gen(ninc)
```

or

```
. gen ninc = inc
. replace ninc = 10000 if ninc<10000
. replace ninc = 200000 if ninc>200000 & !missing(ninc)
```

or

```
. gen ninc = max(min(inc,200000),10000)
```

or

```
. gen ninc = clip(inc,10000,200000)
```

Recoding missing values

You can also set up rules in terms of missing values, either as "from-values" or as "to-values". Here recode mimics the functionality of mvdecode and mvencode (see [D] **mvencode**), although these specialized commands execute much faster.

Say that we want to change missing (.) to 9, storing the data in X

```
. recode x (.=9), gen(X)
```

or

```
. gen X = cond(x==., 9, x)
```

or

```
. mvencode x, mv(.=9) gen(X)
```

We want to change 9 to .a and 8 to ., storing the data in z.

```
. recode x (9=.a) (8=.), gen(z)
```

or

```
. gen z = cond(x==9, .a, cond(x==8, ., x))
```

or

```
. mvdecode x, mv(9=.a, 8=.) gen(z)
```

Recoding subsets of the data

We want to swap in x the values 1 and 2 only for those observations for which age>40, leaving all other values unchanged. We issue the commands

```
. recode x (1=2) (2=1) if age>40, gen(y)
```

or

```
. gen y = cond(x==1,2,cond(x==2,1,x)) if age>40
```

We are in for a surprise. y is missing for observations that do not satisfy the if condition. This outcome is in accordance with how Stata's data-manipulation commands usually work. However, it may not be what you intend. The copyrest option specifies that x be copied into y for all nonselected observations:

```
. recode x (1=2) (2=1) if age>40, gen(y) copy
```

or

```
. gen y = x
. recode y (1=2) (2=1) if age>40
```

or

```
. gen y = cond(age>40,cond(x==1,2,cond(x==2,1,x)),x)
```

Otherwise rules

In all our examples so far, recode had an implicit rule that specified that values that did not meet the conditions of any of the rules were to be left unchanged. recode also allows you to use an "otherwise rule" to specify how untransformed values are to be transformed. recode supports three kinds of otherwise conditions:

nonmissing	all nonmissing not yet transformed
missing	all missing values not yet transformed
else	all values, missing or nonmissing, not yet transformed

The otherwise rules are to be specified *after* the standard transformation rules. nonmissing and missing may be combined with each other, but not with else.

Consider a recode that swaps the values 1 and 2, transforms all other nonmissing values to 3, and transforms all missing values (i.e., sysmiss and the extended missing values) to . (sysmiss). We could type

```
. recode x (1=2) (2=1) (nonmissing=3) (missing=.), gen(z)
```

or

```
. gen z = cond(x==1,2,cond(x==2,1,cond(!missing(x),3),.))
```

As a variation, if we had decided to recode all extended missing values to .a but to keep sysmiss . distinct at ., we could have typed

```
. recode x (1=2) (2=1) (.=.) (nonmissing=3) (missing=.a), gen(z)
```

Test for overlapping rules

recode evaluates the rules from left to right. Once a value has been transformed, it will not be transformed again. Thus if rules "overlap", the first matching rule is applied, and further matches are ignored. A common form of overlapping is illustrated in the following example:

    ... (1/5 = 1) (5/10 = 2)

Here 5 occurs in the condition parts of both rules. Because rules are matched left to right, 5 matches the first rule, and the second rule will not be tested for 5, unless recode is instructed to test for rule overlap with the test option.

Other instances of overlapping rules usually arise because you mistyped the rules. For instance, you are recoding voting intentions for parties in elections into three groups of parties (left, center, right), and type

    ... (1/5 = 1) ... (3 = 2)

Party 3 matches the conditions (1/5) and (3). Since recode applies the first matching rule, party 3 will be mapped into party category 1. The second matching rule is ignored. It is not clear what was wrong in this example. You may have included party 3 in the range 1/5 or mistyped 3 in the second rule. Either way, recode did not notice the problem and your data analysis is in jeopardy. The test option specifies that recode display a warning message if values are matched by more than one rule. With the test option specified, recode also tests whether all rules were applied at least once and displays a warning message otherwise. Rules that never matched any data may indicate that you mistyped a rule, although some conditions may not have applied to (a selection of) your data.

Methods and Formulas

recode is implemented as an ado-file.

Acknowledgment

This version of recode was written by Jeroen Weesie, Department of Sociology, Utrecht University, The Netherlands.

Also See

[D] **generate** — Create or change contents of variable

[D] **mvencode** — Change missing values to numeric values and vice versa

Title

> **rename** — Rename variable

Syntax

Rename variable

> <u>ren</u>ame *old_varname new_varname*

Change or remove prefix on variables

> renpfix *old_stub* [*new_stub*]

Description

rename changes the name of existing variable *old_varname* to *new_varname*; the contents of the variable are unchanged.

renpfix changes the prefix *old_stub* to *new_stub* for all variable names that start with *old_stub*. If *new_stub* is not specified, the *old_stub* prefix is removed.

Remarks

▷ Example 1

rename allows you to change variable names freely. Say that we have labor market data for siblings.

```
. use http://www.stata-press.com/data/r10/renamexmpl
. describe
Contains data from http://www.stata-press.com/data/r10/renamexmpl.dta
  obs:            277
  vars:             6                          9 Jan 2007 11:57
  size:         7,756 (99.3% of memory free)
```

variable name	storage type	display format	value label	variable label
famid	float	%9.0g		
edu	float	%9.0g		
exp	float	%9.0g		
promo	float	%9.0g		
sex	float	%9.0g	sex	
inc	float	%9.0g		

```
Sorted by:  famid
```

We decide to rename the `exp` and `inc` variables.

```
. rename exp experience
. rename inc income
. describe
Contains data from http://www.stata-press.com/data/r10/renamexmpl.dta
  obs:           277
 vars:             6                          9 Jan 2007 11:57
 size:         7,756 (99.3% of memory free)
```

variable name	storage type	display format	value label	variable label
famid	float	%9.0g		
edu	float	%9.0g		
experience	float	%9.0g		
promo	float	%9.0g		
sex	float	%9.0g	sex	
income	float	%9.0g		

```
Sorted by:  famid
     Note:  dataset has changed since last saved
```

Variable `exp` is now called `experience`, and variable `inc` is now called `income`.

◁

▷ Example 2

We have several variables with names that start with the prefix `income` (the variables are `income86`, `income87`, and `income88`). We want them to have names that begin with the prefix `inc`:

```
. use http://www.stata-press.com/data/r10/renamexmpl2, clear
. describe
Contains data from http://www.stata-press.com/data/r10/renamexmpl2.dta
  obs:            20
 vars:             5                          9 Jan 2007 12:12
 size:           480 (99.9% of memory free)
```

variable name	storage type	display format	value label	variable label
id	float	%9.0g		
sex	float	%9.0g		
income86	float	%9.0gc		
income87	float	%9.0gc		
income88	float	%9.0gc		

```
Sorted by:  id
. renpfix income inc
```

```
. describe
Contains data from http://www.stata-press.com/data/r10/renamexmpl2.dta
  obs:            20
  vars:            5                              9 Jan 2007 12:12
  size:          480 (99.9% of memory free)
```

variable name	storage type	display format	value label	variable label
id	float	%9.0g		
sex	float	%9.0g		
inc86	float	%9.0gc		
inc87	float	%9.0gc		
inc88	float	%9.0gc		

```
Sorted by:  id
     Note:  dataset has changed since last saved
```

The variables are now named `inc86`, `inc87`, and `inc88`.

◁

Methods and Formulas

`renpfix` is implemented as an ado-file.

References

Cox, N. J., and J. Weesie. 2001. dm88: Renaming variables, multiply and systematically. *Stata Technical Bulletin* 60: 4–6. Reprinted in *Stata Technical Bulletin Reprints*, vol. 10, pp. 41–44.

Jenkins, S. P., and N. J. Cox. 2001. dm83: Renaming variables: Changing suffixes. *Stata Technical Bulletin* 59: 5–6. Reprinted in *Stata Technical Bulletin Reprints*, vol. 10, pp. 34–35.

Also See

[D] **generate** — Create or change contents of variable

Title

> **reshape** — Convert data from wide to long form and vice versa

Syntax

Basic syntax

> **reshape long** *stubnames* , **i(**varlist**)** [*options*]

> **reshape wide** *stubnames* , **i(**varlist**)** [*options*]

> **reshape long**

> **reshape wide**

> **reshape error**

Advanced syntax

> **reshape i** *varlist*

> **reshape j** *varname* [*values*] [, **string**]

> **reshape xij** *fvarnames* [, **atwl(**chars**)**]

> **reshape xi** [*varlist*]

> **reshape** [**query**]

> **reshape clear**

options	description
Main	
* **i(**varlist**)**	use *varlist* as the ID variables
j(varname [values]**)**	use *stubnames* for X_{ij} variables; optionally, and on **Advanced** tab, use *values* from *varname* to identify subobservations
Advanced	
string	allow the subobservation identifier to include strings
atwl(chars**)**	substitute *chars* for the @ character when converting to long form

* **i(**varlist**)** is required.

where *values* is $\#[-\#]$ $[\#[-\#] \dots]$

and *fvarnames* are either variable names, variable names with @ characters, or a mix of the two. The @ character denotes where the # *j* suffix appears.

Description

reshape converts data from *wide* to *long* form and vice versa.

Options

i(*varlist*) specifies the variables whose unique values denote a logical observation. i() is required.

j(*varname* [*values*]) specifies the variable whose unique values denote a subobservation. *values* lists the unique values to be used from *varname*, which typically are not explicitly stated since reshape will determine them automatically from the data.

string specifies that j() may contain string values.

atwl(*chars*) specifies that *chars* be substituted for the @ character when converting the data to the long form.

Remarks

Remarks are presented under the following headings:

> *Description of basic syntax*
> *Wide and long data forms*
> *Avoiding and correcting mistakes*
> *reshape long and reshape wide without arguments*
> *Missing variables*
> *Advanced issues with basic syntax: i()*
> *Advanced issues with basic syntax: j()*
> *Advanced issues with basic syntax: xij*
> *Advanced issues with basic syntax: the atwl() option*
> *Advanced issues with basic syntax: string identifiers for j()*
> *Advanced issues with basic syntax: second-level nesting*
> *Description of advanced syntax*

Description of basic syntax

Before using reshape, you need to determine whether the data are in long or wide form. You also must determine the logical observation (i) and the subobservation (j) by which to organize the data. Suppose that you had the following data, which could be organized in wide or long form as follows:

i		...... X_{ij}				i	j		X_{ij}
id	sex	inc80	inc81	inc82		id	year	sex	inc
1	0	5000	5500	6000		1	80	0	5000
2	1	2000	2200	3300		1	81	0	5500
3	0	3000	2000	1000		1	82	0	6000
						2	80	1	2000
						2	81	1	2200
						2	82	1	3300
						3	80	0	3000
						3	81	0	2000
						3	82	0	1000

Given these data, you could use `reshape` to convert from one form to the other:

```
. reshape long inc, i(id) j(year)          /* goes from left form to right */
. reshape wide inc, i(id) j(year)          /* goes from right form to left */
```

Since we did not specify `sex` in the command, Stata assumes that it is constant within the logical observation, here `id`.

Wide and long data forms

Think of the data as a collection of observations X_{ij}, where i is the logical observation, or group identifier, and j is the subobservation, or within-group identifier.

Wide-form data are organized by logical observation, storing all the data on a particular observation in one row. Long-form data are organized by subobservation, storing the data in multiple rows.

▷ Example 1

For example, we might have data on a person's ID, gender, and annual income over the years 1980–1982. We have two X_{ij} variables with the data in wide form:

```
. use http://www.stata-press.com/data/r10/reshape1
. list
```

	id	sex	inc80	inc81	inc82	ue80	ue81	ue82
1.	1	0	5000	5500	6000	0	1	0
2.	2	1	2000	2200	3300	1	0	0
3.	3	0	3000	2000	1000	0	0	1

To convert these data to the long form, we type

```
. reshape long inc ue, i(id) j(year)
(note: j = 80 81 82)
```

Data	wide	->	long
Number of obs.	3	->	9
Number of variables	8	->	5
j variable (3 values)		->	year
xij variables:			
	inc80 inc81 inc82	->	inc
	ue80 ue81 ue82	->	ue

There is no variable named `year` in our original, wide-form dataset. `year` will be a new variable in our long dataset. After this conversion, we have

```
. list, sep(3)
```

	id	year	sex	inc	ue
1.	1	80	0	5000	0
2.	1	81	0	5500	1
3.	1	82	0	6000	0
4.	2	80	1	2000	1
5.	2	81	1	2200	0
6.	2	82	1	3300	0
7.	3	80	0	3000	0
8.	3	81	0	2000	0
9.	3	82	0	1000	1

We can return to our original, wide-form dataset by using `reshape wide`.

```
. reshape wide inc ue, i(id) j(year)
(note:  j = 80 81 82)
```

Data	long	->	wide
Number of obs.	9	->	3
Number of variables	5	->	8
j variable (3 values)	year	->	(dropped)
xij variables:			
	inc	->	inc80 inc81 inc82
	ue	->	ue80 ue81 ue82

```
. list
```

	id	inc80	ue80	inc81	ue81	inc82	ue82	sex
1.	1	5000	0	5500	1	6000	0	0
2.	2	2000	1	2200	0	3300	0	1
3.	3	3000	0	2000	0	1000	1	0

Converting from wide to long creates the j (`year`) variable. Converting back from long to wide drops the j (`year`) variable.   ◁

❑ Technical Note

If your data are in wide form and you do not have a group identifier variable (the i(*varlist*) required option), you can create one easily by using `generate`; see [D] **generate**. For instance, in the last example, if we did not have the `id` variable in our dataset, we could have created it by typing

```
. generate id = _n
```
❑

Avoiding and correcting mistakes

`reshape` often detects when the data are not suitable for reshaping; an error is issued, and the data remain unchanged.

▷ Example 2

The following wide data contain a mistake:

```
. use http://www.stata-press.com/data/r10/reshape2, clear
. list
```

	id	sex	inc80	inc81	inc82
1.	1	0	5000	5500	6000
2.	2	1	2000	2200	3300
3.	3	0	3000	2000	1000
4.	2	0	2400	2500	2400

```
. reshape long inc, i(id) j(year)
(note:   j = 80 81 82)
i=id does not uniquely identify the observations;
there are multiple observations with the same value of id.
Type "reshape error" for a listing of the problem observations.
r(9);
```

The i variable must be unique when the data are in the wide form; we typed i(id), yet we have 2 observations for which id is 2. (Is person 2 a male or female?)

◁

▷ Example 3

It is not a mistake when the i variable is repeated when the data are in long form, but the following data have a similar mistake:

```
. use http://www.stata-press.com/data/r10/reshapexp1
. list
```

	id	year	sex	inc
1.	1	80	0	5000
2.	1	81	0	5500
3.	1	81	0	5400
4.	1	82	0	6000

```
. reshape wide inc, i(id) j(year)
(note:   j = 80 81 82)
year not unique within id;
there are multiple observations at the same year within id.
Type "reshape error" for a listing of the problem observations.
r(9);
```

In the long form, i(id) does not have to be unique, but j(year) must be unique within i; otherwise, what is the value of inc in 1981 for which id==1?

reshape told us to type reshape error to view the problem observations.

```
. reshape error
(note: j = 80 81 82)

i (id) indicates the top-level grouping such as subject id.
j (year) indicates the subgrouping such as time.
The data are in the long form;  j should be unique within i.

There are multiple observations on the same year within id.

The following 2 of 4 observations have repeated year values:
```

	id	year
2.	1	81
3.	1	81

```
(data now sorted by id year)
```
                                                                    ◁

▷ Example 4

Consider some long-form data that have no mistakes. We list the first 4 observations.

```
. list in 1/4
```

	id	year	sex	inc	ue
1.	1	80	0	5000	0
2.	1	81	0	5500	1
3.	1	82	0	6000	0
4.	2	80	1	2000	1

Say that when converting the data to wide form, however, we forget to mention the ue variable (which varies within person).

```
. reshape wide inc, i(id) j(year)
(note:  j = 80 81 82)
ue not constant within id
Type "reshape error" for a listing of the problem observations.
r(9);
```

Here reshape observed that ue was not constant within i and so could not restructure the data so that there were single observations on i. We should have typed

```
. reshape wide inc ue, i(id) j(year)
```
                                                                    ◁

In summary, there are three cases in which reshape will refuse to convert the data:

1. The data are in wide form and i is not unique

2. The data are in long form and j is not unique within i

3. The data are in long form and an unmentioned variable is not constant within i

▷ Example 5

With some mistakes, reshape will probably convert the data and produce a surprising result. Suppose that we forget to mention that variable ue varies within id in the following wide data:

```
. use http://www.stata-press.com/data/r10/reshape1
. list
```

	id	sex	inc80	inc81	inc82	ue80	ue81	ue82
1.	1	0	5000	5500	6000	0	1	0
2.	2	1	2000	2200	3300	1	0	0
3.	3	0	3000	2000	1000	0	0	1

```
. reshape long inc, i(id) j(year)
(note: j = 80 81 82)
```

Data		wide	->	long
Number of obs.		3	->	9
Number of variables		8	->	7
j variable (3 values)			->	year
xij variables:				
	inc80 inc81 inc82		->	inc

```
. list, sep(3)
```

	id	year	sex	inc	ue80	ue81	ue82
1.	1	80	0	5000	0	1	0
2.	1	81	0	5500	0	1	0
3.	1	82	0	6000	0	1	0
4.	2	80	1	2000	1	0	0
5.	2	81	1	2200	1	0	0
6.	2	82	1	3300	1	0	0
7.	3	80	0	3000	0	0	1
8.	3	81	0	2000	0	0	1
9.	3	82	0	1000	0	0	1

We did not state that ue varied within i, so the variables ue80, ue81, and ue82 were left as is. reshape did not complain. There is no real problem here because no information has been lost. In fact, this may actually be the result we wanted. Probably, however, we simply forgot to include ue among the X_{ij} variables.

If you obtain an unexpected result, here is how to undo it:

1. If you typed reshape long ... to produce the result, type reshape wide (without arguments) to undo it.

2. If you typed reshape wide ... to produce the result, type reshape long (without arguments) to undo it.

So, we can type

```
. reshape wide
```

to get back to our original, wide-form data and then type the reshape long command that we intended:

```
. reshape long inc ue, i(id) j(year)
```

◁

reshape long and reshape wide without arguments

Whenever you type a `reshape long` or `reshape wide` command with arguments, `reshape` remembers it. Thus you might

```
. reshape long inc ue, i(id) j(year)
```

and work with the data like that. You could then type

```
. reshape wide
```

to convert the data back to the wide form. Then later you could type

```
. reshape long
```

to convert them back to the long form. If you save the data, you can even continue using `reshape wide` and `reshape long` without arguments during a future Stata session.

Be careful. If you create new X_{ij} variables, you must tell `reshape` about them by typing the full `reshape` command, although no real damage will be done if you forget. If you are converting from long to wide form, `reshape` will catch your error and refuse to make the conversion. If you are converting from wide to long, `reshape` will convert the data, but the result will be surprising: remember what happened when we forgot to mention variable ue and ended up with ue80, ue81, and ue82 in our long data; see *Example 5*. You can `reshape long` to undo the unwanted change and then try again.

Missing variables

When converting data from wide form to long form, `reshape` does not demand that all the variables exist. Missing variables are treated as variables with missing observations.

▷ Example 6

Let's drop ue81 from the wide form of the data.

```
. use http://www.stata-press.com/data/r10/reshape1, clear
. drop ue81
. list
```

	id	sex	inc80	inc81	inc82	ue80	ue82
1.	1	0	5000	5500	6000	0	0
2.	2	1	2000	2200	3300	1	0
3.	3	0	3000	2000	1000	0	1

```
. reshape long inc ue, i(id) j(year)
(note:  j = 80 81 82)
(note: ue81 not found)
```

Data	wide	->	long
Number of obs.	3	->	9
Number of variables	7	->	5
j variable (3 values)		->	year
xij variables:			
inc80 inc81 inc82		->	inc
ue80 ue81 ue82		->	ue

```
. list, sep(3)
```

	id	year	sex	inc	ue
1.	1	80	0	5000	0
2.	1	81	0	5500	.
3.	1	82	0	6000	0
4.	2	80	1	2000	1
5.	2	81	1	2200	.
6.	2	82	1	3300	0
7.	3	80	0	3000	0
8.	3	81	0	2000	.
9.	3	82	0	1000	1

reshape placed missing values where ue81 values were unavailable. If we reshaped these data back to wide form by typing

```
. reshape wide inc ue, i(id) j(year)
```

the variable ue81 would be created and would contain all missing values.

◁

Advanced issues with basic syntax: i()

The i() option can indicate one i variable (as our past examples have illustrated) or multiple variables. An example of multiple i variables would be hospital ID and patient ID within each hospital.

```
. reshape ... , i(hid pid)
```

Unique pairs of values for hid and pid in the data define the grouping variable for reshape.

Advanced issues with basic syntax: j()

The j() option takes a variable name (as our past examples have illustrated) or a variable name and a list of values. When the values are not provided, reshape deduces them from the data. Specifying the values with the j() option is rarely needed.

reshape never makes a mistake when the data are in long form and you type reshape wide. The values are easily obtained by tabulating the j variable.

reshape can make a mistake when the data are in wide form and you type reshape long if your variables are poorly named. Say that you have the variables inc80, inc81, and inc82, recording income in each of the indicated years, and you have a variable named inc2, which is not income but indicates when the area was reincorporated. You type

```
. reshape long inc, i(id) j(year)
```

reshape sees the variables inc2, inc80, inc81, and inc82 and decides that there are four groups in which j = 2, 80, 81, and 82.

The easiest way to solve the problem is to rename the inc2 variable to something other than "inc" followed by a number; see [D] **rename**.

You can also keep the name and specify the j values. To perform the reshape, you can type

```
. reshape long inc, i(id) j(year 80-82)
```

or

```
. reshape long inc, i(id) j(year 80 81 82)
```

You can mix the dash notation for value ranges with individual numbers. `reshape` would understand 80 82-87 89 91-95 as a valid values specification.

At the other extreme, you can omit the j() option altogether with `reshape long`. If you do, the j variable will be named _j.

Advanced issues with basic syntax: xij

When specifying variable names, you may include @ characters to indicate where the numbers go.

▷ Example 7

Let's reshape the following data from wide to long:

```
. use http://www.stata-press.com/data/r10/reshape3, clear
. list
```

	id	sex	inc80r	inc81r	inc82r	ue80	ue81	ue82
1.	1	0	5000	5500	6000	0	1	0
2.	2	1	2000	2200	3300	1	0	0
3.	3	0	3000	2000	1000	0	0	1

```
. reshape long inc@r ue, i(id) j(year)
(note:  j = 80 81 82)
```

Data	wide	->	long
Number of obs.	3	->	9
Number of variables	8	->	5
j variable (3 values)		->	year
xij variables:			
	inc80r inc81r inc82r	->	incr
	ue80 ue81 ue82	->	ue

```
. list, sep(3)
```

	id	year	sex	incr	ue
1.	1	80	0	5000	0
2.	1	81	0	5500	1
3.	1	82	0	6000	0
4.	2	80	1	2000	1
5.	2	81	1	2200	0
6.	2	82	1	3300	0
7.	3	80	0	3000	0
8.	3	81	0	2000	0
9.	3	82	0	1000	1

At most one @ character may appear in each name. If no @ character appears, results are as if the @ character appeared at the end of the name. So, the equivalent `reshape` command to the one above is

```
. reshape long inc@r ue@, i(id) j(year)
```

inc@r specifies variables named inc#r in the wide form and incr in the long form. The @ notation may similarly be used for converting data from long to wide:

```
. reshape wide inc@r ue, i(id) j(year)
```

◁

Advanced issues with basic syntax: the atwl() option

Option `atwl()` is for use when @ characters are also specified. `atwl` stands for at-when-long. When you specify a name such as `inc@r` or `ue@`, in the long form the name becomes `incr` and `ue`, and the @ character is ignored. `atwl()` allows you to change @ into something.

If you specify `atwl(X)`, the long-form names become `incXr` and `ueX`. If you specify `atwl(yr)`, the long-form names become `incyrr` and `ueyr`.

Advanced issues with basic syntax: string identifiers for j()

The `string` option allows j to take on string values.

▷ Example 8

Consider the following wide data on husbands and wives. In these data, `incm` is the income of the man and `incf` is the income of the woman.

```
. use http://www.stata-press.com/data/r10/reshape4, clear
. list
```

	id	kids	incm	incf
1.	1	0	5000	5500
2.	2	1	2000	2200
3.	3	2	3000	2000

These data can be reshaped into separate observations for males and females by typing

```
. reshape long inc, i(id) j(sex) string
(note:  j = f m)
```

Data		wide	->	long
Number of obs.		3	->	6
Number of variables		4	->	4
j variable (2 values)			->	sex
xij variables:				
		incf incm	->	inc

The `string` option specifies that j take on nonnumeric values. The result is

```
. list, sep(2)
```

	id	sex	kids	inc
1.	1	f	0	5500
2.	1	m	0	5000
3.	2	f	1	2200
4.	2	m	1	2000
5.	3	f	2	2000
6.	3	m	2	3000

sex will be a string variable. Similarly, these data can be converted from long to wide by typing

```
. reshape wide inc, i(id) j(sex) string
```

◁

Strings are not limited to being single characters or even having the same length. You can specify the location of the string identifier in the variable name by using the @ notation.

▷ Example 9

Suppose that our variables are named id, kids, incmale, and incfem.

```
. use http://www.stata-press.com/data/r10/reshapexp2, clear
. list
```

	id	kids	incmale	incfem
1.	1	0	5000	5500
2.	2	1	2000	2200
3.	3	2	3000	2000

```
. reshape long inc, i(id) j(sex) string
(note:  j = fem male)
```

Data	wide	->	long
Number of obs.	3	->	6
Number of variables	4	->	4
j variable (2 values)		->	sex
xij variables:			
	incfem incmale	->	inc

```
. list, sep(2)
```

	id	sex	kids	inc
1.	1	fem	0	5500
2.	1	male	0	5000
3.	2	fem	1	2200
4.	2	male	1	2000
5.	3	fem	2	2000
6.	3	male	2	3000

If the wide data had variables named minc and finc, the appropriate reshape command would have been

    . reshape long @inc, i(id) j(sex) string

The resulting variable in the long form would be named inc.

We can also place strings in the middle of the variable names. If the variables were named incMome and incFome, the reshape command would be

    . reshape long inc@ome, i(id) j(sex) string

Be careful with string identifiers because it is easy to be surprised by the result. Say that we have wide data having variables named incm, incf, uem, uef, agem, and agef. To make the data long, we might type

    . reshape long inc ue age, i(id) j(sex) string

Along with these variables, we also have the variable agenda. reshape will decide that the sexes are m, f, and nda. This would not happen without the string option if the variables were named inc0, inc1, ue0, ue1, age0, and age1, even with variable agenda present in the data.

◁

Advanced issues with basic syntax: second-level nesting

Sometimes the data may have more than one possible j variable for reshaping; suppose that your data have both a year variable and a sex variable. One logical observation in the data might be represented in any of the following four forms:

    . list in 1/4   // The long-long form

	hid	sex	year	inc
1.	1	f	90	3200
2.	1	f	91	4700
3.	1	m	90	4500
4.	1	m	91	4600

    . list in 1/2   // The long-year wide-sex form

	hid	year	minc	finc
1.	1	90	4500	3200
2.	1	91	4600	4700

    . list in 1/2   // The wide-year long-sex form

	hid	sex	inc90	inc91
1.	1	f	3200	4700
2.	1	m	4500	4600

```
. list in 1       // The wide-wide form
```

	hid	minc90	minc91	finc90	finc91
1.	1	4500	4600	3200	4700

reshape can convert any of these forms to any other. Converting data from the long–long form to the wide–wide form (or any of the other forms) takes two reshape commands. Here is how we would do it:

From		To		Command
year	sex	year	sex	
long	long	long	wide	reshape wide @inc, i(hid year) j(sex) string
long	wide	long	long	reshape long @inc, i(hid year) j(sex) string
long	long	wide	long	reshape wide inc, i(hid sex) j(year)
wide	long	long	long	reshape long inc, i(hid sex) j(year)
long	wide	wide	wide	reshape wide minc finc, i(hid) j(year)
wide	wide	long	wide	reshape long minc finc, i(hid) j(year)
wide	long	wide	wide	reshape wide @inc90 @inc91, i(hid) j(sex) string
wide	wide	wide	long	reshape long @inc90 @inc91, i(hid) j(sex) string

Description of advanced syntax

The advanced syntax is simply a different way of specifying the reshape command, and it has one seldom-used feature that provides extra control. Rather than typing one reshape command to describe the data and perform the conversion, such as

```
. reshape long inc, i(id) j(year)
```

you type a sequence of reshape commands. The initial commands describe the data, and the last command performs the conversion:

```
. reshape i id
. reshape j year
. reshape xij inc
. reshape long
```

reshape i corresponds to i() in the basic syntax.

reshape j corresponds to j() in the basic syntax.

reshape xij corresponds to the variables specified in the basic syntax.

There is also one more specification, which has no counterpart in the basic syntax:

```
. reshape xi varlist
```

In the basic syntax, Stata assumes that all unspecified variables are constant within i. The advanced syntax works the same way, unless you specify the reshape xi command, which names the constant-within-i variables. If you specify reshape xi, any variables that you do not explicitly specify are dropped from the data during the conversion.

As a practical matter, you should explicitly drop the unwanted variables before conversion. For instance, suppose that the data have variables inc80, inc81, inc82, sex, age, and age2 and that you no longer want the age2 variable. You could specify

```
. reshape xi sex age
```

or

```
. drop age2
```

and leave `reshape xi` unspecified.

`reshape xi` does have one minor advantage. It saves `reshape` the work of determining which variables are unspecified. This saves a relatively small amount of computer time.

Another advanced-syntax feature is `reshape query`, which is equivalent to typing `reshape` by itself. `reshape query` reports which `reshape` parameters have been defined. `reshape i`, `reshape j`, `reshape xij`, and `reshape xi` specifications may be given in any order and may be repeated to change or correct what has been specified.

Finally, `reshape clear` clears the definitions. `reshape` definitions are stored with the dataset when you save it. `reshape clear` allows you to erase these definitions.

The basic syntax of `reshape` is implemented in terms of the advanced syntax, so you can mix basic and advanced syntaxes.

Saved Results

`reshape` stores the following characteristics with the data (see [P] **char**):

_dta[ReS_i]	i variable names
_dta[ReS_j]	j variable name
_dta[ReS_jv]	j values if specified
_dta[ReS_Xij]	X_{ij} variable names
_dta[ReS_Xi]	X_i variable names if specified
_dta[ReS_atwl]	atwl() value if specified
_dta[ReS_str]	1 if option `string` specified; 0 otherwise

Methods and Formulas

`reshape` is implemented as an ado-file.

Acknowledgment

This version of `reshape` was based, in part, on the work of Jeroen Weesie from Utrecht University, The Netherlands (Weesie 1997).

References

Gould, W. W. 1997. stata48: Updated reshape. *Stata Technical Bulletin* 39: 4–16. Reprinted in *Stata Technical Bulletin Reprints*, vol. 7, pp. 5–20.

Weesie, J. 1997. dm48: An enhancement of reshape. *Stata Technical Bulletin* 38: 2–4. Reprinted in *Stata Technical Bulletin Reprints*, vol. 7, pp. 40–43.

——. 1998. dm58: A package for the analysis of husband-wife data. *Stata Technical Bulletin* 43: 9–13. Reprinted in *Stata Technical Bulletin Reprints*, vol. 8, pp. 13–20.

Also See

Title

> **rmdir** — Remove directory

Syntax

rmdir *directory_name*

Double quotes may be used to enclose the directory name, and the quotes must be used if the directory name contains embedded blanks.

Description

rmdir removes an empty directory (folder).

Remarks

Examples:

Windows:

```
. rmdir myproj
. rmdir c:\projects\myproj
. rmdir "c:\My Projects\Project 1"
```

Macintosh and Unix

```
. rmdir myproj
. rmdir ~/projects/myproj
```

Also See

[D] **cd** — Change directory

[D] **copy** — Copy file from disk or URL

[D] **dir** — Display filenames

[D] **erase** — Erase a disk file

[D] **shell** — Temporarily invoke operating system

[D] **type** — Display contents of a file

[D] **mkdir** — Create directory

Title

> **sample** — Draw random sample

Syntax

> sample # $\left[\textit{if}\right]$ $\left[\textit{in}\right]$ $\left[\, ,\, \underline{\text{c}}\text{ount} \text{ by}(\textit{groupvars})\right]$

by is allowed; see [D] **by**.

Description

sample draws random samples of the data in memory. "Sampling" here is defined as drawing observations without replacement; see [R] **bsample** for sampling with replacement.

The size of the sample to be drawn can be specified as a percentage or as a count:

> sample without the count option draws a #% pseudorandom sample of the data in memory, thus discarding $(100 - \#)\%$ of the observations.

> sample with the count option draws a #-observation pseudorandom sample of the data in memory, thus discarding _N − # observations. # can be larger than _N, in which case all observations are kept.

In either case, observations not meeting the optional if and in criteria are kept (sampled at 100%).

If you are interested in reproducing results, you must first set the random-number seed; see [D] **generate**.

Options

count specifies that # in sample # be interpreted as an observation count rather than as a percentage. Typing sample 5 without the count option means that a 5% sample be drawn; typing sample 5, count, however, would draw a sample of 5 observations.

Specifying # as greater than the number of observations in the dataset is not considered an error.

by(*groupvars*) specifies that a #% sample be drawn within each set of values of *groupvars*, thus maintaining the proportion of each group.

count may be combined with by(). For example, typing sample 50, count by(sex) would draw a sample of size 50 for men and 50 for women.

Specifying by *varlist*: sample # is equivalent to specifying sample #, by(*varlist*); use whichever syntax you prefer.

(*Continued on next page*)

Remarks

▷ Example 1

We have NLSY data on young women aged 14–26 years in 1968 and wish to draw a 10% sample of the data in memory.

```
. use http://www.stata-press.com/data/r10/nlswork
(National Longitudinal Survey.  Young Women 14-26 years of age in 1968)

. describe, short

Contains data from http://www.stata-press.com/data/r10/nlswork.dta
  obs:        28,534                          National Longitudinal Survey.
                                              Young Women 14-26 years of age
                                              in 1968
 vars:            21                          7 Dec 2006 17:02
 size:     1,055,758 (88.8% of memory free)
Sorted by:  idcode   year

. sample 10
(25681 observations deleted)

. describe, short

Contains data from http://www.stata-press.com/data/r10/nlswork.dta
  obs:         2,853                          National Longitudinal Survey.
                                              Young Women 14-26 years of age
                                              in 1968
 vars:            21                          7 Dec 2006 17:02
 size:       105,561 (98.9% of memory free)
Sorted by:
    Note:  dataset has changed since last saved
```

Our original dataset had 28,534 observations. The sample-10 dataset has 2.853 observations, which is the nearest number to $.10 \times 28534$.

◁

▷ Example 2

Among the variables in our data is race; race $= 1$ denotes whites, race $= 2$ denotes blacks, and race $= 3$ denotes other. We want to keep 100% of the nonwhite women but only 10% of the white women.

```
. use http://www.stata-press.com/data/r10/nlswork, clear
(National Longitudinal Survey.  Young Women 14-26 years of age in 1968)

. tab race

 1=white,
 2=black,
  3=other |     Freq.     Percent        Cum.
----------+-----------------------------------
        1 |    20,180       70.72       70.72
        2 |     8,051       28.22       98.94
        3 |       303        1.06      100.00
----------+-----------------------------------
    Total |    28,534      100.00

. sample 10 if race == 1
(18162 observations deleted)
```

```
. describe, short
Contains data from http://www.stata-press.com/data/r10/nlswork.dta
  obs:          10,372                          National Longitudinal Survey.
                                                Young Women 14-26 years of age
                                                in 1968
  vars:             21                          7 Dec 2006 17:02
  size:        383,764 (95.9% of memory free)
Sorted by:
    Note:   dataset has changed since last saved
. display .10*20180 + 8051 + 303
10372
```

◁

▷ Example 3

Now let's suppose that we want to keep 10% of each of the three categories of `race`.

```
. use http://www.stata-press.com/data/r10/nlswork, clear
(National Longitudinal Survey.  Young Women 14-26 years of age in 1968)
. sample 10, by(race)
(25681 observations deleted)
. tab race
   1=white,
   2=black,
    3=other |     Freq.     Percent       Cum.
------------+-----------------------------------
          1 |     2,018       70.73      70.73
          2 |       805       28.22      98.95
          3 |        30        1.05     100.00
------------+-----------------------------------
      Total |     2,853      100.00
```

This differs from simply typing `sample 10` in that with `by()`, `sample` holds constant the percentages of white, black, and other women.

◁

❏ Technical Note

We have a large dataset on disk containing 125,235 observations. We wish to draw a 10% sample of this dataset without loading the entire dataset (perhaps because the dataset will not fit in memory). `sample` will not solve this problem—the dataset must be loaded first—but it is rather easy to solve it ourself. Say that `bigdata.dct` contains the dictionary for this dataset; see [D] **infile**. One solution is to type

```
. infile using bigdata if uniform()<=.1
dictionary {
    etc.
}
(12,580 observations read)
```

The `if` modifier on the end of `infile` drew uniformly distributed random numbers over the interval 0 and 1 and kept each observation if the random number was less than or equal to 0.1. This, however, did not draw an exact 10% sample—the sample was expected to contain only 10% of the observations, and here we obtained just more than 10%. This is probably a reasonable solution.

If the sample must contain precisely 12,524 observations, however, after getting too many observations, we could type

```
. generate u=uniform()
. sort u
. keep in 1/12524
(56 observations deleted)
```

That is, we put the resulting sample in random order and keep the first 12,524 observations. Now our only problem is making sure that, at the first step, we have more than 12,524 observations. Here we were lucky, but half the time we will not be so lucky—after typing `infile ... if uniform()<=.1`, we will have less than a 10% sample. The solution, of course, is to draw more than a 10% sample initially and then cut it back to 10%.

How much more than 10% do we need? That depends on the number of records in the original dataset, which in our example is 125,235.

A little experimentation with `bitesti` (see [R] **bitest**) provides the answer:

```
. bitesti 125235 12524 .102
```

N	Observed k	Expected k	Assumed p	Observed p
125235	12524	12773.97	0.10200	0.10000

```
Pr(k >= 12524)              = 0.990466  (one-sided test)
Pr(k <= 12524)              = 0.009777  (one-sided test)
Pr(k <= 12524 or k >= 13025) = 0.019584  (two-sided test)
```

Initially drawing a 10.2% sample will yield a sample larger than 10% 99 times of 100. If we draw a 10.4% sample, we are virtually assured of having enough observations (type `bitesti 125235 12524 .104` for yourself).

❑

Methods and Formulas

`sample` is implemented as an ado-file.

References

Cox, N. J. 2001. dm86: Sampling without replacement: Absolute sample sizes and keeping all observations. *Stata Technical Bulletin* 59: 8–9. Reprinted in *Stata Technical Bulletin Reprints*, vol. 10, pp. 38–39.

Weesie, J. 1997. dm46: Enhancement to the sample command. *Stata Technical Bulletin* 37: 6–7. Reprinted in *Stata Technical Bulletin Reprints*, vol. 7, pp. 37–38.

Also See

[R] **bsample** — Sampling with replacement

Title

> **save** — Save datasets

Syntax

Save data in memory to file

> <u>sa</u>ve [*filename*] [, *save_options*]

Save data in memory to file in Stata 8/Stata 9 format

> saveold *filename* [, *saveold_options*]

save_options	description
<u>nol</u>abel	omit value labels from the saved dataset
replace	overwrite existing dataset
all	save e(sample) with the dataset; programmer's option
<u>orphans</u>	save all value labels
emptyok	save dataset even if zero observations and zero variables

saveold_options	description
<u>nol</u>abel	omit value labels from the saved dataset
replace	overwrite existing dataset
all	save e(sample) with the dataset; programmer's option

Description

save stores the dataset currently in memory on disk under the name *filename*. If *filename* is not specified, the name under which the data were last known to Stata (c(filename)) is used. If *filename* is specified without an extension, .dta is used. If your *filename* contains embedded spaces, remember to enclose it in double quotes.

saveold saves the dataset currently in memory on disk under the name *filename* in Stata 8/Stata 9 format. (Stata 8 and Stata 9 share the same dataset format.)

If you are using Stata 10 and want to save a file so that it may be read by someone using Stata 8 or Stata 9, simply use the saveold command. Stata 9 allows value labels to be up to 32,000 characters long. If Stata 8 tries to read a Stata 9 dataset with value labels that exceed the Stata 8 limit (244 for Stata/SE; 80 for Stata/IC), Stata 8 will ignore those labels and read the rest of the dataset.

Options for save

nolabel omits value labels from the saved dataset. The associations between variables and value label names, however, are saved along with the dataset label and the variable labels.

replace permits save or saveold to overwrite an existing dataset.

all is for use by programmers. If specified, e(sample) will be saved with the dataset. You could run a regression; save mydata, all; drop _all; use mydata; and predict yhat if e(sample).

orphans saves all value labels, including those not attached to any variable.

emptyok is a programmer's option. It specifies that the dataset be saved, even if it contains zero observations and zero variables. If emptyok is not specified and the dataset is empty, save responds with the message "no variables defined".

Options for saveold

nolabel omits value labels from the saved dataset. The associations between variables and value label names, however, are saved along with the dataset label and the variable labels.

replace permits save or saveold to overwrite an existing dataset.

all is for use by programmers. If specified, e(sample) will be saved with the dataset. You could run a regression; save mydata, all; drop _all; use mydata; and predict yhat if e(sample).

Remarks

Stata keeps the data on which you are currently working in your computer's memory. You put the data there in the first place by using the input, infile, insheet, or infix command; see [U] **21 Inputting data**. Thereafter, you can save the dataset on disk so that you can use it easily in the future. Stata stores your data on disk in a compressed format that only Stata understands. This does not mean, however, that you are locked into using only Stata. Any time you wish, you can use the outfile or outsheet commands to create an ASCII-format dataset that all software packages understand; see [D] **outfile** and [D] **outsheet**.

Stata goes to a lot of trouble to keep you from accidentally losing your data. When you attempt to leave Stata by typing exit, Stata checks that your data have been safely stored on disk. If not, Stata refuses to let you leave. (You can tell Stata that you want to leave anyway by typing exit, clear.) Similarly, when you save your data in a disk file, Stata ensures that the disk file does not already exist. If it does exist, Stata refuses to save it. You can use the replace option to tell Stata that it is okay to overwrite an existing file.

▷ Example 1

We have entered data into Stata for the first time. We have the following data:

```
. describe
Contains data
    obs:             39                      Minnesota Highway Data, 1973
    vars:             5
    size:           936 (99.6% of memory free)

              storage  display   value
variable name  type    format    label     variable label

acc_rate       float   %9.0g               Accident rate
spdlimit       float   %9.0g               Speed limit
acc_pts        float   %9.0g               Access points per mile
rate           float   %9.0g      rcat     Accident rate per million
                                              vehicle miles
spdcat         float   %9.0g      scat     Speed limit category

Sorted by:
    Note:  dataset has changed since last saved
```

We have a dataset containing 39 observations on five variables, and, evidently, we have gone to a lot of trouble to prepare this dataset. We have used the label data command to label the data Minnesota Highway Data, the label variable command to label all the variables, and the label define and label values commands to attach value labels to the last two variables. (See [U] **12.6.3 Value labels** for information about doing this.)

At the end of the describe, Stata notes that the "dataset has changed since last saved". This is Stata's way of gently reminding us that these data need to be saved. Let's save our data:

```
. save hiway
file hiway.dta saved
```

We type save hiway, and Stata stores the data in a file named hiway.dta. (Stata automatically added the .dta suffix.) Now when we describe our data, we no longer get the warning that our dataset has not been saved; instead, we are told the name of the file in which the data are saved:

```
. describe
Contains data from hiway.dta
    obs:             39                      Minnesota Highway Data, 1973
    vars:             5
    size:           936 (99.7% of memory free)    18 Jan 2007 11:42

              storage  display   value
variable name  type    format    label     variable label

acc_rate       float   %9.0g               Accident rate
spdlimit       float   %9.0g               Speed limit
acc_pts        float   %9.0g               Access points per mile
rate           float   %9.0g      rcat     Accident rate per million
                                              vehicle miles
spdcat         float   %9.0g      scat     Speed limit category

Sorted by:
```

Just to prove to you that the data have really been saved, let's eliminate the copy of the data in memory by typing drop _all:

```
. drop _all

. describe
Contains data
  obs:            0
  vars:           0
  size:           0 (100.0% of memory free)
Sorted by:
```

We now have no data in memory. Since we saved our dataset, we can retrieve it by typing use hiway:

```
. use hiway
(Minnesota Highway Data, 1973)

. describe
Contains data from hiway.dta
  obs:           39                          Minnesota Highway Data, 1973
  vars:           5                          18 Jan 2007 11:42
  size:         936 (99.7% of memory free)
```

variable name	storage type	display format	value label	variable label
acc_rate	float	%9.0g		Accident rate
spdlimit	float	%9.0g		Speed limit
acc_pts	float	%9.0g		Access points per mile
rate	float	%9.0g	rcat	Accident rate per million vehicle miles
spdcat	float	%9.0g	scat	Speed limit category

```
Sorted by:
```

◁

▷ Example 2

Continuing with our previous example, we have saved our data in the file hiway.dta. We continue to work with our data and discover an error; we made a mistake when we typed one of the values for the variable spdlimit:

```
. list in 1/3
```

	acc_rate	spdlimit	acc_pts	rate	spdcat
1.	1.61	50	2.2	Below 4	Above 60
2.	1.81	60	6.8	Below 4	55 to 60
3.	1.84	55	14	Below 4	55 to 60

In the first observation, the variable spdlimit is 50, whereas the spdcat variable indicates that the speed limit is more than 60 miles per hour. We check our original copy of the data and discover that the spdlimit variable ought to be 70. We can fix it with the replace command:

```
. replace spdlimit=70 in 1
(1 real change made)
```

If we were to `describe` our data now, Stata would warn us that our data have now changed since they were last saved:

```
. describe
Contains data from hiway.dta
  obs:            39                          Minnesota Highway Data, 1973
  vars:            5                          18 Jan 2007 11:42
  size:          936 (99.7% of memory free)

              storage  display    value
variable name   type   format     label      variable label

acc_rate        float  %9.0g                 Accident rate
spdlimit        float  %9.0g                 Speed limit
acc_pts         float  %9.0g                 Access points per mile
rate            float  %9.0g      rcat       Accident rate per million
                                                vehicle miles
spdcat          float  %9.0g      scat       Speed limit category

Sorted by:
     Note:  dataset has changed since last saved
```

We take our cue and attempt to `save` the data again:

```
. save hiway
file hiway.dta already exists
r(602);
```

Stata refuses to honor our request, telling us instead that "file hiway.dta already exists". Stata will not let us accidentally overwrite an existing dataset. To `replace` the data, we must do so explicitly by typing `save hiway, replace`. If we want to save the file under the same name as it was last known to Stata, we can omit the filename:

```
. save, replace
file hiway.dta saved
```

Now our data are saved.

◁

Methods and Formulas

`saveold` is implemented as an ado-file.

Also See

[D] **compress** — Compress data in memory

[D] **fdasave** — Save and use datasets in FDA (SAS XPORT) format

[D] **outfile** — Write ASCII-format dataset

[D] **outsheet** — Write spreadsheet-style dataset

[D] **use** — Use Stata dataset

[U] **11.6 File-naming conventions**

Title

---

separate — Create separate variables

Syntax

separate *varname* [*if*] [*in*] , by(*byvar* | *exp*) [*options*]

options	description
Main	
* by(*byvar*)	categorize observations into groups defined by *byvar*
* by(*exp*)	categorize observations into two groups defined by *exp*
Options	
<u>g</u>enerate(*stubname*)	name new variables by suffixing values to *stubname*; default is to use *varname* as prefix
<u>seq</u>uential	use as name suffix categories numbered sequentially from 1
<u>miss</u>ing	create variables for the missing values
<u>short</u>label	create shorter variable labels

* Either by(*byvar*) or by(*exp*) must be specified.

Description

separate creates new variables containing values from *varname*.

Options

> Main

by(*byvar* | *exp*) specifies one variable defining the categories or a logical expression that categorizes the observations into two groups.

If by(*byvar*) is specified, *byvar* may be a numeric or string variable taking on any values.

If by(*exp*) is specified, the expression must evaluate to true (1), false (0), or missing.

by() is required.

> Options

generate(*stubname*) specifies how the new variables are to be named. If generate() is not specified, separate uses the name of the original variable, shortening it if necessary. If generate() is specified, separate uses *stubname*. If any of the resulting names is too long when the values are suffixed, it is not shortened and an error message is issued.

sequential specifies that categories be numbered sequentially from 1. By default, separate uses the actual values recorded in the original variable, if possible, and sequential numbers otherwise. separate can use the original values if they are all nonnegative integers smaller than 10,000.

missing also creates a variable for the category *missing* if missing occurs (*byvar* takes on the value missing or *exp* evaluates to missing). The resulting variable is named in the usual manner but with an appended underscore; e.g., bp_. By default, separate creates no such variable. The contents of the other variables are unaffected by whether missing is specified.

shortlabel creates a variable label that is shorter than the default. By default, when separate generates the new variable labels, it includes the name of the variable being separated. shortlabel specifies that the variable name be omitted from the new variable labels.

Remarks

▷ Example 1

We have data on the miles per gallon (mpg) and country of manufacture of 74 automobiles. We want to compare the distributions of mpg for domestic and foreign automobiles by plotting the quantiles of the two distributions (see [R] **diagnostic plots**).

```
. use http://www.stata-press.com/data/r10/auto
(1978 Automobile Data)

. separate mpg, by(foreign)
```

variable name	storage type	display format	value label	variable label
mpg0	byte	%8.0g		mpg, foreign == Domestic
mpg1	byte	%8.0g		mpg, foreign == Foreign

```
. list mpg* foreign
```

	mpg	mpg0	mpg1	foreign
1.	22	22	.	Domestic
2.	17	17	.	Domestic
3.	22	22	.	Domestic
		(*output omitted*)		
22.	16	16	.	Domestic
23.	17	17	.	Domestic
24.	28	28	.	Domestic
		(*output omitted*)		
73.	25	.	25	Foreign
74.	17	.	17	Foreign

(*Continued on next page*)

. qqplot mpg0 mpg1

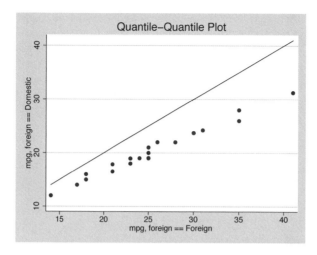

In our auto dataset, the foreign cars have better gas mileage.

◁

Saved Results

separate saves the following in r():

Macros
r(varlist)   names of the newly created variables

Methods and Formulas

separate is implemented as an ado-file.

Acknowledgment

separate was originally written by Nicholas J. Cox of Durham University.

Also See

[R] **tabulate oneway** — One-way tables of frequencies

[R] **tabulate twoway** — Two-way tables of frequencies

Title

shell — Temporarily invoke operating system

Syntax

$\{$ <u>she</u>ll | ! $\}$ $\big[$ *operating_system_command* $\big]$

winexec *program_name* $\big[$ *program_args* $\big]$

$\{$ <u>xshe</u>ll | !! $\}$ $\big[$ *operating_system_command* $\big]$

Command availability:

| | Stata for ... | | | |
Command	Windows	Macintosh	Unix(GUI)	Unix(console)
shell	X	X	X	X
winexec	X	X	X	–
xshell	–	–	X	–

Description

shell (synonym: "!") allows you to send commands to your operating system or to enter your operating system for interactive use. In Stata for Macintosh, the *operating_system_command* is required.

winexec allows you to start other programs (such as browsers) from Stata's command line and without Stata waiting for the program to complete before continuing.

xshell (Stata for Unix(GUI) only) brings up an **xterm** in which the command is to be executed.

Remarks

Remarks are presented under the following headings:

Stata for Windows
Stata for Macintosh
Stata for Unix(GUI)
Stata for Unix(console)

Stata for Windows

shell, without arguments, preserves your session and invokes the operating system. The Command window will disappear, and a DOS window will appear, indicating that you may not continue in Stata until you exit the DOS shell. To reenter Stata, type **exit** at your operating system's prompt. Your Stata session is reestablished just as if you had never left.

489

Say that you are using Stata for Windows and you suddenly realize you need to do two things. You need to enter your operating system for a few minutes. Rather than exiting Stata, doing what you have to do, and then restarting Stata, you type shell in the Command window. A DOS window appears:

```
Microsoft Windows [Version 6.0.6000]
Copyright (c) 2006 Microsoft Corporation.  All rights reserved.
C:\data>
```

You can now do whatever you need to do in DOS, and Stata will wait until you exit the DOS window before continuing.

Experienced Stata users seldom type out the word shell. They type "!". Also you do not have to enter your operating system, issue a command, and then exit back to Stata. If you want to execute one command, you can type the command right after the word shell or the exclamation point:

. !rename try15.dta final.dta

If you do this, the DOS window will open and close as the command is executed. If the DOS window does not close as it should, see the technical note at the end of this section.

Stata for Windows users can also use the winexec command, which allows you to launch any Windows application from within Stata. You can think of it as a shortcut for clicking on the Windows **Start** button, choosing **Run...**, and typing a command.

Assume that you are working in Stata and decide that you want to run a text editor:

. winexec notepad

(The Windows application Notepad will start and run at the same time as Stata)

You could even pass a filename to your text editor:

. winexec notepad c:\docs\myfile.txt

You may need to specify a complete path to the executable that you wish to launch:

. winexec c:\windows\notepad c:\docs\myfile.txt

The important difference between winexec and shell is that Stata does not wait for whatever program winexec launches to complete before continuing. Stata will wait for the program shell launches to complete before performing any further commands.

❏ Technical Note

If, when you use shell to issue a command, the DOS window does not close properly, Windows has the preference set that prevents the window from closing after executing a command. For example, if you typed !dir, a DOS window would appear containing a directory listing, but the window would not go away and return you to Stata until you closed the window manually.

You may view this as desirable behavior, but if you do not, you can change it. Under Windows, navigate to C:\Windows. You can do this by double-clicking on the **My Computer** icon on your desktop, then double-clicking on the **C:** drive in the **My Computer** window, and then double-clicking on the **Windows** folder in the **C:** drive window.

In the **Windows** folder, you will find an icon named command.com. Right-click this icon, and a menu will appear. Choose **Properties** from that menu, and a tabbed dialog box will appear. Click the **Program** tab, and you will see a *Close on exit* checkbox near the bottom of the dialog box. Check this box if you want your DOS shells to close automatically after they complete any commands that you issue with the shell command in Stata for Windows.

❏

❏ Technical Note

Although we do not recommend it, Stata for Windows users can change the shell that Stata calls. By default, Stata for Windows calls the program `command.com` for a DOS shell when running under Windows ME or 98 and calls `cmd.exe` when running under Windows Vista, XP, 2000, or NT.

To change the shell that Stata calls, set the global macro `$S_SHELL` to contain the name of the executable program that you want Stata to use for the shell.

❏

Stata for Macintosh

`shell`, with arguments, invokes your operating system, executes one command, and redirects the output to the Results window. The command must complete before you can enter another command in the Command window.

Say that you are using Stata for Macintosh and suddenly realize that there are two things you have to do. You need to switch to the Finder or enter commands from a terminal for a few minutes. Rather than exiting Stata, doing what you have to do, and then switching back to Stata, you type `shell` and the command in the Command window to execute one command. You then repeat this step for each command that you want to execute from the shell.

Experienced Stata users seldom type out the word `shell`. They type "`!`".

    . !mv try15.dta final.dta

Be careful not to execute commands, such as `vi`, that require interaction from you. Since all output is redirected to Stata's Results window, you will not be able to interact with the command from Stata. This will effectively lock up Stata because the command will never complete.

Stata for Macintosh users can also use the `winexec` command, which allows you to launch any native Mach-O application from within Stata. You may, however, have to specify the absolute path to the application. If the application you wish to launch is a Mac OS X application bundle, you must specify an absolute path to the executable in the bundle.

Assume that you are working in Stata and decide that you want to run a text editor:

    . winexec /Applications/TextEdit.app/Contents/MacOS/TextEdit
            (The OS X application TextEdit will start and run at the same time as Stata)

You could even pass a filename to your text editor:

    . winexec /Applications/TextEdit.app/Contents/MacOS/TextEdit /Users/cnguyen/myfile.do

If you specify a file path as an argument to the program to be launched, you must specify an absolute path. Also using ~ in the path will not resolve to a home directory. `winexec` cannot launch PEF binaries such as those from Mac OS 9 and some Carbon applications. If an application cannot be launched from a terminal window, it cannot be launched by `winexec`.

The important difference between `winexec` and `shell` is that Stata does not wait for whatever program `winexec` launches to complete before continuing. Stata will wait for the program `shell` launches to complete before performing any further commands. `shell` is appropriate for executing shell commands; `winexec` is appropriate for launching applications.

Stata for Unix(GUI)

shell, without arguments, preserves your session and invokes the operating system. The Command window will disappear, and an xterm window will appear, indicating that you may not do anything in Stata until you exit the xterm window. To reenter Stata, type exit at the Unix prompt. Your Stata session is reestablished just as if you had never left.

Say that you are using Stata for Unix(GUI) and suddenly realize that you need to do two things. You need to enter your operating system for a few minutes. Rather than exiting Stata, doing what you have to do, and then restarting Stata, you type shell in the Command window. An xterm window will appear:

       mycomputer$ _

You can now do whatever you need to do, and Stata will wait until you exit the window before continuing.

Experienced Stata users seldom type out the word shell. They type "!". Also you do not have to enter your operating system, issue a command, and then exit back to Stata. If you want to execute one command, you can type the command right after the word shell or the exclamation point:

       . !mv try15.dta final.dta

Be careful because sometimes you will want to type

       . !!vi myfile.do

and, in other cases,

       . winexec xedit myfile.do

!! is a synonym for xshell—a command different from, but related to, shell—and winexec is a different and related command, too.

Before we get into this, understand that if all you want is a shell from which you can issue Unix commands, type shell or !:

       . !
       mycomputer$ _

When you are through, type exit to the Unix prompt, and you will return to Stata:

       mycomputer$ exit

       . _

If, on the other hand, you want to specify in Stata the Unix command that you want to execute, you need to decide whether you want to use shell, xshell, or winexec. The answer depends on whether the command you want to execute requires a terminal window or is an X application:

... does not need a terminal window:	use shell ... (synonym: !...)
... needs a terminal window:	use xshell ... (synonym: !!...)
... is an X application:	use winexec ... (no synonym)

When you type shell mv try15.dta final.dta, Stata invokes your shell (/bin/sh, /bin/csh, etc.) and executes the specified command (mv here), routing the standard output and standard error back to Stata. Typing !mv try15.dta final.dta is the same as typing shell mv try15.dta final.dta.

When you type xshell vi myfile.do, Stata invokes an xterm window (which in turn invokes a shell) and executes the command there. Typing !!vi myfile.do is equivalent to typing xshell vi myfile.do.

When you type `winexec xedit myfile.do`, Stata directly invokes the command specified (`xedit` here). No `xterm` window is brought up nor is a shell invoked because, here, `xterm` does not need it. `xterm` is an X application that will create its own window in which to run. You could have typed `!!xedit myfile.do`. That would have brought up an unnecessary `xterm` window from which `xedit` would have been executed, and that would not matter. You could even have typed `!xedit myfile.do`. That would have invoked an unnecessary shell from which `xedit` would have been executed, and that would not matter, either. The important difference, however, is that `shell` and `xshell` wait until the process completes before allowing Stata to continue, and `winexec` does not.

❏ Technical Note

You can set Stata global macros to control the behavior of `shell` and `xshell`. The macros are

`$S_SHELL`	defines the shell to be used by `shell` when you type a command following `shell`. The default is something like "`/bin/sh -c`", although this can vary, depending on how your Unix environment variables are set.
`$S_XSHELL`	defines shell to be used by `shell` and `xshell` when they are typed without arguments. The default is "`xterm`".
`$S_XSHELL2`	defines shell to be used by `xshell` when it is typed with arguments. The default is "`xterm -e`".

For instance, if you type in Stata

```
. global S_XSHELL2 "/usr/X11R6/bin/xterm -e"
```

and then later type

```
. !!vi myfile.do
```

Stata would issue the command `/usr/X11R6/bin/xterm -e vi myfile.do` to Unix.

If you do make changes, we recommend that you record the changes in your `profile.do` file.

❏

Stata for Unix(console)

`shell`, without arguments, preserves your session and then invokes your operating system. Your Stata session will be suspended until you `exit` the shell, at which point your Stata session is reestablished just as if you had never left.

Say that you are using Stata and you suddenly realize that you need to do two things. You need to enter your operating system for a few minutes. Rather than exiting Stata, doing what you have to do, and then restarting Stata, you type `shell`. A Unix prompt appears:

```
. shell
(Type exit to return to Stata)
$ _
```

You can now do whatever you need to do and type `exit` when you finish. You will return to Stata just as if you had never left.

Experienced Stata users seldom type out the word shell. They type '!'. Also you do not have to enter your operating system, issue a command, and then exit back to Stata. If you want to execute one command, you can type the command right after the word shell or the exclamation point. If you want to edit the file myfile.do, and if vi is the name of your favorite editor, you could type

```
. !vi myfile.do
```
        Stata opens your editor.
        When you exit your editor:
```
. _
```

Also See

[R] **query** — Display system parameters

[D] **cd** — Change directory

[D] **copy** — Copy file from disk or URL

[D] **dir** — Display filenames

[D] **erase** — Erase a disk file

[D] **mkdir** — Create directory

[D] **rmdir** — Remove directory

[D] **type** — Display contents of a file

Title

sort — Sort data

Syntax

<u>so</u>rt *varlist* $\left[\,in\,\right]$ $\left[\,,\,\text{stable}\right]$

Description

sort arranges the observations of the current data into ascending order based on the values of the variables in *varlist*. There is no limit to the number of variables in the *varlist*. Missing numeric values are interpreted as being larger than any other number, so they are placed last with $. < .a < .b \cdots < .z$. When you sort on a string variable, however, null strings are placed first. The dataset is marked as being sorted by *varlist* unless in *range* is specified. If in *range* is specified, only those observations are rearranged. The unspecified observations remain in the same place.

Option

stable specifies that observations with the same values of the variables in *varlist* keep the same relative order in the sorted data that they had previously. For instance, consider the following data:

```
x b
3 1
1 2
1 1
1 3
2 4
```

Typing sort x without the stable option produces one of the following six orderings:

x b	x b	x b	x b	x b	x b
1 2	1 2	1 1	1 1	1 3	1 3
1 1	1 3	1 3	1 2	1 1	1 2
1 3	1 1	1 2	1 3	1 2	1 1
2 4	2 4	2 4	2 4	2 4	2 4
3 1	3 1	3 1	3 1	3 1	3 1

Without the stable option, the ordering of observations with equal values of *varlist* is randomized. With sort x, stable, you will always get the first ordering and never the other five.

If your intent is to have the observations sorted first on x and then on b within tied values of x (the fourth ordering above), you should type sort x b rather than sort x, stable.

stable is seldom used and, when specified, causes sort to execute more slowly.

Remarks

Sorting data is one of the more common tasks involved in processing data. Sometimes, before Stata can perform some task, the data must be in a specific order. For example, if you want to use the by *varlist*: prefix, the data must be sorted in order of *varlist*. Match-merging two datasets with the merge command places similar requirements on both datasets. You use the sort command to fulfill these requirements.

▷ Example 1

Sorting data can also be informative. Suppose that we have data on automobiles, and each car's make and mileage rating (called make and mpg) are included among the variables in the data. We want to list the five cars with the lowest mileage rating in our data:

```
. use http://www.stata-press.com/data/r10/auto
(1978 Automobile Data)

. keep make mpg weight

. sort mpg, stable

. list make mpg in 1/5
```

	make	mpg
1.	Linc. Continental	12
2.	Linc. Mark V	12
3.	Cad. Deville	14
4.	Cad. Eldorado	14
5.	Linc. Versailles	14

◁

▷ Example 2: Tracking the sort order

Stata keeps track of the order of your data. For instance, we just sorted the above data on mpg. When we ask Stata to describe the data in memory, it tells us how the dataset is sorted:

```
. describe
Contains data from http://www.stata-press.com/data/r10/auto.dta
  obs:            74                          1978 Automobile Data
  vars:            3                          13 Apr 2007 17:45
  size:        1,924 (99.7% of memory free)  (_dta has notes)
```

variable name	storage type	display format	value label	variable label
make	str18	%-18s		Make and Model
mpg	int	%8.0g		Mileage (mpg)
weight	int	%8.0gc		Weight (lbs.)

```
Sorted by:  mpg
     Note:  dataset has changed since last saved
```

Stata keeps track of changes in sort order. If we were to make a change to the mpg variable, Stata would know that the data are no longer sorted. Remember that the first observation in our data has mpg equal to 12, as does the second. Let's change the value of the first observation:

```
. replace mpg=13 in 1
(1 real change made)
```

```
. describe
Contains data from http://www.stata-press.com/data/r10/auto.dta
  obs:            74                          1978 Automobile Data
  vars:            3                          13 Apr 2007 17:45
  size:        1,924 (99.7% of memory free)  (_dta has notes)
```

variable name	storage type	display format	value label	variable label
make	str18	%-18s		Make and Model
mpg	int	%8.0g		Mileage (mpg)
weight	int	%8.0gc		Weight (lbs.)

```
Sorted by:
    Note:  dataset has changed since last saved
```

After making the change, Stata indicates that our dataset is "Sorted by:" nothing. Let's put the dataset back as it was:

```
. replace mpg=12 in 1
(1 real change made)
. sort mpg                                                                   ◁
```

❏ Technical Note

Stata does not track changes in the sort order and will sometimes decide that a dataset is not sorted when, in fact, it is. For instance, if we were to change the first observation of our auto dataset from 12 miles per gallon to 10, Stata would decide that the dataset is "Sorted by:" nothing, just as it did above when we changed mpg from 12 to 13. Our change in example 2 did change the order of the data, so Stata was correct. Changing mpg from 12 to 10, however, does not really affect the sort order.

As far as Stata is concerned, any change to the variables on which the data are sorted means that the data are no longer sorted, even if the change actually leaves the order unchanged. Stata may be dumb, but it is also fast. It sorts already-sorted datasets instantly, so Stata's ignorance costs us little.

❏

▷ Example 3: Sorting on multiple variables

Data can be sorted by more than one variable, and in such cases, the sort order is lexicographic. If we sort the data by two variables, for instance, the data are placed in ascending order of the first variable, and then observations that share the same value of the first variable are placed in ascending order of the second variable. Let's order our automobile data by mpg and within mpg by weight:

(Continued on next page)

```
. sort mpg weight
. list in 1/8, sep(4)
```

	make	mpg	weight
1.	Linc. Mark V	12	4,720
2.	Linc. Continental	12	4,840
3.	Peugeot 604	14	3,420
4.	Linc. Versailles	14	3,830
5.	Cad. Eldorado	14	3,900
6.	Merc. Cougar	14	4,060
7.	Merc. XR-7	14	4,130
8.	Cad. Deville	14	4,330

The data are in ascending order of mpg, and, within each mpg category, the data are in ascending order of weight. The lightest car that achieves 14 miles per gallon in our data is the Peugeot 604.

◁

❏ Technical Note

The sorting technique used by Stata is fast, but the order of variables not included in the *varlist* is not maintained. If you wish to maintain the order of additional variables, include them at the end of the *varlist*. There is no limit to the number of variables by which you may sort.

❏

▷ Example 4: Descending sorts

Sometimes you may want to order a dataset by descending sequence of something. Perhaps we wish to obtain a list of the five cars achieving the best mileage rating. The sort command orders the data only into ascending sequences. Another command, gsort, orders the data in ascending or descending sequences; see [D] **gsort**. You can also create the negative of a variable and achieve the desired result:

```
. generate negmpg = -mpg
. sort negmpg
. list in 1/5
```

	make	mpg	weight	negmpg
1.	VW Diesel	41	2,040	-41
2.	Subaru	35	2,050	-35
3.	Datsun 210	35	2,020	-35
4.	Plym. Champ	34	1,800	-34
5.	Toyota Corolla	31	2,200	-31

We find that the VW Diesel tops our list.

◁

▷ Example 5: Sorting on string variables

sort may also be used on string variables. The data are sorted alphabetically:

. sort make

. list in 1/5

	make	mpg	weight	negmpg
1.	AMC Concord	22	2,930	-22
2.	AMC Pacer	17	3,350	-17
3.	AMC Spirit	22	2,640	-22
4.	Audi 5000	17	2,830	-17
5.	Audi Fox	23	2,070	-23

◁

❑ Technical Note

Bear in mind that Stata takes "alphabetically" to mean that all uppercase letters come before lowercase letters. As far as Stata is concerned, the following list is sorted alphabetically:

. list, sep(0)

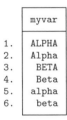

	myvar
1.	ALPHA
2.	Alpha
3.	BETA
4.	Beta
5.	alpha
6.	beta

❑

References

Royston, P. 2001. Sort a list of items. *Stata Journal* 1: 105–106.

Schumm, L. P. 2006. Stata tip 28: Precise control of dataset sort order. *Stata Journal* 6: 144–146.

Also See

[D] **describe** — Describe data in memory or in file

[D] **gsort** — Ascending and descending sort

[U] **11 Language syntax**

Title

---
split — Split string variables into parts

Syntax

split *strvar* $[$ *if* $]$ $[$ *in* $]$ $[$, *options* $]$

options	description
Main	
g̲enerate(*stub*)	begin new variable names with *stub*; default is *strvar*
p̲arse(*parse_strings*)	parse on specified strings; default is to parse on spaces
l̲imit(*#*)	create a maximum of *#* new variables
no̲trim	do not trim leading or trailing spaces of original variable
Destring	
destring	apply destring to new string variables, replacing initial string variables with numeric variables where possible
i̲gnore("*chars*")	remove specified nonnumeric characters
force	convert nonnumeric strings to missing values
float	generate numeric variables as type float
percent	convert percent variables to fractional form

Description

split splits the contents of a string variable *strvar* into one or more parts, using one or more *parse_strings* (by default, blank spaces), so that new string variables are generated. Thus split is useful for separating "words" or other parts of a string variable. *strvar* itself is not modified.

Options

◯─┐ Main └──

generate(*stub*) specifies the beginning characters of the new variable names so that new variables *stub*1, *stub*2, etc., are produced. *stub* defaults to *strvar*.

parse(*parse_strings*) specifies that, instead of using spaces, parsing use one or more *parse_strings*. Most commonly, one string that is one punctuation character will be specified. For example, if parse(,) is specified, "1,2,3" is split into "1", "2", and "3".

You can also specify (1) two or more strings that are alternative separators of "words" and (2) strings that consist of two or more characters. Alternative strings should be separated by spaces. Strings that include spaces should be bound by " ". Thus if parse(, " ") is specified, "1,2 3" is also split into "1", "2", and "3". Note particularly the difference between, say, parse(a b) and parse(ab): with the first, a and b are both acceptable as separators, whereas with the second, only the string ab is acceptable.

limit(*#*) specifies an upper limit to the number of new variables to be created. Thus limit(2) specifies that, at most, two new variables be created.

500

notrim specifies that the original string variable not be trimmed of leading and trailing spaces before being parsed. notrim is not compatible with parsing on spaces, as the latter implies that spaces in a string are to be discarded. You can either specify a parsing character or, by default, allow a trim.

⌐ Destring ⌐

destring applies destring to the new string variables, replacing the variables initially created as strings by numeric variables where possible. See [D] **destring**.

ignore(), force, float, percent; see [D] **destring**.

Remarks

split is used to split string variable into two or more component parts, for example, "words". You might need to correct a mistake, or the string variable might be a genuine composite that you wish to subdivide before doing more analysis.

The basic steps applied by split are, given one or more separators, to find those separators within the string and then to generate one or more new string variables, each containing a part of the original. The separators could be, for example, spaces or other punctuation symbols, but they can in turn be strings containing several characters. The default separator is a space.

The key string functions for subdividing string variables and, indeed, strings in general, are strpos(), which finds the position of separators, and substr(), which extracts parts of the string. (See [D] **functions**.) split is based on the use of those functions.

If your problem is not defined by splitting on separators, you will probably want to use substr() directly. Suppose that you have a string variable date containing dates in the form "21011952", so that the last four characters define a year. This string contains no separators. To extract the year, you would use substr(date,-4,4). Again suppose that each woman's obstetric history over the last 12 months was recorded by a str12 variable containing values such as "npppppppbnn", where p, b, and n denote months of pregnancy, birth, and nonpregnancy. Once more, there are no separators, so you would use substr() to subdivide the string.

split discards the separators, as it presumes that they are irrelevant to further analysis or that you could restore them at will. If this is not what you want, you might use substr() (and possibly strpos()).

Finally, before we turn to examples, compare split with the egen function ends(), which produces the head, the tail, or the last part of a string. This function, like all egen functions, produces just one new variable as a result. In contrast, split typically produces several new variables as the result of one command. For more details and discussion, including comments on the special problem of recognizing personal names, see [D] **egen**.

split can be useful when input to Stata is somehow misread as one string variable. If you copy and paste into the Data Editor, say, under Windows by using the clipboard, but data are space separated, what you regard as separate variables will be combined because the Data Editor expects comma- or tab-separated data. If some parts of your composite variable are numeric characters that should be put into numeric variables, you could use destring at the same time; see [D] **destring**.

```
. split var1, destring
```

Here no generate() option was specified, so the new variables will have names var11, var12, and so forth. You may now wish to use rename to produce more informative variable names. See [D] **rename**.

You can also use `split` to subdivide genuine composites. For example, email addresses such as `tech-support@stata.com` may be split at "@":

```
. split address, p(@)
```

This sequence yields two new variables: `address1`, containing the part of the email address before the "@", such as "tech-support", and `address2`, containing the part after the "@", such as "stata.com". The separator itself, "@", is discarded. Because `generate()` was not specified, the name `address` was used as a stub in naming the new variables. `split` displays the names of new variables created, so you will see quickly whether the number created matches your expectations.

If the details of individuals were of no interest and you wanted only machine names, either

```
. egen machinename = ends(address), tail p(@)
```

or

```
. generate machinename = substr(address, strpos(address,"@") + 1,.)
```

would be more direct.

Next, suppose that a string variable holds names of legal cases that should be split into variables for plaintiff and defendant. The separators could be " V ", " V. ", " VS ", and " VS. ". (We assume that any inconsistency in the use of uppercase and lowercase has been dealt with by the string function `upper()`; see [D] **functions**.) Note particularly the leading and trailing spaces in our detailing of separators: the first separator is " V ", for example, not "V", which would incorrectly split "GOLIATH V DAVID" into "GOLIATH ", " DA", and "ID". The alternative separators are given as the argument to `parse()`:

```
. split case, p(" V " " V. " " VS " " VS. ")
```

Again with default naming of variables and recalling that separators are discarded, we expect new variables `case1` and `case2`, with no creation of `case3` or further new variables. Whenever none of the separators specified were found, `case2` would have empty values, so we can check:

```
. list case if case2 == ""
```

Suppose that a string variable contains fields separated by tabs. For example, `insheet` leaves tabs unchanged. Knowing that a tab is `char(9)`, we can type

```
. split data, p(`=char(9)') destring
```

`p(char(9))` would not work. The argument to `parse()` is taken literally, but evaluation of functions on the fly can be forced as part of macro substitution.

Finally, suppose that a string variable contains substrings bound in parentheses, such as (1 2 3) (4 5 6). Here we can split on the right parentheses and, if desired, those afterward. For example,

```
. split data, p(")")
. foreach v in `r(varlist)' {
        replace `v' = `v' + ")"
. }
```

Saved Results

`split` saves the following in `r()`:

Scalars
    `r(nvars)`   number of new variables created
    `r(varlist)`   names of the newly created variables

Methods and Formulas

split is implemented as an ado-file.

Acknowledgments

split was written by Nicholas J. Cox, Durham University, who, in turn, thanks Michael Blasnik, M. Blasnick & Associates, for ideas contributed to an earlier jointly written program.

Also See

[D] **rename** — Rename variable

[D] **separate** — Create separate variables

[D] **destring** — Convert string variables to numeric variables and vice versa

[D] **egen** — Extensions to generate

[D] **functions** — Functions

Title

stack — Stack data

Syntax

stack *varlist* [*if*] [*in*], { into(*newvars*) | group(*#*) } [*options*]

options	description
Main	
* into(*newvars*)	identify names of new variables to be created
* group(*#*)	stack *#* groups of variables in *varlist*
clear	clear dataset from memory
wide	keep variables in *varlist* that are not specified in *newvars*

* Either into(*newvars*) or group(*#*) is required.

Description

stack stacks the variables in *varlist* vertically, resulting in a dataset with variables *newvars* and $\_N \cdot (N_v/N_n)$ observations, where N_v is the number of variables in *varlist* and N_n is the number in *newvars*. stack creates the new variable _stack identifying the groups.

Options

<u>Main</u>

into(*newvars*) identifies the names of the new variables to be created. into() may be specified using variable ranges (e.g., into(v1-v3)). Either into() or group(), but not both, must be specified.

group(*#*) specifies the number of groups of variables in *varlist* to be stacked. The created variables will be named according to the first group in *varlist*. Either into() or group(), but not both, must be specified.

clear indicates that it is okay to clear the dataset in memory. If you do not specify this option, you will be asked to confirm your intentions.

wide includes any of the original variables in *varlist* that are not specified in *newvars* in the resulting data.

Remarks

▷ Example 1

This command is best understood by examples. We begin with artificial but informative examples and end with useful examples:

```
. use http://www.stata-press.com/data/r10/stackxmpl
. list
```

	a	b	c	d
1.	1	2	3	4
2.	5	6	7	8

```
. stack  a b  c d, into(e f) clear
. list
```

	_stack	e	f
1.	1	1	2
2.	1	5	6
3.	2	3	4
4.	2	7	8

We formed the new variable e by stacking a and c, and we formed the new variable f by stacking b and d. _stack is automatically created and set equal to 1 for the first (a, b) group and equal to 2 for the second (c, d) group. (When _stack==1, the new data e and f contain the values from a and b. When _stack==2, e and f contain values from c and d.)

There are two groups because we specified four variables in the *varlist* and two variables in the into list, and $4/2 = 2$. If there were six variables in the *varlist*, there would be $6/2 = 3$ groups. If there were also three variables in the into list, there would be $6/3 = 2$ groups. Specifying six variables in the *varlist* and four variables in the into list would result in an error because $6/4$ is not an integer.

◁

▷ Example 2

Variables may be repeated in the *varlist*, and the *varlist* need not contain all the variables:

```
. use http://www.stata-press.com/data/r10/stackxmpl, clear
. list
```

	a	b	c	d
1.	1	2	3	4
2.	5	6	7	8

```
. stack  a b  a c, into(a bc) clear
. list
```

	_stack	a	bc
1.	1	1	2
2.	1	5	6
3.	2	1	3
4.	2	5	7

a was stacked on a and called a, whereas b was stacked on c and called bc.

If we had wanted the resulting variables to be called simply a and b, we could have used

```
. stack  a b  a c, group(2) clear
```

which is equivalent to

```
. stack  a b  a c, into(a b) clear
```

◁

▷ Example 3

In this artificial but informative example, the `wide` option includes the variables in the original dataset that were specified in *varlist* in the output dataset:

```
. use http://www.stata-press.com/data/r10/stackxmpl, clear
. list
```

	a	b	c	d
1.	1	2	3	4
2.	5	6	7	8

```
. stack  a b  c d, into(e f) clear wide
. list
```

	_stack	e	f	a	b	c	d
1.	1	1	2	1	2	.	.
2.	1	5	6	5	6	.	.
3.	2	3	4	.	.	3	4
4.	2	7	8	.	.	7	8

In addition to the stacked e and f variables, the original a, b, c, and d variables are included. They are set to missing where their values are not appropriate.

◁

▷ Example 4

This is the last artificial example. When you specify the `wide` option and repeat the same variable name in both the *varlist* and the `into` list, the variable will contain the stacked values:

```
. use http://www.stata-press.com/data/r10/stackxmpl, clear
. list
```

	a	b	c	d
1.	1	2	3	4
2.	5	6	7	8

```
. stack a b  a c, into(a bc) clear wide
. list
```

	_stack	a	bc	b	c
1.	1	1	2	2	.
2.	1	5	6	6	.
3.	2	1	3	.	3
4.	2	5	7	.	7

◁

▷ Example 5

We want one graph of y against x1 and y against x2. We might be tempted to type scatter y x1 x2, but that would graph y against x2 and x1 against x2. One solution is to type

```
. save mydata
. stack  y x1  y x2, into(yy x12) clear
. generate y1 = yy if _stack==1
. generate y2 = yy if _stack==2
. scatter y1 y2 x12
. use mydata, clear
```

The names yy and x12 are supposed to suggest the contents of the variables. yy contains (y,y), and x12 contains (x1,x2). We then make y1 defined at the x1 points but missing at the x2 points—graphing y1 against x12 is the same as graphing y against x1 in the original dataset. Similarly, y2 is defined at the x2 points but missing at x1—graphing y2 against x12 is the same as graphing y against x2 in the original dataset. Therefore, scatter y1 y2 x12 produces the desired graph.

◁

▷ Example 6

We wish to graph y1 against x1 and y2 against x2 on the same graph. The logic is the same as above, but let's go through it. Perhaps we have constructed two cumulative distributions by using cumul (see [R] **cumul**):

```
. use http://www.stata-press.com/data/r10/citytemp
(City Temperature Data)
. cumul tempjan, gen(cjan)
. cumul tempjuly, gen(cjuly)
```

We want to graph both cumulatives in the same graph; that is, we want to graph cjan against tempjan and cjuly against tempjuly. Remember that we could graph the tempjan cumulative by typing

```
. scatter cjan tempjan, c(l) m(o) sort
  (output omitted )
```

We can graph the tempjuly cumulative similarly. To obtain both on the same graph, we must stack the data:

```
. stack  cjuly tempjuly  cjan tempjan, into(c temp) clear
. generate cjan  = c if _stack==1
(958 missing values generated)
. generate cjuly = c if _stack==2
(958 missing values generated)
. scatter cjan cjuly temp, c(l l) m(o o) sort
  (output omitted )
```

Alternatively, if we specify the wide option, we do not have to regenerate cjan and cjuly because they will be created automatically:

```
. use http://www.stata-press.com/data/r10/citytemp, clear
(City Temperature Data)
. cumul tempjan, gen(cjan)
. cumul tempjuly, gen(cjuly)
. stack  cjuly tempjuly  cjan tempjan, into(c temp) clear wide
```

```
. scatter cjan cjuly temp, c(l l) m(o o) sort
(output omitted )
```

◁

❏ Technical Note

There is a third way, not using the `wide` option, that is exceedingly tricky but is sometimes useful:

```
. use http://www.stata-press.com/data/r10/citytemp, clear
(City Temperature Data)
. cumul tempjan, gen(cjan)
. cumul tempjuly, gen(cjuly)
. stack cjuly tempjuly  cjan tempjan, into(c temp) clear
. sort _stack temp
. scatter c temp, c(L) m(o)
(output omitted )
```

Note the use of `connect`'s capital L rather than lowercase l option. `c(L)` connects points only from left to right; because the data are sorted by `_stack temp`, `temp` increases within the first group (`cjuly` vs. `tempjuly`) and then starts again for the second (`cjan` vs. `tempjan`); see [G] *connectstyle*.

❏

Methods and Formulas

`stack` is implemented as an ado-file.

Also See

[D] **contract** — Make dataset of frequencies and percentages

[D] **reshape** — Convert data from wide to long form and vice versa

[D] **xpose** — Interchange observations and variables

Title

> **statsby** — Collect statistics for a command across a by list

Syntax

statsby [*exp_list*] [, *options*]: *command*

options	description
Main	
*by(*varlist* [, missing])	equivalent to interactive use of **by** *varlist*:
Options	
clear	replace data in memory with results
saving(*filename*, ...)	save results to *filename*; save statistics in double precision; save results to *filename* every # replications
total	include results for the entire dataset
subsets	include all combinations of subsets of groups
Reporting	
nodots	suppress the replication dots
noisily	display any output from *command*
trace	trace the *command*
nolegend	suppress table legend
verbose	display the full table legend
Advanced	
basepop(*exp*)	restrict initializing sample to *exp*; seldom used
force	do not check for **svy** commands; seldom used
forcedrop	retain only observations in **by()** groups when calling *command*; seldom used

* by(*varlist*) is required on the dialog box because **statsby** is useful to the interactive user only when using **by()**. All weight types supported by *command* are allowed except **pweights**; see [U] **11.1.6 weight**.

exp_list contains	(*name*: *elist*)
	elist
	eexp
elist contains	*newvarname* = (*exp*)
	(*exp*)
eexp is	*specname*
	[*eqno*]*specname*
specname is	_b
	_b[]
	_se
	_se[]
eqno is	##
	name

509

exp is a standard Stata expression; see [U] **13 Functions and expressions**.

Distinguish between [ ], which are to be typed, and [ ], which indicate optional arguments.

Description

statsby collects statistics from *command* across a by() list. Typing

. statsby *exp_list*, by(*varname*): *command*

executes *command* for each group identified by *varname*, building a dataset of the associated values from the expressions in *exp_list*. The resulting dataset replaces the current dataset, unless the saving() option is supplied. *varname* can refer to a numeric or a string variable.

command defines the statistical command to be executed. Most Stata commands and user-written programs can be used with statsby, as long as they follow standard Stata syntax and allow the *if* qualifier; see [U] **11 Language syntax**. The by prefix cannot be part of *command*.

exp_list specifies the statistics to be collected from the execution of *command*. If no expressions are given, *exp_list* assumes a default depending upon whether *command* changes results in e() and r(). If *command* changes results in e(), the default is _b. If *command* changes results in r() (but not e()), the default is all the scalars posted to r(). It is an error not to specify an expression in *exp_list* otherwise.

Options

    ⌐ Main ⌐

by(*varlist* [, missing]) specifies a list of existing variables that would normally appear in the by *varlist*: section of the command if you were to issue the command interactively. By default, statsby ignores groups in which one or more of the by() variables is missing. Alternatively, missing causes missing values to be treated like any other values in the by() groups, and results from the entire dataset are included with use of the subsets option. If by() is not specified, *command* will be run on the entire dataset. *varlist* can contain both numeric and string variables.

    ⌐ Options ⌐

clear specifies that it is okay to replace the data in memory, even though the current data have not been saved to disk.

saving(*filename* [, *suboptions*]) creates a Stata data file (.dta file) consisting of, for each statistic in *exp_list*, a variable containing the replicates.

    double specifies that the results for each replication be stored as doubles, meaning 8-byte reals. By default, they are stored as floats, meaning 4-byte reals.

    every(#) specifies that results be written to disk every #th replication. every() should be specified in conjunction with saving() only when *command* takes a long time for each replication. This will allow recovery of partial results should your computer crash. See [P] **postfile**.

total specifies that *command* be run on the entire dataset, in addition to the groups specified in the by() option.

subsets specifies that *command* be run for each group defined by any combination of the variables in the by() option.

┌─ Reporting ┐

`nodots` suppresses display of the replication dots. By default, one dot character is printed for each `by()` group. A red 'x' is printed if *command* returns with an error or if one of the values in *exp_list* is missing.

`noisily` causes the output of *command* to be displayed for each `by` group. This option implies the `nodots` option.

`trace` causes a trace of the execution of *command* to be displayed. This option implies the `noisily` option.

`nolegend` suppresses the display of the table legend, which identifies the rows of the table with the expressions they represent.

`verbose` requests that the full table legend be displayed. By default, coefficients and standard errors are not displayed.

┌─ Advanced ┐

`basepop(`*exp*`)` specifies a base population that `statsby` uses to evaluate the *command* and to set up for collecting statistics. The default base population is the entire dataset, or the dataset specified by any *if* or *in* conditions specified on the *command*.

One situation where `basepop()` is useful is collecting statistics over the panels of a panel dataset by using an estimator that works for time series, but not panel data, e.g.,

> `. statsby, by(mypanels) basepop(mypanels==2): arima ...`

`force` suppresses the restriction that *command* not be a `svy` command. `statsby` does not perform subpopulation estimation for survey data, so it should not be used with `svy`. `statsby` reports an error when it encounters `svy` in *command* if the `force` option is not specified. This option is seldom used, so use it only if you know what you are doing.

`forcedrop` forces `statsby` to drop all observations except those in each `by()` group before calling *command* for the group. This allows `statsby` to work with user-written commands that completely ignore *if* and *in* but do not return an error when either is specified. `forcedrop` is seldom used.

Remarks

Remarks are presented under the following headings:

> *Collecting coefficients and standard errors*
> *Collecting saved results*
> *All subsets*

Collecting coefficients and standard errors

▷ Example 1

We begin with an example using the `auto.dta` dataset. In this example, we want to collect the coefficients from a regression in which we model the price of a car on its weight, length, and mpg. We want to run this model for both domestic and foreign cars. We can do this easily by using `statsby` with the extended expression _b.

```
. use http://www.stata-press.com/data/r10/auto
(1978 Automobile Data)
. statsby _b, by(foreign) verbose nodots: regress price weight length mpg
        command:  regress price weight length mpg
      _b_weight:  _b[weight]
      _b_length:  _b[length]
         _b_mpg:  _b[mpg]
        _b_cons:  _b[_cons]
             by:  foreign
. list
```

	foreign	_b_wei~t	_b_length	_b_mpg	_b_cons
1.	Domestic	6.767233	-109.9518	142.7663	2359.475
2.	Foreign	4.784841	13.39052	-18.4072	-6497.49

If we were interested only in the coefficient of a particular variable, such as mpg, we would specify that particular coefficient; see [U] **13.5 Accessing coefficients and standard errors**.

```
. use http://www.stata-press.com/data/r10/auto, clear
(1978 Automobile Data)
. statsby mpg=_b[mpg], by(foreign) nodots: regress price weight length mpg
        command:  regress price weight length mpg
            mpg:  _b[mpg]
             by:  foreign
. list
```

	foreign	mpg
1.	Domestic	142.7663
2.	Foreign	-18.4072

The extended expression _se indicates that we want standard errors.

```
. use http://www.stata-press.com/data/r10/auto, clear
(1978 Automobile Data)
. statsby _se, by(foreign) verbose nodots: regress price weight length mpg
        command:  regress price weight length mpg
     _se_weight:  _se[weight]
     _se_length:  _se[length]
        _se_mpg:  _se[mpg]
       _se_cons:  _se[_cons]
             by:  foreign
. list
```

	foreign	_se_we~t	_se_le~h	_se_mpg	_se_cons
1.	Domestic	1.226326	39.48193	134.7221	7770.131
2.	Foreign	1.670006	50.70229	59.37442	6337.952

◁

▷ Example 2

For multiple-equation estimations, we can use [*eqno*]_b ([*eqno*]_se) to get the coefficients (standard errors) of a specific equation or use _b (_se) to get the coefficients (standard errors) of all the equations. To demonstrate, we use heckman and a slightly different dataset.

```
. use http://www.stata-press.com/data/r10/statsby, clear
. statsby _b, by(group) verbose nodots: heckman price mpg, sel(trunk)
        command:  heckman price mpg, sel(trunk)
    price_b_mpg:  [price]_b[mpg]
   price_b_cons:  [price]_b[_cons]
  select_b_tr~k:  [select]_b[trunk]
   select_b_cons:  [select]_b[_cons]
   athrho_b_cons:  [athrho]_b[_cons]
   lnsigma_b_c~s:  [lnsigma]_b[_cons]
             by:  group
. list, compress noobs
```

group	price_b~g	price_~s	select_~k	select~s	athrho_~s	lnsigm~s
1	-253.9293	11836.33	-.0122223	1.248342	-.31078	7.895351
2	-242.5759	11906.46	-.0488969	1.943078	-1.399222	8.000272
3	-172.6499	9813.357	-.0190373	1.452783	-.3282423	7.876059
4	-250.7318	10677.31	.0525965	.3502012	.6133645	7.96349

To collect the coefficients of the first equation only, we would specify [price]_b instead of _b.

```
. use http://www.stata-press.com/data/r10/statsby, clear
. statsby [price]_b, by(group) verbose nodots: heckman price mpg, sel(trunk)
        command:  heckman price mpg , sel(trunk)
    price_b_mpg:  [price]_b[mpg]
   price_b_cons:  [price]_b[_cons]
             by:  group
. list
```

	group	price_b~g	price_~s
1.	1	-253.9293	11836.33
2.	2	-242.5759	11906.46
3.	3	-172.6499	9813.357
4.	4	-250.7318	10677.31

◁

❑ Technical Note

If *command* fails on one or more groups, statsby will capture the error messages and ignore those groups.

❑

(*Continued on next page*)

Collecting saved results

Many Stata commands save results of calculations; see [U] **13.6 Accessing results from Stata commands**. statsby can collect the saved results and expressions involving these saved results, too. Expressions must be bound in parentheses.

▷ Example 3

Suppose that we want to collect the mean and the median of price, as well as their ratios, and we want to collect them for both domestic and foreign cars. We might type

```
. use http://www.stata-press.com/data/r10/auto, clear
(1978 Automobile Data)
. statsby mean=r(mean) median=r(p50) ratio=(r(mean)/r(p50)), by(foreign) nodots:
> summarize price, detail
      command:  summarize price, detail
         mean:  r(mean)
       median:  r(p50)
        ratio:  r(mean)/r(p50)
           by:  foreign
. list
```

	foreign	mean	median	ratio
1.	Domestic	6072.423	4782.5	1.269717
2.	Foreign	6384.682	5759	1.108644

◁

❏ Technical Note

In *exp_list*, *newvarname* is not required. If no new variable name is specified, statsby names the new variables _stat_1, _stat_2, and so forth.

❏

All subsets

▷ Example 4

When there are two or more variables in the by(*varlist*), we can execute *command* for any combination, or subset, of the variables in the by() option by specifying the subsets option.

```
. use http://www.stata-press.com/data/r10/auto, clear
(1978 Automobile Data)
. statsby mean=r(mean) median=r(p50) n=r(N), by(foreign rep78) subsets nodots:
> summarize price, detail
      command:  summarize price, detail
         mean:  r(mean)
       median:  r(p50)
            n:  r(N)
           by:  foreign rep78
```

```
. list
```

	foreign	rep78	mean	median	n
1.	Domestic	1	4564.5	4564.5	2
2.	Domestic	2	5967.625	4638	8
3.	Domestic	3	6607.074	4749	27
4.	Domestic	4	5881.556	5705	9
5.	Domestic	5	4204.5	4204.5	2
6.	Domestic	.	6179.25	4853	48
7.	Foreign	3	4828.667	4296	3
8.	Foreign	4	6261.444	6229	9
9.	Foreign	5	6292.667	5719	9
10.	Foreign	.	6070.143	5719	21
11.	.	1	4564.5	4564.5	2
12.	.	2	5967.625	4638	8
13.	.	3	6429.233	4741	30
14.	.	4	6071.5	5751.5	18
15.	.	5	5913	5397	11
16.	.	.	6165.257	5006.5	74

In the above dataset, observation 6 is for domestic cars, regardless of the repair record; observation 10 is for foreign cars, regardless of the repair record; observation 11 is for both foreign cars and domestic cars given that the repair record is 1; and the last observation is for the entire dataset.

◁

❏ Technical Note

To see the output from *command* for each group identified in the by() option, we can use the noisily option.

```
. use http://www.stata-press.com/data/r10/auto, clear
(1978 Automobile Data)
. statsby mean=r(mean) se=(r(sd)/sqrt(r(N))), by(foreign) noisily nodots:
> summarize price
statsby: First call to summarize with data as is:
. summarize price
```

Variable	Obs	Mean	Std. Dev.	Min	Max
price	74	6165.257	2949.496	3291	15906

```
statsby legend:
    command:  summarize price
       mean:  r(mean)
         se:  r(sd)/sqrt(r(N))
         by:  foreign
Statsby groups
running (summarize price) on group 1
. summarize price
```

Variable	Obs	Mean	Std. Dev.	Min	Max
price	52	6072.423	3097.104	3291	15906

```
running (summarize price) on group 2
```

```
. summarize price
    Variable |       Obs        Mean   Std. Dev.       Min        Max
       price |        22    6384.682    2621.915       3748      12990
. list
```

	foreign	mean	se
1.	Domestic	6072.423	429.4911
2.	Foreign	6384.682	558.9942

Methods and Formulas

statsby is implemented as an ado-file.

Acknowledgment

Speed improvements in statsby were based on code written by Michael Blasnik, MBlasnik & Associates.

References

Hardin, J. W. 1996. dm42: Accrue statistics for a command across a by list. *Stata Technical Bulletin* 32: 5–9. Reprinted in *Stata Technical Bulletin Reprints*, vol. 6, pp. 13–18.

Newson, R. 1999. dm65: A program for saving a model fit as a dataset. *Stata Technical Bulletin* 49: 2–5. Reprinted in *Stata Technical Bulletin Reprints*, vol. 9, pp. 19–23.

——. 2000. dm65.1: Update to a program for saving a model fit as a dataset. *Stata Technical Bulletin* 58: 25. Reprinted in *Stata Technical Bulletin Reprints*, vol. 10, p. 7.

Also See

[P] **postfile** — Save results in Stata dataset

[D] **collapse** — Make dataset of summary statistics

[R] **bootstrap** — Bootstrap sampling and estimation

[R] **jackknife** — Jackknife estimation

[R] **permute** — Monte Carlo permutation tests

[D] **by** — Repeat Stata command on subsets of the data

Title

> **sysuse** — Use shipped dataset

Syntax

Use example dataset installed with Stata

> sysuse ["]*filename*["] [, clear]

List example Stata datasets installed with Stata

> sysuse dir [, all]

Description

sysuse *filename* loads the specified Stata-format dataset that was shipped with Stata or that is stored along the adopath. If *filename* is specified without a suffix, .dta is assumed.

sysuse dir lists the names of the datasets shipped with Stata plus any other datasets stored along the adopath.

Options

clear specifies that it is okay to replace the data in memory, even though the current data have not been saved to disk.

all specifies that all datasets be listed, even those that include an underscore (_) in their name. By default, such datasets are not listed.

Remarks

Remarks are presented under the following headings:

> *Typical use*
> *A note concerning shipped datasets*
> *Using user-installed datasets*
> *How sysuse works*

Typical use

A few datasets are included with Stata and are stored in the system directories. These datasets are often used in the help files to demonstrate a certain feature.

Typing

```
. sysuse dir
```

lists the names of those datasets. One such dataset is lifeexp.dta. If you simply type use lifeexp, you will see

```
. use lifeexp
file lifeexp.dta not found
r(601);
```

Type `sysuse`, however, and the dataset is loaded:

```
. sysuse lifeexp
(Life expectancy, 1998)
```

The datasets shipped with Stata are stored in different folders (directories) so that they do not become confused with your datasets.

A note concerning shipped datasets

Not all the datasets used in the manuals are shipped with Stata. To obtain the other datasets, see [D] **webuse**.

The datasets used to demonstrate Stata are often fictional. If you want to know whether a dataset is real or fictional, and its history, load the dataset and type

```
. notes
```

A few datasets have no notes. This means that the datasets are believed to be real, but that they were created so long ago that information about their original source has been lost. Treat such datasets as if they were fictional.

Using user-installed datasets

Any datasets you have installed using `net` or `ssc` (see [R] **net** and [R] **ssc**) can be listed by typing `sysuse dir` and can be loaded using `sysuse` *filename*.

Any datasets you store in your personal ado folder (see [P] **sysdir**) are also listed by `sysuse dir` and can be loaded using `sysuse` *filename*.

How sysuse works

`sysuse` simply looks across the adopath for `.dta` files; see [P] **sysdir**.

By default, `sysuse dir` does not list a dataset that contains an underscore (_) in its name. By convention, such datasets are used by ado-files to achieve their ends and probably are not of interest to you. If you type `sysuse dir, all` all datasets are listed.

Saved Results

`sysuse dir` saves in the macro `r(files)` the list of dataset names.

`sysuse` *filename* saves in the macro `r(fn)` the *filename*, including the full path specification.

Methods and Formulas

`sysuse` is implemented as an ado-file.

Also See

[D] **webuse** — Use dataset from Stata web site

[D] **use** — Use Stata dataset

[P] **findfile** — Find file in path

[P] **sysdir** — Query and set system directories

[R] **net** — Install and manage user-written additions from the Internet

[R] **ssc** — Install and uninstall packages from SSC

Title

type — Display contents of a file

Syntax

type ["]*filename*["] [, *options*]

Note: double quotes must be used to enclose *filename* if the name contains blanks.

options	description
asis	show file as is; default is to display files with suffixes .smcl or .sthlp as SMCL
smcl	display file as SMCL; default for files with suffixes .smcl or .sthlp
showtabs	display tabs as <T> rather than being expanded
starbang	list lines in the file that begin with "*!"

Description

type lists the contents of a file stored on disk. This command is similar to the Windows command window type and Unix more(1) or pg(1) commands.

In Stata for Unix, cat is a synonym for type.

Options

asis specifies that the file be shown exactly as it is. The default is to display files with suffixes .smcl or .sthlp as SMCL, meaning that the SMCL directives are interpreted and properly rendered. Thus type can be used to look at files created by the log using command.

smcl specifies that the file be displayed as SMCL, meaning that the SMCL directives are interpreted and properly rendered. This is the default for files with suffixes .smcl or .sthlp.

showtabs requests that any tabs be displayed as <T> rather than being expanded.

starbang lists only the lines in the specified file that begin with the characters "*!". Such comment lines are typically used to indicate the version number of ado-files, class files, etc. starbang may not be used with SMCL files.

Remarks

▷ Example 1

We have raw data containing the level of Lake Victoria Nyanza and the number of sunspots during the years 1902–1921 stored in a file called sunspots.raw. We want to read this dataset into Stata by using infile, but we cannot remember the order in which we entered the variables. We can find out by using the type command:

```
. type sunspots.raw
1902 -10   5    1903  13 24    1904  18 42
1905  15  63    1906  29 54    1907  21 62
1908  10  49    1909   8 44    1910   1 19
1911  -7   6    1912 -11  4    1913  -3  1
1914  -2  10    1915   4 47    1916  15 57
1917  35 104    1918  27 81    1919   8 64
1920   3  38    1921  -5 25
```

Looking at this output, we now remember that the variables are entered year, level, and number of sunspots. We can read this dataset by typing `infile year level spots using sunspots`.

If we wanted to see the tabs in `sunspots.raw`, we could type

```
. type sunspots.raw, showtabs
1902 -10    5<T>1903  13 24<T>1904  18 42
1905  15   63<T>1906  29 54<T>1907  21 62
1908  10   49<T>1909   8 44<T>1910   1 19
1911  -7    6<T>1912 -11  4<T>1913  -3  1
1914  -2   10<T>1915   4 47<T>1916  15 57
1917  35  104<T>1918  27 81<T>1919   8 64
1920   3   38<T>1921  -5 25
```

◁

▷ Example 2

In a previous Stata session, we typed `log using myres` and created `myres.smcl`, containing our results. We can use `type` to list the log:

```
. type myres.smcl

      log:  /work/peb/dof/myres.smcl
 log type:  smcl
opened on:  20 Jan 2007, 15:37:48
. use lbw
(Hosmer & Lemeshow data)
. xi: logistic low age lwt i.race smoke ptl ht ui
i.race          _Irace_1-3        (naturally coded; _Irace_1 omitted)
Logistic regression                      Number of obs  =        189
                                         LR chi2(8)     =      33.22
                                         Prob > chi2    =     0.0001
Log likelihood =   -100.724              Pseudo R2      =     0.1416
  (output omitted )
. estat gof
Logistic model for low, goodness-of-fit test
  (output omitted )
. log close
      log:  /work/peb/dof/myres.smcl
 log type:  smcl
closed on:  20 Jan 2007, 15:38:30
```

We could also use `view` to look at the log; see [R] **view**.

◁

Also See

[R] **translate** — Print and translate logs

[R] **view** — View files and logs

[D] **cd** — Change directory

[D] **copy** — Copy file from disk or URL

[D] **dir** — Display filenames

[D] **erase** — Erase a disk file

[D] **mkdir** — Create directory

[D] **rmdir** — Remove directory

[D] **shell** — Temporarily invoke operating system

[U] **11.6 File-naming conventions**

Title

> **use** — Use Stata dataset

Syntax

Load Stata-format dataset

> <u>use</u> *filename* [, clear <u>nol</u>abel]

Load subset of Stata-format dataset

> <u>use</u> [*varlist*] [*if*] [*in*] using *filename* [, clear <u>nol</u>abel]

Description

use loads a Stata-format dataset previously saved by save into memory. If *filename* is specified without an extension, .dta is assumed. If your *filename* contains embedded spaces, remember to enclose it in double quotes.

In the second syntax for use, a subset of the data may be read.

Options

clear specifies that it is okay to replace the data in memory, even though the current data have not been saved to disk.

nolabel prevents value labels in the saved data from being loaded. It is unlikely that you will ever want to specify this option.

Remarks

▷ Example 1

We have no data in memory. In a previous session, we issued the command save hiway to save the Minnesota Highway Data that we had been analyzing. We retrieve it now:

```
. use hiway
(Minnesota Highway Data, 1973)
```

Stata loads the data into memory and shows us that the dataset is labeled "Minnesota Highway Data, 1973".

◁

▷ Example 2

We continue to work with our `hiway` data and find an error in our data that needs correcting:

```
. replace spdlimit=70 in 1
(1 real change made)
```

We remember that we need to forward some information from another dataset to a colleague. We use that other dataset:

```
. use accident
no; data in memory would be lost
r(4);
```

Stata refuses to load the data because we have not saved the `hiway` data since we changed it.

```
. save hiway, replace
file hiway.dta saved
. use accident
(Minnesota Accident Data)
```

After we save our `hiway` data, Stata lets us load our `accident` dataset. If we had not cared whether our changed `hiway` dataset were saved, we could have typed `use accident, clear` to tell Stata to load the accident data without saving the changed dataset in memory.

◁

❏ Technical Note

In example 2, you saved a revised `hiway.dta` dataset, which you forward to your colleague. Your colleague issues the command

```
. use hiway
```

and gets the message

```
file hiway.dta not Stata format
r(610);
```

Your colleague is using a version of Stata older than Stata 8. If your colleague is using Stata 7, you can save the dataset in Stata 7 format by using the `saveold` command; see [D] **save**.

Newer versions of Stata can always read datasets created by older versions of Stata. Stata/MP and Stata/SE can read datasets created by Stata/IC. Stata/IC can read datasets created by Stata/MP and Stata/SE if those datasets conform to Stata/IC's limits; type `help limits`.

❏

❏ Technical Note

If you are working with really large datasets, you might use your data and get the following result:

```
. use nlswork
insufficient memory
r(950);
```

If this occurs, you must increase the amount of memory allocated to Stata; see [U] **6 Setting the size of memory**.

❏

▷ Example 3

In the previous technical note, the dataset we were trying to use was too large for the memory we had allocated, so we had to increase the amount of memory. An alternative would be to load only some of the variables:

```
. use ln_wage grade age tenure race using nlswork
(National Longitudinal Survey.  Young Women 14-26 years of age in 1968)
. describe
Contains data from nlswork.dta
  obs:        28,534                        National Longitudinal Survey.
                                            Young Women 14-26 years of age
                                            in 1968
  vars:            5                        7 Dec 2006 17:02
  size:      428,010 (89.8% of memory free)

              storage  display    value
variable name   type   format     label     variable label

age             byte   %8.0g                age in current year
race            byte   %8.0g                1=white, 2=black, 3=other
grade           byte   %8.0g                current grade completed
tenure          float  %9.0g                job tenure, in years
ln_wage         float  %9.0g                ln(wage/GNP deflator)

Sorted by:
```

Stata successfully loaded the five variables.

◁

▷ Example 4

You are new to Stata and want to try working with a Stata dataset that was used in an example in [XT] **xtlogit**. You load the dataset:

```
. use http://www.stata-press.com/data/r10/union
(NLS Women 14-24 in 1968)
```

The dataset is successfully loaded, but it would have been shorter to type

```
. webuse union
(NLS Women 14-24 in 1968)
```

webuse is a synonym for use http://www.stata-press.com/data/r10/; see [D] **webuse**.

◁

Also See

[D] **compress** — Compress data in memory

[D] **datasignature** — Determine whether data have changed

[D] **fdasave** — Save and use datasets in FDA (SAS XPORT) format

[D] **infile (fixed format)** — Read ASCII (text) data in fixed format with a dictionary

[D] **infile (free format)** — Read unformatted ASCII (text) data

[D] **infix (fixed format)** — Read ASCII (text) data in fixed format

[D] **insheet** — Read ASCII (text) data created by a spreadsheet

[D] **odbc** — Load, write, or view data from ODBC sources

[D] **save** — Save datasets

[D] **sysuse** — Use shipped dataset

[D] **webuse** — Use dataset from Stata web site

[TS] **haver** — Load data from Haver Analytics database

[U] **11.6 File-naming conventions**

[U] **21 Inputting data**

Title

> **webuse** — Use dataset from Stata web site

Syntax

Load dataset over the web

> webuse $\left[" \right]$ *filename* $\left[" \right]$ [, clear]

Report URL from which datasets will be obtained

> webuse query

Specifying URL from which dataset will be obtained

> webuse set $\left[\text{http://} \right]$ *url* $\left[/ \right]$

Reset URL to default

> webuse set

Description

webuse *filename* loads the specified dataset, obtaining it over the web. By default, datasets are obtained from http://www.stata-press.com/data/r10/. If *filename* is specified without a suffix, .dta is assumed.

webuse query reports the URL from which datasets will be obtained.

webuse set allows you to specify the URL to be used as the source for datasets. webuse set without arguments resets the source to http://www.stata-press.com/data/r10/.

Option

clear specifies that it is okay to replace the data in memory, even though the current data have not been saved to disk.

Remarks

Remarks are presented under the following headings:

> Typical use
> A note concerning example datasets
> Redirecting the source

Typical use

In the examples in the Stata manuals, we see things such as

> . use http://www.stata-press.com/data/r10/lifeexp

The above is used to load—in this instance—the dataset `lifeexp.dta`. You can type that, and it will work:

```
. use http://www.stata-press.com/data/r10/lifeexp
(Life expectancy, 1998)
```

Or you may simply type

```
. webuse lifeexp
(Life expectancy, 1998)
```

`webuse` is a synonym for use `http://www.stata-press.com/data/r10/`.

A note concerning example datasets

The datasets used to demonstrate Stata are often fictional. If you want to know whether a dataset is real or fictional, and its history, load the dataset and type

```
. notes
```

A few datasets have no notes. This means that the datasets are believed to be real but that they were created so long ago that information about their original source has been lost. Treat such datasets as if they were fictional.

Redirecting the source

By default, `webuse` obtains datasets from http://www.stata-press.com/data/r10/, but you can change that. Say that the site http://www.zzz.edu/users/sue/ has several datasets that you wish to explore. You can type

```
. webuse set http://www.zzz.edu/users/~sue
```

`webuse` will become a synonym for use `http://www.zzz.edu/users/~sue/` for the rest of the session or until you give another `webuse` command.

When you set the URL, you may omit the trailing slash (as we did above), or you may include it:

```
. webuse set http://www.zzz.edu/users/~sue/
```

You may also omit `http://`:

```
. webuse set www.zzz.edu/users/~sue
```

If you type `webuse set` without arguments, the URL will be reset to the default, http://www.stata-press.com/data/r10/:

```
. webuse set
```

Methods and Formulas

`webuse` is implemented as an ado-file.

Also See

[D] **sysuse** — Use shipped dataset

[D] **use** — Use Stata dataset

Title

> **xmlsave** — Save and use datasets in XML format

Syntax

Save data in memory to XML-format dataset

> <u>xmlsav</u>e *filename* [*if*] [*in*] [, *xmlsave_options*]

Save subset of data in memory to XML-format dataset

> <u>xmlsav</u>e *varlist* using *filename* [*if*] [*in*] [, *xmlsave_options*]

Use XML-format dataset

> xmluse *filename* [, *xmluse_options*]

xmlsave_options	description
Main	
<u>doc</u>type(dta)	save XML file by using Stata's .dta format
<u>doc</u>type(excel)	save XML file by using Excel XML format
dtd	include Stata DTD in XML file
<u>leg</u>ible	format XML to be more legible
<u>rep</u>lace	overwrite existing *filename*

xmluse_options	description
<u>doc</u>type(dta)	load XML file by using Stata's .dta format
<u>doc</u>type(excel)	load XML file by using Excel XML format
<u>sh</u>eet("*sheetname*")	Excel worksheet to load
<u>cell</u>s(*upper-left:lower-right*)	Excel cell range to load
<u>dat</u>estring	import Excel dates as strings
<u>alls</u>tring	import all Excel data as strings
<u>first</u>row	treat first row of Excel data as variable names
<u>miss</u>ing	treat inconsistent Excel types as missing
<u>nocom</u>press	do not compress Excel data
clear	replace data in memory

Description

xmlsave and xmluse allow datasets to be saved or used in XML file formats for Stata's .dta and Microsoft Excel's SpreadsheetML format. XML files are advantageous because they are structured text files that are highly portable between applications that understand XML.

xmlsave saves the data in memory in the dta XML format by default. To save the data, type

> . xmlsave *filename*

528

although sometimes you will want to explicitly specify which document type definition (DTD) to use by typing

> . xmlsave *filename*, doctype(dta)

xmluse can read either an Excel-format XML or a Stata-format XML file into Stata. You type

> . xmluse *filename*

Stata will read into memory the XML file *filename*.xml, containing the data after determining whether the file is of document type dta or excel. As with the xmlsave command, the document type can also be explicitly specified with the doctype() option.

> . xmluse *filename*, doctype(dta)

It never hurts to specify the document type; it is actually recommended since there is no guarantee that Stata will be able to determine the document type from the content of the XML file. Whenever the doctype() option is omitted, a note will be displayed that identifies the document type Stata used to load the dataset.

If *filename* is specified without an extension, .xml is assumed.

Options for xmlsave

⌐ Main ⌐

doctype(dta | excel) specifies the document type definition (DTD) to use when saving the dataset.

doctype(dta), the default, specifies that an XML file will be saved using Stata's .dta format (see [P] **file formats .dta**). This is analogous to Stata's binary dta format for datasets. All data that can normally be represented in a normal dta file will be represented by this document type.

doctype(excel) specifies that an XML file will be saved using Microsoft's SpreadsheetML DTD. SpreadsheetML is the term given by Microsoft to the Excel XML format. Specifying this document type produces a generic spreadsheet with variable names as the first row, followed by data. It can be imported by any version of Microsoft Excel that supports Microsoft's SpreadsheetML format.

dtd when combined with doctype(dta) embeds the necessary document type definition into the XML file so that a validating parser of another application can verify the dta XML format. This option is rarely used, however, because it increases file size with information that is purely optional.

legible adds indents and other optional formatting to the XML file, making it more legible for a person to read. This extra formatting, however, is unnecessary and in larger datasets can significantly increase the file size.

replace permits xmlsave to overwrite existing *filename*.xml.

Options for xmluse

doctype(dta | excel) specifies the document type definition (DTD) to use when loading data from *filename*.xml. Although it is optional, use of doctype() is encouraged. If this option is omitted with xmluse, the document type of *filename*.xml will be determined automatically. When this occurs, a note will display the document type used to translate *filename*.xml. This automatic determination of document type is not guaranteed, and the use of this option is encouraged to prevent ambiguity between various XML formats. Specifying the document type explicitly also improves speed, as the data are only passed over once to load, instead of twice to determine the document type. In larger datasets, this advantage can be noticeable.

doctype(dta) specifies that an XML file will be loaded using Stata's dta format. This document type follows closely Stata's binary .dta format (see [P] **file formats .dta**).

doctype(excel) specifies that an XML file will be loaded using Microsoft's SpreadsheetML document type definition. SpreadsheetML is the term given by Microsoft to the Excel XML format.

sheet("*sheetname*") imports the worksheet named *sheetname*. Excel files can contain multiple worksheets within one document, so using the sheet() option specifies which of these to load. The default is to import the first worksheet to occur within *filename*.xml.

cells(*upper-left*:*lower-right*) specifies a cell range within an Excel worksheet to load. The default range is the entire range of the worksheet, even if portions are empty. Often the use of cells() is necessary because data are offset within a spreadsheet, or only some of the data need to be loaded. Cell-range notation follows the letter-for-column and number-for-row convention that is popular within all spreadsheet applications. The following are valid examples:

. xmluse *filename*, doctype(excel) cells(A1:D100)

. xmluse *filename*, doctype(excel) cells(C23:AA100)

datestring forces all Excel SpreadsheetML date formats to be imported as strings to retain time information that would otherwise be lost if automatically converted to Stata's date format. With this option, time information can be parsed from the string after loading it.

allstring forces Stata to import all Excel SpreadsheetML data as string data. Although data type information is dictated by SpreadsheetML, there are no constraints to keep types consistent within columns. When such inconsistent use of data types occurs in SpreadsheetML, the only way to resolve inconsistencies is to import data as string data.

firstrow specifies that the first row of data in an Excel worksheet consist of variable names. The default behavior is to generate generic names. If any name is not a valid Stata variable name, a generic name will be substituted in its place.

missing forces any inconsistent data types within SpreadsheetML columns to be imported as missing data. This can be necessary for various reasons but often will occur when a formula for a particular cell results in an error, thus inserting a cell of type ERROR into a column that was predominantly of a NUMERIC type.

nocompress specifies that data not be compressed after loading from an Excel SpreadsheetML file. Because data type information in SpreadsheetML can be ambiguous, Stata initially imports with broad data types and, after all data are loaded, performs a compress (see [D] **compress**) to reduce data types to a more appropriate size. The following table shows the data type conversion used before compression and the data types that would result from using nocompress:

SpreadsheetML type	Initial Stata type
String	str244
Number	double
Boolean	double
DateTime	double
Error	str244

clear clears data in memory before loading from *filename*.xml.

Remarks

XML stands for Extensible Markup Language and is a highly adaptable text format derived from SGML. The World Wide Web Consortium is responsible for maintaining the XML language standards. See http://www.w3.org/XML/ for information regarding the XML language, as well as a thorough definition of its syntax.

The document type dta, used by both xmlsave and xmluse, represents Stata's own document type definition (DTD) for representing Stata .dta files in XML. Stata reserves the right to modify the specification for this DTD at any time, although this is unlikely to be a frequent event.

The document type excel, used by both xmlsave and xmluse, corresponds to the DTD developed by Microsoft for use in modern versions of Microsoft Excel spreadsheets. This product may incorporate intellectual property owned by Microsoft Corporation. The terms and conditions under which Microsoft is licensing such intellectual property may be found at *http://msdn.microsoft.com/library/en-us/odcXMLRef/html/odcXMLRefLegalNotice.asp*. For more information about Microsoft Office and XML, see http://www.microsoft.com/office/xml/.

❏ Technical Note

When you import data from Excel to Stata, a common hurdle is handling Excel's use of inconsistent data types within columns. Numbers, strings, and other types can be mixed freely within a column of Excel data. Stata, however, requires that all data in a variable be of one consistent type. This can cause problems when a column of data from Excel is imported into Stata and the data types vary across rows.

By default, xmluse attempts to import Excel data by using the data type information stored in the XML file. If an error due to data type inconsistencies is encountered, you can use the options firstrow, missing, and cells() to isolate the problem while retaining as much of the data-type information as possible.

However, identifying the problem and determining which option to apply can sometimes be difficult. Often you may not care what format the data are imported into Stata, as long as you can import them. The quick solution for these situations is to use the allstring option to guarantee that all the data are imported as strings, assuming that the XML file itself was valid. Often converting the data back into numeric form after they are imported into Stata is easier, given Stata's vast data-management commands.

❏

▷ Example 1: Saving XML files

To save the current Stata dataset to a file, auto.xml, type

    . xmlsave auto

To overwrite an existing XML dataset with a new file containing the variables make, mpg, and weight, type

    . xmlsave make mpg weight using auto, replace

To save the dataset to an XML file for use with Microsoft Excel, type

    . xmlsave auto, doctype(excel) replace

◁

▷ Example 2: Using XML files

Assuming that we have a file named `auto.xml` saved using the `doctype(dta)` option of `xmlsave`, we can read in this dataset with the command

```
. xmluse auto, doctype(dta) clear
```

If the file was saved from Microsoft Excel to a file called `auto.xml` that contained the worksheet `Rollover Data`, with the first row representing column headers (or variable names), we could import the worksheet by typing

```
. xmluse auto, doctype(excel) sheet("Rollover Data") firstrow clear
```

Continuing with the previous example: if we wanted just the first column of data in that worksheet, and we knew that there were only 75 rows, including one for the variable name, we could have typed

```
. xmluse auto, doc(excel) sheet("Rollover Data") cells(A1:A75) first clear
```

◁

Also See

[D] **compress** — Compress data in memory

[D] **fdasave** — Save and use datasets in FDA (SAS XPORT) format

[D] **infile** — Overview of reading data into Stata

[D] **odbc** — Load, write, or view data from ODBC sources

[D] **outfile** — Write ASCII-format dataset

[D] **outsheet** — Write spreadsheet-style dataset

[D] **save** — Save datasets

[P] **file formats .dta** — Description of .dta file format

Title

xpose — Interchange observations and variables

Syntax

xpose , clear [*options*]

options	description
* clear	reminder that untransposed data will be lost if not previously saved
format	use largest numeric display format from untransposed data
format(%*fmt*)	apply specified format to all variables in transposed data
varname	add variable _varname containing original variable names
promote	use the most compact data type that preserves numeric accuracy

* clear is required.

Description

xpose transposes the data, changing variables into observations and observations into variables. All new variables—that is, those created by the transposition—are made the default storage type. Thus any original variables that were strings will result in observations containing missing values. (If you transpose the data twice, you will lose the contents of string variables.)

Options

clear is required and is supposed to remind you that the untransposed data will be lost (unless you have saved the data previously).

format specifies that the largest numeric display format from your untransposed data be applied to the transposed data.

format(%*fmt*) specifies that the specified numeric display format be applied to all variables in the transposed data.

varname adds the new variable _varname to the transposed data containing the original variable names. Also, with or without the varname option, if the variable _varname exists in the dataset before transposition, those names will be used to name the variables after transposition. Thus transposing the data twice will (almost) yield the original dataset.

promote specifies that the transposed data use the most compact numeric data type that preserves the original data accuracy.

If your data contain any variables of type double, all variables in the transposed data will be of type double.

If variables of type float are present, but there are no variables of type double or long, the transposed variables will be of type float. If variables of type long are present, but there are no variables of type double or float, the transposed variables will be of type long.

Remarks

▷ Example 1

We have a dataset on something by county and year that contains

```
. use http://www.stata-press.com/data/r10/xposexmpl
. list
```

	county	year1	year2	year3
1.	1	57.2	11.3	19.5
2.	2	12.5	8.2	28.9
3.	3	18	14.2	33.2

Each observation reflects a county. To change this dataset so that each observation reflects a year, type

```
. xpose, clear varname
. list
```

	v1	v2	v3	_varname
1.	1	2	3	county
2.	57.2	12.5	18	year1
3.	11.3	8.2	14.2	year2
4.	19.5	28.9	33.2	year3

We would now have to drop the first observation (corresponding to the previous county variable) to make each observation correspond to one year. Had we not specified the varname option, the variable _varname would not have been created. The _varname variable is useful, however, if we want to transpose the dataset back to its original form.

```
. xpose, clear
. list
```

	county	year1	year2	year3
1.	1	57.2	11.3	19.5
2.	2	12.5	8.2	28.9
3.	3	18	14.2	33.2

◁

Methods and Formulas

xpose is implemented as an ado-file.

See Hamilton (2004, chap. 2) for an introduction to Stata's data-management features.

Reference

Hamilton, L. C. 2006. *Statistics with Stata (Updated for version 9)*. Belmont, CA: Duxbury.

Also See

[D] **reshape** — Convert data from wide to long form and vice versa

[D] **stack** — Stack data

Subject and author index

This is the subject and author index for the *Stata Data Management Reference Manual*. Readers interested in topics other than data management should see the combined subject index (and the combined author index) in the *Stata Quick Reference and Index*. The combined index indexes the *Getting Started* manuals, the *User's Guide*, and all the Reference manuals except the *Mata Reference Manual*.

Semicolons set off the most important entries from the rest. Sometimes no entry will be set off with semicolons, meaning that all entries are equally important.

D

W

X

Y